T0299983

NUMERICAL METHODS IN ASTROPHYSICS
An Introduction

Series in Astronomy and Astrophysics

The Series in Astronomy and Astrophysics includes books on all aspects of theoretical and experimental astronomy and astrophysics. Books in the series range in level from textbooks and handbooks to more advanced expositions of current research.

Other books in the series

Very High Energy Gamma-Ray Astronomy
T C Weekes

The Physics of Interstellar Dust
E Krügel

Dust in the Galactic Environment, 2nd Edition
D C B Whittet

Dark Sky, Dark Matter
J M Overduin and P S Wesson

An Introduction to the Science of Cosmology
D J Raine and E G Thomas

The Origin and Evolution of the Solar System
M M Woolfson

The Physics of the Interstellar Medium, 2nd Edition
J E Dyson and D A Williams

Optical Astronomical Spectroscopy
C R Kitchin

Dust and Chemistry in Astronomy
T J Millar and D A Williams (eds)

Stellar Astrophysics
R J Tayler (ed)

Series in Astronomy and Astrophysics

NUMERICAL METHODS IN ASTROPHYSICS

An Introduction

Peter Bodenheimer
University of California
Santa Cruz, USA

Gregory P. Laughlin
University of California
Santa Cruz, USA

Michał Różyczka
Nicolaus Copernicus Astronomical Center
Warsaw, Poland

Harold W. Yorke
Jet Propulsion Laboratory
Pasadena, California, USA

The software mentioned in this book is now available for download on our Web site at: http://www.crcpress.com/e_products/downloads/default.asp

CRC Press
Taylor & Francis Group
6000 Broken Sound Parkway NW, Suite 300
Boca Raton, FL 33487-2742

Printed in the United States of America on acid-free paper
10 9 8 7 6 5 4 3 2

International Standard Book Number-10: 0-7503-0883-4 (Hardcover)
International Standard Book Number-13: 978-0-7503-0883-0 (Hardcover)

Library of Congress Cataloging-in-Publication Data

Numerical methods in astrophysics : an introduction / Peter Bodenheimer ... [et al.].
p. cm. -- (Series in astronomy and astrophysics)
Includes bibliographical references and index.
ISBN 0-7503-0883-4 (alk. paper)
1. Astrophysics--Mathematics. 2. Hydrodynamics. 3. Radiative transfer. 4. Differential equations, Partial--Numerical solutions. I. Bodenheimer, Peter. II. Title. III. Series.

QB462.3.I58 2006
523.0101'51--dc22 2006050481

Visit the Taylor & Francis Web site at
http://www.taylorandfrancis.com

and the CRC Press Web site at
http://www.crcpress.com

Contents

Introduction

Observations of astrophysical phenomena on every scale — ranging from the dynamics of Saturn's rings to the orbits of planets around nearby stars, out to distant supernovae and colliding galaxies, and even of the Big Bang itself — lead inexorably to a single conclusion: hydrodynamical flows and orbital mechanics are absolutely vital elements for understanding how the universe works. Furthermore, radiation arising from astrophysical processes continually bathes the Earth in subtly flickering light. This faint illumination is the sole link between the events themselves and our instruments; nevertheless, when it is properly spatially resolved or decomposed into the various frequencies of radiation, it is rich with information. The theory of radiation transport thus forms a third, equally indispensable element in our interpretation of these phenomena.

A theoretical interpretation of observational data often progresses through several stages of refinement, including, first, very approximate "back of the envelope" calculations based on the fundamental laws of physics; second, analytic or semi-analytic calculations often based on linear perturbation theory; and, third, full-scale numerical simulations. As new observational capabilities are developed (over a wide range of wavelengths and with increasing spectral and spatial resolution), the inherent complexity of astrophysical phenomena grows more and more evident.

Theoretical estimates also indicate that the explanation of some kinds of objects and phenomena requires the exploration of physical regimes where the laws of physics are not well known. The regions that we must probe in this manner include the interiors of neutron stars and giant planets, or turbulence in the interstellar medium. Even outside such extreme cases, the analysis of observations requires the consideration of additional physical effects beyond hydrodynamics and radiation transport, such as chemistry, nuclear physics, magnetic fields, atomic physics, and the properties of matter, including the equation of state. The broad scope of physical effects does require, in some situations, consideration of a rather complex theoretical model.

Analytical methods invariably involve approximations in the laws of physics; under some circumstances such approximations are reasonable, but in others they are not. In some cases, the wide variety of processes that must be considered makes the overall problem intractable by purely analytical methods. Such methods are, of course, necessary to formulate individual segments of the problem, as Chapter 1 illustrates, but there comes a point beyond which analytical methods will not go, and one must resort to a numerical approach. Nevertheless, once the numerical results have been obtained, the order-of-magnitude estimates are extremely useful to compare various time scales and to determine whether the results are physically reasonable. The analytical solutions also are indispensable as test cases against which the accuracy of numerical solutions can be determined.

Indeed, when one considers some of the main unsolved problems of astrophysics, such as the formation of clusters of galaxies and clusters of stars or the origin of binary stars and planetary systems, it appears that only a limited amount of progress can be made through analytical methods, and the use of numerical simulations is absolutely necessary. With the rapid development of computer power and availability of storage, such simulations have become increasingly sophisticated, and graphical representations of the results have led to spectacular pictures. One must, however, always be somewhat skeptical of such detailed numerical results, because even in the most comprehensive numerical treatments there are still physical approximations whose validity should be evaluated. Also, there are numerous nonphysical numerical effects, which can degrade the accuracy of a solution. In hydrodynamical problems involving three space dimensions or problems in frequency-dependent radiation transfer, the solution is limited as well by the finite numerical resolution that can be achieved. The results must be judged, for example, on the basis of the physically necessary spatial resolution compared with the actual resolution. It is the purpose of this book to outline some of the basic numerical methods available to solve the equations of gravitational dynamics, hydrodynamics, and radiation transport; to indicate which methods are most suitable for particular problems; to show what the accuracy requirements are in numerical simulations; and to suggest ways to test for and reduce the inevitable numerical effects.

While most of the book is devoted to the numerical methods themselves, with emphasis on practical application rather than extensive theoretical development, the first chapter focuses on the fundamental equations and provides some basic derivations. The second chapter discusses the basic method of finite differences on a grid, with application mainly to hydrodynamic problems in one space dimension and time. The third chapter deals with the problem of the motion of N point-mass particles under the influence only of their mutual gravitational interactions. Chapter 4 combines the elements of the N-body dynamics problem with certain properties of fluid dynamics in a treatment of the method of "smoothed particle hydrodynamics," extensively used in astrophysics problems, especially when three space dimensions are required. Discussed in Chapter 5 is one of the fundamental sets of equations in astrophysics: the equations of stellar structure and evolution, which form a special case of a simplified form of the equations of hydrodynamics with radiation transport. The chapter covers the physical approximations generally used as well as the basic numerical solution technique. Chapter 6 deals with more advanced techniques in grid-based hydrodynamics, with applications in particular to problems in two space dimensions. Chapter 7 summarizes some of the basic techniques for calculating the gravitational forces in an astrophysical system. Chapter 8 is a brief introduction to the extension of the numerical techniques of Chapter 6 to problems where magnetic fields are important. Chapter 9 discusses some specific problems in grid-based methods for radiation transfer, in some cases, combined with hydrodynamics.

Chapter 10 contains brief "user manuals" for the numerical codes that are provided. These codes allow the user to set up his or her own problems in orbital dynamics, smoothed particle hydrodynamics, grid-based hydrodynamics, radiation transfer, and stellar evolution. Sample input data are provided and, in some cases, a graph of sample results is shown. The codes are compatible with the compilers f77 and f90. In a few cases, the codes require subroutines that are published in *Numerical Recipes in Fortran* (Press et al., 1992). Because of copyright concerns, these subroutines are not provided here, but it is clearly indicated in the codes which programs are needed, and the user is free to obtain these on his or her own.

A large number of numerical techniques have been developed to solve astrophysical problems, and this book is not intended to be a comprehensive encyclopedia, which encompasses all of them. As an introduction to the subject, it does cover basic particle methods and basic grid methods and gives the reader the general impression of the overall landscape, while going into the full, necessary detail on only a limited set of problems. Some of the methods discussed here are also treated in *Numerical Recipes in Fortran*, especially in Chapter 16 (Ordinary Differential Equations) and Chapter 19 (Partial Differential Equations). We overlap with that excellent work only to the extent necessary to provide a self-contained treatment; in some cases, we simply summarize *Numerical Recipes in Fortran* procedures and refer the reader to it for details.

In summary, the following sets of equations, commonly used in astrophysics, are discussed analytically, mostly in Chapter 1, and numerically in the rest of the book:

- Eulerian equations of hydrodynamics (1.42)
- Lagrangian equations of hydrodynamics (1.62)
- Navier–Stokes equations (1.73)
- Equation of radiative transfer (1.90) or, in simpler form (1.91)
- Equations of magnetohydrodynamics (1.148)
- Stellar structure equations (5.3), (5.4), (5.7), and (5.11)
- Equations of radiation hydrodynamics (9.63), (9.64), and (9.65)

We express our gratitude to Richard Wünsch and Artur Gawryszczak for setting up the ZEUS program in a form suitable for readers of this book. We give our sincere thanks to our colleagues who have given permission for the use of their codes: Lars Hernquist, Michael Norman, John Chambers, and Martin Duncan. We also thank those authors who have given us permission for the use of their published figures.

REFERENCES

Press, W. H., Teukolsky, S. A., Vetterling, W. T., and Flannery, B. P. (1992) *Numerical Recipes in Fortran,* 2nd ed. (Cambridge: Cambridge University Press).

Supplementary Resources Disclaimer

Additional resources were previously made available for this title on CD. However, as CD has become a less accessible format, all resources have been moved to a more convenient online download option.

You can find these resources available here: https://www.routledge.com/9780750308830

Please note: Where this title mentions the associated disc, please use the downloadable resources instead.

1 Basic Equations

The equations discussed in this chapter include those of pure hydrodynamics, pure hydrodynamics plus viscosity (Navier–Stokes equations), pure hydrodynamics plus radiation transfer (radiation hydrodynamics), and pure hydrodynamics plus magnetic fields (magnetohydrodynamics). The reader interested in proceeding directly to the practical applications can skip many of the details of this chapter, but should develop an appreciation of its basic results.

Both the hydrodynamics equations and the radiation transfer equation can be derived from a fundamental statistical relation known as the *Boltzmann equation*. This chapter discusses the equation itself, shows how the equations of hydrodynamics and radiation transfer can be generated from it, introduces some of the basic physical variables and notations that will be used in the rest of the book, and extends the physical discussion to include cases where viscosity or magnetic fields must be included. The equations of hydrodynamics are first derived in Eulerian form, where the coordinate system is fixed in space, and then in Lagrangian (*co-moving*) form, in which the coordinate system is attached to moving mass elements. The equations are then generalized to include viscosity. Next, the equation of radiation transfer is derived and given in its most general form as well as in some special cases. Finally, the Maxwell equations are presented and used to derive the equations of magnetohydrodynamics. Clearly this chapter provides only a brief introduction to these fundamental equations; a much more detailed treatment can be found in the two-volume series by Frank Shu (1991, 1992).

1.1 THE BOLTZMANN EQUATION

Gases and liquids are ensembles of atoms and molecules. Ideally, we would want to know the position $\mathbf{x}_i$, velocity $\mathbf{u}_i$, and force per unit mass $\mathbf{F}_i$, acting on each particle i in a gas or liquid. From this knowledge — at least according to classical mechanics — we can determine the future position and velocity from

$$\frac{d\mathbf{x}_i}{dt} = \mathbf{u}_i \qquad \frac{d\mathbf{u}_i}{dt} = \mathbf{F}_i(\mathbf{x}_j, \mathbf{u}_j, t) \quad \text{for all } j, \tag{1.1}$$

where it is implicitly assumed that the force $\mathbf{F}_i$ on i can be determined by a known function of the positions and velocities of all particles. Of course, integrating these *N-body* equations in time is only occasionally possible in practice. Even considering a modest amount of gas (say, 1 mole) requires following the trajectories of 6×10^{23} particles, a feat that is far beyond the capabilities of present-day computers.

However, in general, we do not need to know the exact positions and exact velocities of all particles at any particular time. It is sufficient to consider identical particles with nearly identical positions and nearly identical velocities in a statistical manner; the evolution of such ensembles is governed by the equations of statistical physics. In fact, in many problems of astrophysics, they can be regarded as continuous media.

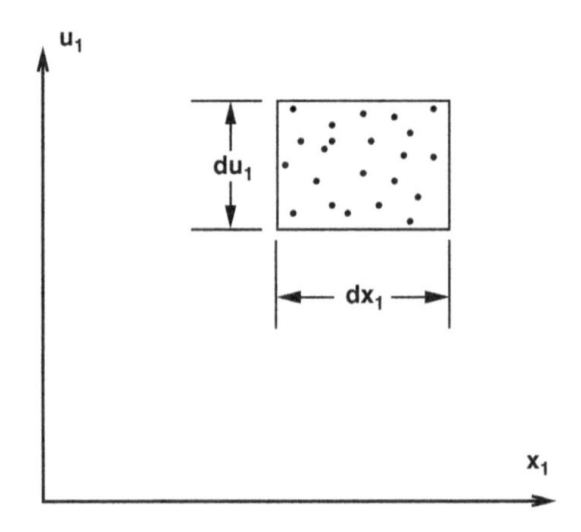

FIGURE 1.1 Particles in cell of 6D phase space projected onto the $x_1 - u_1$ plane.

Such media are described by much simpler *equations of hydrodynamics*. The basic statistical equation from which the equations of hydrodynamics can be derived is the Boltzmann equation. We shall discuss it briefly, referring readers interested in a more advanced approach to Huang (1987) or Lifshitz and Pitaevskii (1980).

Consider an ensemble of identical particles without internal degrees of freedom (such particles are incapable of any internal motions, such as vibration or rotation). Assume that quantum effects may be neglected in their motions relative to each other. Let $\mathbf{x} = (x_1, x_2, x_3)$ give the position of a particle in three-dimensional space, $\mathbf{u} = (u_1, u_2, u_3)$ its velocity, and $\mathbf{q} = (q_1, q_2, q_3)$ its momentum, such that $\mathbf{q} = m\mathbf{u}$, where m is the mass of the particle. Note that $\mathbf{u}$ is the actual velocity of an individual particle, as distinct from the mean fluid velocity $\mathbf{v}$, which will be used in most of the book. The *distribution function* $f(x_1, x_2, x_3, u_1, u_2, u_3, t)$, where t is the time, is defined such that

$$dN = f(\mathbf{x}, \mathbf{u}, t)\,\mathbf{dx}\,\mathbf{du}, \tag{1.2}$$

where dN is the number of particles that at time t are positioned between $\mathbf{x}$ and $\mathbf{x}+\mathbf{dx}$ and have velocities in the range $\mathbf{u}$ to $\mathbf{u} + \mathbf{du}$.

Assume that the particles are subject to an external force field $\mathbf{F}$, which does not change appreciably over a distance comparable to the mean interparticle separation. Let the field be normalized such that $\mathbf{F} = (F_1, F_2, F_3)$ is the force per unit mass. Consider a volume element between $\mathbf{x}$ and $\mathbf{x} + \mathbf{dx}$, occupied by particles with velocities between $\mathbf{u}$ and $\mathbf{u} + \mathbf{du}$. This volume element has six dimensions; its projection onto the (x_1, u_1) plane is shown in Figure 1.1. As long as this volume element contains a sufficiently large number of particles, a statistical approach will be valid. For the moment, neglect the collisions between particles. In time dt, the momentum $\mathbf{q}$ of any particle will change to $\mathbf{q} + m\mathbf{F}dt$, its velocity to $\mathbf{u} + \mathbf{F}dt$, and its position $\mathbf{x}$ to $\mathbf{x} + \mathbf{u}dt$. Thus, the number of particles $f(\mathbf{x}, \mathbf{u}, t)\,\mathbf{dx}\,\mathbf{du}$ is equal to the number of particles $f(\mathbf{x} + \mathbf{u}dt, \mathbf{u} + \mathbf{F}dt, t + dt)\,\mathbf{dx}\,\mathbf{du}$, that is:

$$f(\mathbf{x} + \mathbf{u}dt, \mathbf{u} + \mathbf{F}dt, t + dt) - f(\mathbf{x}, \mathbf{u}, t) = 0. \tag{1.3}$$

If collisions are allowed for, then the right-hand side of Equation (1.3) will be different from zero

$$f(\mathbf{x}+\mathbf{u}dt, \mathbf{u}+\mathbf{F}dt, t+dt) - f(\mathbf{x}, \mathbf{u}, t) = [\Delta f]_{\text{coll}}, \tag{1.4}$$

where $[\Delta f]_{\text{coll}}$ is the change in f in time dt due to collisions. Since the left-hand side represents the change in f during a time interval dt, we may write

$$\begin{aligned} &\frac{\partial f}{\partial t} + u_1 \cdot \frac{\partial f}{\partial x_1} + u_2 \cdot \frac{\partial f}{\partial x_2} + u_3 \cdot \frac{\partial f}{\partial x_3} \\ &\quad + F_1 \cdot \frac{\partial f}{\partial u_1} + F_2 \cdot \frac{\partial f}{\partial u_2} + F_3 \cdot \frac{\partial f}{\partial u_3} = \left[\frac{\partial f}{\partial t}\right]_{\text{coll}}, \end{aligned} \tag{1.5}$$

or, in equivalent standard notation (implied summation over repeated indices)

$$\frac{\partial f}{\partial t} + u_i \frac{\partial f}{\partial x_i} + F_i \frac{\partial f}{\partial u_i} = \left[\frac{\partial f}{\partial t}\right]_{\text{coll}}. \tag{1.6}$$

This is Boltzmann's transport equation (hereafter just "the Boltzmann equation"), which describes the evolution of the distribution function in the six-dimensional space $(\mathbf{x}, \mathbf{u})$ called *phase space*. It expresses the fact that the change in the number of particles within a given phase space volume element $(\mathbf{dx}\ \mathbf{du})$ is equal to the net number of particles, which enter or leave that element (Figure 1.2 and Figure 1.3).

Now, at location $\mathbf{x}$, note that $f(\mathbf{x}, \mathbf{u}, t)\,\mathbf{du}$ is the number of particles per unit volume in the interval $\mathbf{u}$ to $\mathbf{u}+\mathbf{du}$. Thus, the total number of particles per unit volume is:

$$n(\mathbf{x}, t) \equiv \int f(\mathbf{x}, \mathbf{u}, t)\,\mathbf{du}. \tag{1.7}$$

Associated with n is the approximate *mean particle separation*, given by

$$l = n^{1/3}; \tag{1.8}$$

for the statistical treatment to be valid, the interval dx_1 in Figure 1.1 must be much larger than l. The mass density (hereafter just "density") is defined by

$$\rho(\mathbf{x}, t) \equiv \int m f(\mathbf{x}, \mathbf{u}, t)\,\mathbf{du} \tag{1.9}$$

and the mean or *bulk velocity* of the medium $\mathbf{v}$ (hereafter just "velocity") by

$$\mathbf{v}(\mathbf{x}, t) \equiv \frac{1}{\rho} \int \mathbf{u} m f(\mathbf{x}, \mathbf{u}, t)\,\mathbf{du}, \tag{1.10}$$

where the integration domain is the whole velocity space. When the collisions are elastic (i.e., conserving momentum and kinetic energy of the colliding particles) and the density of the medium is low enough for collisions involving more than two particles to be neglected, then, in the absence of external forces, f is obtained from

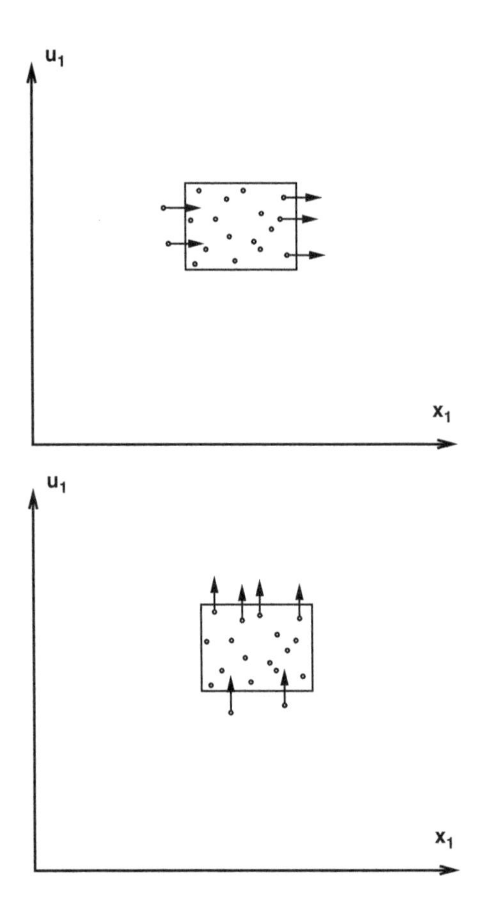

FIGURE 1.2 The *left-hand side* of the Boltzmann equation. The number of particles in a phase space volume element **dx du** changes as particles of velocity **u** change their position (top) or particles experiencing a force **F** change their velocity (bottom).

statistical mechanics (Lifshitz and Pitaevskii, 1980), and is given by the *Maxwellian velocity distribution*

$$f(\mathbf{x}, \mathbf{u}, t)\mathbf{du} = n(\mathbf{x}, t)\left[\frac{m}{2\pi k_B T(\mathbf{x}, t)}\right]^{\frac{3}{2}} \exp\left[-\frac{m(\mathbf{u}-\mathbf{v})^2}{2k_B T(\mathbf{x}, t)}\right] \mathbf{du}, \tag{1.11}$$

where k_B is the Boltzmann constant and T is the temperature. Here, $f\mathbf{du}$ represents the number of particles per unit volume with velocities between $\mathbf{u}$ and $\mathbf{u} + \mathbf{du}$ in an *equilibrium state*, which is obtained in the limit $t \to \infty$. In this state, $\partial f/\partial t = 0$ and the number of particles entering the volume **dx du** by collisions is equal to the number leaving it by collisions.

Let

$$\tilde{\mathbf{u}} \equiv \mathbf{u} - \mathbf{v} \tag{1.12}$$

be the random *peculiar* velocity at which a particle moves with respect to the bulk flow. The *specific internal energy*, i.e., internal energy per unit mass, ϵ, associated

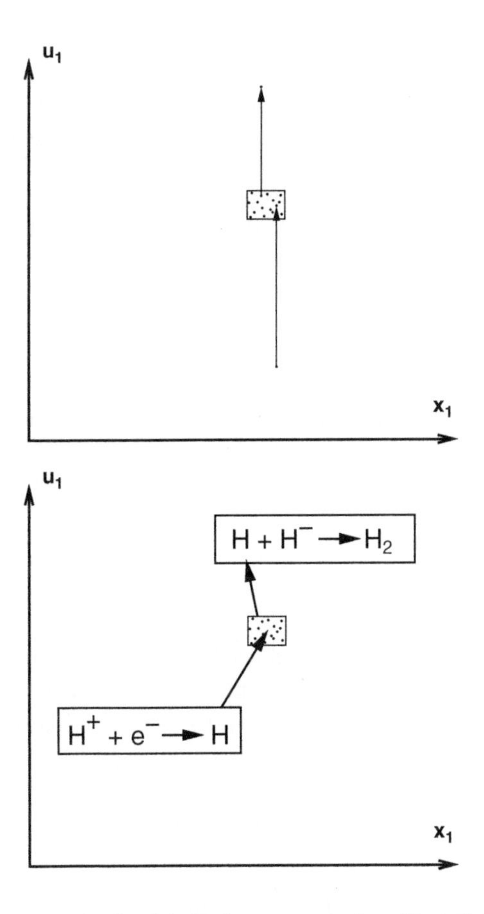

FIGURE 1.3 The *right hand side* of the Boltzmann equation. (Top) Collisions between particles can discontinuously change the particles' velocities. (Bottom) Examples of physical processes that can "create" or "destroy" hydrogen particles.

with the peculiar motions of particles, is defined by

$$\rho\epsilon(\mathbf{x}, t) \equiv \frac{1}{2}\int m\tilde{u}^2 f(\mathbf{x}, \mathbf{u}, t)\,\mathrm{d}\mathbf{u}. \tag{1.13}$$

Inserting Equation (1.11) into the definition (1.13) and performing the integration we see that

$$\epsilon = \frac{3}{2}\frac{k_B}{m}T, \tag{1.14}$$

which shows that the temperature is just a measure of the kinetic energy associated with peculiar motions.

Upon concluding this section, let us note that momentum may be used instead of velocity to define phase space, with the corresponding distribution function $f(\mathbf{x}, \mathbf{q}, t)$ $\mathbf{dx\,dq}$. Such a definition is, in fact, more general than the one based on $\mathbf{u}$, as it allows us to include massless particles (photons), which we will deal with in Section 1.6.

1.2 CONSERVATION LAWS OF HYDRODYNAMICS

In this section, we shall derive the equations of hydrodynamics and discuss their physical meaning. The same general set of equations applies to both liquid and gaseous media, which together are referred to as *fluids*. Actual liquids are a real rarity in astrophysical objects; nevertheless, in astrophysical hydrodynamics, one traditionally refers to the gaseous medium as a "fluid." Hereafter we shall follow this tradition. The results of the derivation are summarized in Equation (1.42).

To derive the equations of hydrodynamics, we shall calculate *moments* of the Boltzmann equation with respect to the velocity. In Section 1.6 the method of moments will also be applied to derive the equation of radiative transfer from the Boltzmann equation for photons.

To calculate the kth order moment ($k = 0, 1, 2$) one has to multiply both sides of Equation (1.6) by $\mathcal{U}_k$, where $\mathcal{U}_0 = 1, \mathcal{U}_1 = \mathbf{u}, \mathcal{U}_2 = u^2$ (u being the length of the velocity vector $\mathbf{u}$), and integrate it over the whole velocity space

$$\int \mathcal{U}_k \left[\frac{\partial f}{\partial t} + u_i \frac{\partial f}{\partial x_i} + F_i \frac{\partial f}{\partial u_i} \right] \mathbf{du} = \int \mathcal{U}_k \left[\frac{\partial f}{\partial t} \right]_{\text{coll}} \mathbf{du}. \tag{1.15}$$

To proceed, in principle one would need an explicit expression for the collision term $[\frac{\partial f}{\partial t}]_{\text{coll}}$. However, based on conservation laws for interactions among particles, important general properties of $[\frac{\partial f}{\partial t}]_{\text{coll}}$ can be deduced even if such an expression is not available. If collisions do not destroy or create particles and if every collision conserves mass, momentum, and energy of the colliding particles, then the following integrals over the whole velocity space must vanish

$$\int \left[\frac{\partial f}{\partial t} \right]_{\text{coll}} \mathbf{du} = 0 \tag{1.16}$$

because the number of particles is conserved, and

$$\int \left[\frac{\partial f}{\partial t} \right]_{\text{coll}} u_i \, \mathbf{du} = 0; \quad i = 1, 2, 3 \tag{1.17}$$

because the total momentum vector of particles is conserved, and

$$\int \left[\frac{\partial f}{\partial t} \right]_{\text{coll}} u^2 \, \mathbf{du} = 0 \tag{1.18}$$

because the total energy of particles is conserved. On the other hand, total mass, total momentum, and total energy of the ensemble must be finite, implying that f must satisfy the following necessary condition

$$\lim_{u \to \infty} u^2 f = 0. \tag{1.19}$$

By substituting $\mathbf{du} = du_1 du_2 du_3$ and performing integrations by parts, making use of Equation (1.19), one can show that

$$\int \frac{\partial f}{\partial u_i} \mathbf{du} = 0$$
$$\int u_j \frac{\partial f}{\partial u_i} \mathbf{du} = -\delta_{ij} \frac{\rho}{m}$$
$$\frac{1}{2} \int u^2 \frac{\partial f}{\partial u_i} \mathbf{du} = -\frac{\rho}{m} v_i \,. \tag{1.20}$$

(Here δ_{ij} is the standard "Kronecker" δ, which is taken to be unity when the two indices are the same ($i = j$) and is taken to be zero otherwise.) Equation (1.6), multiplied by m and integrated over the whole velocity space, yields

$$m \int \frac{\partial f}{\partial t} \mathbf{du} + m \int u_i \frac{\partial f}{\partial x_i} \mathbf{du} + m F_i \int \frac{\partial f}{\partial u_i} \mathbf{du} = 0, \tag{1.21}$$

where the collision term on the right-hand side vanishes because of Equation (1.16). Since the third term of the left-hand side of Equation (1.21) also vanishes because of the first line of Equation (1.20), we have

$$m \int \frac{\partial f}{\partial t} \mathbf{du} + m \int \frac{\partial}{\partial x_i} (u_i f) \mathbf{du} = 0 \tag{1.22}$$

or

$$\frac{\partial}{\partial t} \int m f \, \mathbf{du} + \frac{\partial}{\partial x_i} \int u_i m f \, \mathbf{du} = 0 \tag{1.23}$$

(swapping $\frac{\partial}{\partial x_i}$ and u_i is allowed because x_i and u_i are independent coordinates in phase space). Equation (1.23) may now be written in the standard form

$$\frac{\partial \rho}{\partial t} + \frac{\partial}{\partial x_i} (\rho v_i) = 0. \tag{1.24}$$

(Note again that there is implied summation over a repeated index.) This is the first equation of hydrodynamics, known as the *continuity equation*. Assume that the region with $\mathbf{v} \neq 0$ has a finite spatial extent and integrate Equation (1.24) over a domain $\mathcal{V}$, which is so large that $\mathbf{v} = 0$ everywhere on its surface $\mathcal{S}$. From the Gauss theorem, we have

$$\int_{\mathcal{V}} \frac{\partial}{\partial x_i} (\rho v_i) d\mathcal{V} = \int_{\mathcal{S}} \rho v_i \hat{n}_i dA = 0, \tag{1.25}$$

where dA is an area element of $\mathcal{S}$, and $\hat{\mathbf{n}} \equiv (\hat{n}_1, \hat{n}_2, \hat{n}_3)$ is a unit vector normal to dA. We see that

$$\frac{\partial M_{\mathcal{V}}}{\partial t} = 0, \tag{1.26}$$

where $M_{\mathcal{V}}$ is the mass of the medium contained in $\mathcal{V}$. Thus, Equation (1.24) simply expresses the mass conservation law.

Now, multiply Equation (1.6) by $m u_i$ ($i = 1, 2, 3$) and integrate it over the whole velocity space using the second line of Equation (1.20). The result is:

$$\frac{\partial}{\partial t} (\rho v_i) + \frac{\partial}{\partial x_j} \int m u_i u_j f \, \mathbf{du} - \rho F_i = 0. \tag{1.27}$$

(Note that in the rest of this section we use j as the index which is summed over.) Since $\int \tilde{\mathbf{u}} f \, d\mathbf{u} = 0$, we have

$$\int m u_i u_j f \, d\mathbf{u} = \int m v_i v_j f \, d\mathbf{u} + \int m \tilde{u}_i \tilde{u}_j f \, d\mathbf{u} = \rho v_i v_j + \mathrm{P}_{ij}, \tag{1.28}$$

where

$$\mathrm{P}_{ij} \equiv \int m \tilde{u}_i \tilde{u}_j f \, d\mathbf{u} \tag{1.29}$$

is the *pressure tensor*. In most cases of astrophysical relevance, pressure is isotropic, i.e., $\mathrm{P}_{ij} = P\delta_{ij}$, where

$$P \equiv \frac{1}{3} \int m \tilde{u}^2 f \, d\mathbf{u}, \tag{1.30}$$

which is 2/3 of the energy density $\rho\epsilon$. Equation (1.27) reduces to

$$\frac{\partial}{\partial t}(\rho v_i) + \frac{\partial}{\partial x_j}(\rho v_i v_j) = -\frac{\partial P}{\partial x_i} + \rho F_i. \tag{1.31}$$

This is the second equation of hydrodynamics, known as the *momentum equation*. (In fact, it is split into three separate equations, each dealing with one component of the momentum vector.)

The quantity ρv_i is the ith component of the momentum density (momentum per unit volume) vector, while $\rho v_i v_j$ is the momentum flux (momentum per unit area per unit time) carried by the jth component of velocity. When $P = \text{const}$ and $\mathbf{F} = 0$, Equation (1.31) becomes entirely analogous to the continuity equation, expressing the momentum conservation law. When $P \neq \text{const}$ (which is the typical case), a force arising from the pressure gradient appears on its right-hand side, which results from the exchange of energy between the bulk flow and peculiar motions of fluid particles. In a sense, the pressure gradient and external forces (associated, for example, with gravitational or electromagnetic fields) act like sources (or sinks) in the momentum flow, and they are often called *source terms* of the momentum equation. The vector form of Equation (1.31) is:

$$\frac{\partial}{\partial t}(\rho \mathbf{v}) + \nabla \cdot \pi = \rho \mathbf{F}, \tag{1.32}$$

where

$$\pi_{ij} \equiv \rho v_i v_j + P\delta_{ij} \tag{1.33}$$

is the symmetric *momentum flux density tensor*.

Exercise

Assuming that the pressure is isotropic, show that

$$\frac{\partial}{\partial t}\left[\rho\left(\frac{v^2}{2} + \epsilon\right)\right] + \frac{\partial}{\partial x_j}\left[\rho v_j \left(\frac{v^2}{2} + \epsilon\right)\right] = -\frac{\partial h_j}{\partial x_j} - \frac{\partial}{\partial x_j}(P v_j) + \rho v_j F_j, \tag{1.34}$$

where

$$h_j \equiv \int \frac{m}{2} \tilde{u}_j \tilde{u}^2 f \, d\mathbf{u} \tag{1.35}$$

is the *conduction heat flux*. Hint: Multiply Equation (1.6) by $\frac{m}{2}u^2$, integrate it over the whole velocity space, and use the third line of Equation (1.20).

When the *conduction heat flux* (Equation 1.35) can be neglected (which, fortunately, occurs in many cases of astrophysical interest), Equation (1.34) reduces to

$$\frac{\partial \mathcal{E}}{\partial t} + \frac{\partial}{\partial x_j}[(\mathcal{E} + P)v_j] = \rho v_j F_j, \tag{1.36}$$

where

$$\mathcal{E} \equiv \frac{1}{2}\rho v^2 + \rho\epsilon \tag{1.37}$$

is the *total energy density*. Equation (1.36), known as the *total energy equation*, often completes the set of basic equations of hydrodynamics. Alternatively, the *internal energy equation* may be used, in which the *internal energy density*, $e \equiv \rho\epsilon$, appears instead of $\mathcal{E}$. To obtain it, multiply Equation (1.24) by v_i and subtract it from Equation (1.31). The result is:

$$\frac{\partial v_i}{\partial t} + v_j \frac{\partial}{\partial x_j} v_i = -\frac{1}{\rho}\frac{\partial P}{\partial x_i} + F_i. \tag{1.38}$$

(Recall that summation over repeated indices is implied.) Now, combine Equation (1.24) multiplied by $\frac{1}{2}v_j v_j$ and Equation (1.38) multiplied by v_i to obtain the *bulk energy equation*

$$\frac{\partial}{\partial t}\left(\frac{1}{2}\rho v^2\right) + \frac{\partial}{\partial x_j}\left(\frac{1}{2}\rho v^2 v_j\right) = -v_j \frac{\partial P}{\partial x_j} + \rho v_j F_j, \tag{1.39}$$

which governs the evolution of the kinetic energy of the fluid. (Note that $\frac{1}{2}\rho v^2$ is just the density of the kinetic energy associated with the bulk motion of the fluid.) Finally, subtract Equation (1.39) from Equation (1.36). The result is:

$$\frac{\partial e}{\partial t} + \frac{\partial}{\partial x_j}(e v_j) = -P\frac{\partial v_j}{\partial x_j}. \tag{1.40}$$

As opposed to the sourceless continuity equation, the internal energy equation contains a source term $-P\frac{\partial v_j}{\partial x_j}$, which describes the effects of expansion or contraction of the medium. To better illustrate its meaning, integrate it over a domain $\mathcal{V}$ and, assuming $P =$ const, apply the Gauss theorem

$$\int_{\mathcal{V}} \frac{\partial}{\partial x_j}(P v_j) d\mathcal{V} = P \int_{\mathcal{S}} v_j \hat{n}_j dA. \tag{1.41}$$

The last integral is just $\frac{\partial \mathcal{V}}{\partial t}$, i.e., the rate at which the volume of the medium changes. Multiplied by P, it represents the rate of the work that has to be done on the medium to compress it (or the work that is done by the expanding medium).

So far we have derived five equations,

$$\begin{aligned}
\frac{\partial\rho}{\partial t}+\frac{\partial}{\partial x_j}(\rho v_j)&=0\\
\frac{\partial}{\partial t}(\rho v_i)+\frac{\partial}{\partial x_j}(\rho v_i v_j)&=-\frac{\partial P}{\partial x_i}+\rho F_i \quad (i=1,2,3)\\
\frac{\partial e}{\partial t}+\frac{\partial}{\partial x_j}(e v_j)&=-P\frac{\partial v_j}{\partial x_j},
\end{aligned} \tag{1.42}$$

with six unknown functions (ρ, v_1, v_2, v_3, P, and e, where e is the internal energy per unit volume). The quantity F_i is a prescribed external force, such as gravity, per unit mass. For consistency, we have taken j as the index to be summed over in all three equations. To solve Equations (1.42), a relation between the pressure P and the internal energy per unit volume e (or, equivalently, between P and ϵ) is needed apart from boundary and initial conditions. Such a relation is called the *equation of state*. For the ideal monoatomic gas (for which the assumptions listed at the beginning of Section 1.1 apply) the equation of state can be derived from the definitions of pressure (Equation 1.30) and specific internal energy (Equation 1.13)

$$P=\frac{2}{3}\rho\epsilon. \tag{1.43}$$

In this case, the $P-\epsilon$ relation is extremely simple. In general, however, it may be very complicated, involving a number of additional equations (e.g., when the gas is being ionized). Additional complications arise when the internal energy can be transported independently of the flow, i.e., when effects of radiation transfer become important. Then the emission and/or absorption of photons directly modifies the temperature and, according to Equation (1.14), the specific internal energy, which in turn influences the flow via pressure gradients.

The three lines of Equation (1.42) are known as the *Euler equations*. The first term on the left-hand side of each equation is the time derivative of a physical quantity Q, and the second term is the divergence of the flux of that quantity, $\nabla\cdot Q\mathbf{v}$. On the right-hand side, source terms appear, describing the rate at which Q is locally generated. Note that there is a degree of arbitrariness in the choice of source terms, e.g., consistent with Equation (1.32), the pressure term may be moved to the left-hand side of the momentum equation, turning the flux of the ith momentum component into $\rho v_i\mathbf{v}+P$.

1.3 THE VALIDITY OF THE CONTINUOUS MEDIUM APPROXIMATION

Equations (1.42) describe a collection of particles as a continuous medium. For such an approximation to be applicable, several conditions must be fulfilled. First of all, the mean free path of a particle between two consecutive collisions, λ, must be much smaller than the characteristic length scale of the problem, l_{ch}

$$\lambda \ll l_{ch}. \tag{1.44}$$

The above inequality must be really strong because, in order to define local mean values of $\rho(\mathbf{x})$, $\mathbf{v}(\mathbf{x})$, and $T(\mathbf{x})$, a finite-size *element of the medium* must be chosen centered on $\mathbf{x}$ such that its size l_e satisfies the condition

$$\lambda \ll l_e \ll l_{ch}. \tag{1.45}$$

Obviously, the distribution function should not vary over l_e and the number of particles within the element should be large enough for the averaging procedure to be meaningful.

Next, interparticle forces should be short-range, i.e., no interaction should occur between particles separated by distances larger than $\sim l_e$. If this condition is not satisfied, elements of the medium exchange energy and momentum not only with their closest neighbors, but with all remaining elements of the medium. An obvious example of a long-range force is gravity. Gravitational terms in the equations of hydrodynamics are obtained from the Boltzmann equation only if they are from sources external to the medium ($\mathbf{F}$ in Equation 1.3). The effects of self-gravity of the medium itself are not derivable from the Boltzmann equation; in addition, one must invoke the Poisson equation.

Finally, note that peculiar motions of particles close to the boundary of the element carry them out of the element into the adjacent elements with different values of $\rho(\mathbf{x})$, $\mathbf{v}(\mathbf{x})$, and $T(\mathbf{x})$. As a result, forces of friction appear and a microscopic exchange of momentum and internal energy commences between the elements, which is not taken into account by Equations (1.42). It is desirable that these effects, collectively called *diffusion*, be small. When they cannot be neglected, Equations (1.42) should be supplemented with terms describing the phenomena associated with diffusion, which, among others, may involve the conduction heat flux (Equation 1.35). Such terms will be introduced in Section 1.5.

How well is the requirement specified by Equation (1.44) fulfilled under typical astrophysical conditions? In a neutral gas

$$\lambda = \frac{1}{n\sigma}, \tag{1.46}$$

where σ is the *scattering cross section*, typically of the order of $\sim 10^{-15}$ cm^2, and n is the number of particles per unit volume. Interstellar clouds, whose main component is neutral hydrogen (the so-called HI clouds), have densities of a few atoms per cm^3. The corresponding mean free path is $\sim 10^{14}$ cm, or about the distance between the Sun and Jupiter. This may seem large; however, typical sizes of HI clouds are counted in tens of parsecs ($10^{19} - 10^{20}$ cm), and Equation (1.44) is satisfied. We may run into problems when the clouds, which are usually rather smooth, develop strong gradients in ρ and T resulting, for example, from intercloud collisions. A typical outcome of such collisions is shock fronts (see Section 6.1). A shock front is just a few mean free paths wide, and Equation (1.44) is violated. The hydrodynamical approach neglects this violation and treats shocks as discontinuities. This turns out to be entirely acceptable as far as the global evolution of the medium is concerned, but subtle details of the flow across the front, which may be quite important for the observers, are obviously lost.

In intergalactic space, the density is still lower than in HI clouds, but characteristic length and time scales are much longer, and Equations (1.42) are still applicable. In general, on appropriate scales of length and time, Equation (1.42) may be employed to describe the dynamics of nearly all gaseous astrophysical objects. This holds also for objects composed of ionized gas, i.e., of a mixture of ions and free electrons. Electromagnetic forces are long range, but in nearly all situations the charges contained within a fluid element are very effectively screeened from charges residing in other elements. Moreover, because positive and negative charges usually are not systematically displaced, fluid elements may be regarded as neutral. As a result, dynamically important electric fields are generated only in rather unusual circumstances (e.g., close to the surface of a pulsar) and, in most cases, they can be safely neglected. On the other hand, the global neutrality of the medium does not prevent free electrons and ions from moving in response to magnetic fields. This leads to generation of additional magnetic fields, giving rise to very complex phenomena, which we shall deal with in Section 1.7.

1.4 EULERIAN AND LAGRANGIAN FORMULATION OF HYDRODYNAMICS

The Euler equations of hydrodynamics derived in Section 1.2 describe the evolution of the state of the medium at a fixed location in space, $\mathbf{x}$. Such an approach is traditionally called *Eulerian*, and the Eulerian time derivative $\frac{\partial}{\partial t}$ refers to changes occurring as a result of the flow of the medium past $\mathbf{x}$. An alternative approach, traditionally called *Lagrangian*, employs *co-moving* spatial coordinates of a fluid element, defined as the instantaneous components (r_1, r_2, r_3) of its radius vector $\mathbf{r}$. The Lagrangian, or co-moving time derivative $\frac{d}{dt}$ refers to changes occuring *within the element* as it changes its state *and* location. In particular, the velocity of the element is given by $\frac{d\mathbf{r}}{dt}$. Obviously, at the location occupied by the element at an instant t, the "Lagrangian" velocity must be equal to the "Eulerian" velocity with which the element sweeps past $\mathbf{x}$

$$\frac{d\mathbf{r}}{dt} = \mathbf{v}. \tag{1.47}$$

At t_0, let the location of the element be $\mathbf{x}^0 = (x_1^0, x_2^0, x_3^0)$. If the velocity field $\mathbf{v}(\mathbf{x}, t)$ is known for all $\mathbf{x}$ and t, then the radius vector of that element can be found by integration

$$\mathbf{r} = \mathbf{x}^0 + \int_{t_0}^{t} \frac{d\mathbf{r}}{dt}\, dt = \mathbf{r}\left(x_1^0, x_2^0, x_3^0, t\right), \tag{1.48}$$

where the initial coordinates (x_1^0, x_2^0, x_3^0) serve as labels identifying the element. Equation (1.48) may be viewed as a continuous transformation of (x_1^0, x_2^0, x_3^0) into momentary coordinates $[r_1(t), r_2(t), r_3(t)]$. Unfortunately, such a transformation is usually extremely complicated.

According to the definition of the Lagrangian derivative, we may write

$$\frac{dQ}{dt} = \lim_{\delta t \to 0} \frac{Q(\mathbf{x} + \mathbf{v}\delta t, t + \delta t) - Q(\mathbf{x}, t)}{\delta t}, \tag{1.49}$$

where Q stands for any quantity characterizing the medium. Neglecting second and higher-order terms in the Taylor expansion, we have

$$Q(\mathbf{x}+\mathbf{v}\delta t, t+\delta t) = Q(\mathbf{x}, t) + \frac{\partial Q}{\partial t}\delta t + v_j \frac{\partial Q}{\partial x_j}\delta t, \tag{1.50}$$

and, hence,

$$\frac{dQ}{dt} = \frac{\partial Q}{\partial t} + v_j \frac{\partial Q}{\partial x_j}. \tag{1.51}$$

Now, since

$$\frac{\partial}{\partial x_j}(\rho v_j) = v_j \frac{\partial \rho}{\partial x_j} + \rho \frac{\partial v_j}{\partial x_j}, \tag{1.52}$$

we easily arrive at the Lagrangian form of the continuity equation

$$\frac{d\rho}{dt} + \rho \frac{\partial v_j}{\partial x_j} = 0. \tag{1.53}$$

Exercise

Show that the Lagrangian continuity equation is equivalent to

$$\rho \mathcal{V} = \text{const}, \tag{1.54}$$

where $\mathcal{V}$ is the volume of the fluid element. Hint: Follow the derivation of Equation (1.26).

According to Equation (1.54), the Lagrangian continuity equation simply states the fact that the density of the element is inversely proportional to its volume.

From Equation (1.38), the Lagrangian form of the momentum equation may be immediately obtained

$$\frac{dv_i}{dt} = -\frac{1}{\rho}\frac{\partial P}{\partial x_i} + F_i, \tag{1.55}$$

in which we recognize the second law of Newton. It says that the acceleration of the element is equal to the sum of forces acting on that element. (Recall that v_i in Equation (1.55) is differentiated along the element's path.) Finally, the Eulerian energy equation (1.36) directly yields its Lagrangian counterpart

$$\frac{d\mathcal{E}}{dt} + v_j \frac{\partial P}{\partial x_j} + (\mathcal{E} + P)\frac{\partial v_j}{\partial x_j} = \rho v_j F_j. \tag{1.56}$$

Exercise

Show that Equation (1.56) is equivalent to

$$\frac{de}{dt} = \frac{e+P}{\rho}\frac{d\rho}{dt} \tag{1.57}$$

and to

$$\frac{d\epsilon}{dt} - \frac{P}{\rho^2}\frac{d\rho}{dt} = 0. \tag{1.58}$$

Hint: Derive the Lagrangian bulk energy equation and subtract it from Equation (1.56).

Let V be the *specific volume*, i.e., the volume occupied by 1 gram of the medium. We have

$$\frac{dV}{dt} = \frac{d}{dt}\left(\frac{1}{\rho}\right) = -\frac{1}{\rho^2}\frac{d\rho}{dt}, \tag{1.59}$$

and, substituting Equation (1.59) into Equation (1.58), we obtain

$$\frac{d\epsilon}{dt} + P\frac{dV}{dt} = 0, \tag{1.60}$$

which is just the first law of thermodynamics applied to the case when no heating or cooling processes (such as heat conduction) operate, i.e., when the medium evolves adiabatically. The equation for an adiabat

$$P = K\rho^{\frac{5}{3}}. \tag{1.61}$$

where K is the *adiabatic constant*, may be derived from Equation (1.60) by substituting $\epsilon = \frac{3}{2}\frac{P}{\rho}$ (see Equation 1.14).

In analogy to the Eulerian case (Equations 1.42), internal energy density may be used instead of total energy density. The full set of Lagrangian equations takes the form

$$\begin{aligned} \frac{d\rho}{dt} + \rho\frac{\partial v_j}{\partial x_j} &= 0 \\ \rho\frac{dv_i}{dt} &= -\frac{\partial P}{\partial x_i} + \rho F_i \quad (i = 1, 2, 3) \\ \frac{de}{dt} + (e + P)\frac{\partial v_j}{\partial x_j} &= 0, \end{aligned} \tag{1.62}$$

where summation over the repeated index j is implied.

To conclude this section we may say that the Lagrangian equations are mathematically simpler than their Eulerian counterparts, and they give a more straightforward insight into the physics. They can be easily implemented in one-dimensional problems; however, in two- or three-dimensional problems, their practical application requires specialized numerical techniques (see Chapter 4 and Chapter 6, Section 6.2).

1.5 VISCOSITY AND NAVIER–STOKES EQUATIONS

In this section, the Eulerian equations of momentum and energy will be generalized to cases in which the effects associated with the interchange of particles between adjacent fluid elements cannot be neglected. Whenever these effects are significant, we have to deal with the *internal friction* or *viscosity* of the fluid.

Equation (1.31), derived for a nonviscous fluid, may be thought of as describing the *macroscopic* transport of momentum through space as a result of external forces and simple translation of fluid elements. In a viscous fluid, an additional *microscopic* transfer of momentum occurs, caused by friction. To incorporate it into the momentum equation, one has to introduce a modified momentum flux density tensor $\mathbf{\Pi} \equiv \boldsymbol{\pi} - \boldsymbol{\sigma}$, i.e.,

$$\Pi_{ij} \equiv \rho v_i v_j + P\delta_{ij} - \sigma_{ij}, \tag{1.63}$$

where $\boldsymbol{\sigma}$ is the *viscous stress tensor*.

Friction acts whenever adjacent fluid elements slide past each other. This means that σ_{ij} should be composed of spatial derivatives of velocity components. Assuming that spatial variations of the velocity field are not too strong (which should indeed be the case when the fluid is viscous), we may restrict ourselves to the derivatives of the first order. Additionally, we should require σ_{ij} to vanish when the fluid rotates at a constant angular velocity, i.e., like a rigid body, because such motion does not involve relative displacements of fluid elements. A tensor satisfying these conditions must be symmetric and it may be written as a sum of two terms:

$$\sigma_{ij} = \eta \left(\frac{\partial v_i}{\partial x_j} + \frac{\partial v_j}{\partial x_i} - \frac{2}{3}\frac{\partial v_k}{\partial x_k}\delta_{ij} \right) + \zeta \frac{\partial v_k}{\partial x_k}\delta_{ij}, \tag{1.64}$$

where η and ζ are the *shear* and *bulk* viscosity coefficients, respectively. Both η and ζ are positive and do not depend on velocity (Landau and Lifshitz, 1959, Section 15). Note also that the first, traceless term in Equation (1.64) vanishes when the medium contracts uniformly with $\mathbf{v} = v\mathbf{r}$.

Bulk viscosity is related to the energy transfer between translational and internal motions of fluid particles. If it is present, a frictional force opposing changes of the volume of fluid elements is generated. (Note that this effect cannot be observed in an ideal monoatomic gas due to lack of internal degrees of freedom; for such gas $\zeta = 0$.) Also immune to bulk viscosity are incompressible fluids, whose density does not vary with time and whose elements at all times preserve their initial volumes (the velocity field of such fluids is divergence-free and the bulk viscosity term simply vanishes).

In astrophysics one rarely deals with bulk viscosity. Usually, only the shear viscosity coefficient is different from zero, and the term "shear" is frequently omitted. Henceforth, except in cases where it could lead to confusion, we shall refer to shear viscosity simply as viscosity and to η simply as the *dynamic viscosity coefficient*.

With the "viscous" term included, the momentum equation (1.31) takes the form

$$\frac{\partial}{\partial t}(\rho v_i) + \frac{\partial}{\partial x_j}(\rho v_i v_j) = -\frac{\partial P}{\partial x_i} + \frac{\partial}{\partial x_j}\sigma_{ij} + \rho F_i. \tag{1.65}$$

Starting from Equation (1.65) and performing the same transformations as in Section 1.2, we obtain the following "viscous" counterpart of the bulk energy equation (1.39)

$$\frac{\partial}{\partial t}\left(\frac{1}{2}\rho v^2\right) + \frac{\partial}{\partial x_j}\left(\frac{1}{2}\rho v^2 v_j - \sigma_{ij} v_i\right) = -v_j \frac{\partial P}{\partial x_j} + \rho v_j F_j - \sigma_{ij}\frac{\partial v_i}{\partial x_j}. \tag{1.66}$$

To visualize its meaning, imagine a fluid element that at a moment t and a location $\mathbf{x}$ has a velocity $\mathbf{v}$. The expression

$$\frac{1}{2}\rho v^2 v_j - \sigma_{ij} v_i \tag{1.67}$$

represents the bulk energy flux density. Its first term represents the energy *transported* by the element in the direction $\mathbf{v}$, while its second term is the energy *transferred* by friction to other elements around $\mathbf{x}$. Viscosity and friction also cause the kinetic energy of the bulk motion to dissipate into heat (i.e., to *transform* into internal energy). The effect of dissipation is accounted for by the last term on the right-hand side of Equation (1.66). It gives the rate at which the kinetic energy is lost due to friction and, as such, it always must be negative. We have

$$\begin{aligned}\sigma_{ij}\frac{\partial v_i}{\partial x_j} &= \eta\frac{\partial v_i}{\partial x_j}\left(\frac{\partial v_i}{\partial x_j}+\frac{\partial v_j}{\partial x_i}-\frac{2}{3}\frac{\partial v_k}{\partial x_k}\delta_{ij}\right)+\zeta\frac{\partial v_i}{\partial x_j}\frac{\partial v_k}{\partial x_k}\delta_{ij}\\ &= \eta\frac{\partial v_i}{\partial x_j}\tilde{\sigma}_{ij}+\zeta\left(\frac{\partial v_i}{\partial x_i}\right)^2,\end{aligned} \tag{1.68}$$

where the tensor

$$\tilde{\sigma}_{ij} \equiv \frac{\partial v_i}{\partial x_j}+\frac{\partial v_j}{\partial x_i}-\frac{2}{3}\frac{\partial v_k}{\partial x_k}\delta_{ij} \tag{1.69}$$

is traceless and symmetric. Consequently, we may write

$$\tilde{\sigma}_{ij}\frac{\partial v_i}{\partial x_j} = \frac{1}{2}\tilde{\sigma}_{ij}\left(\frac{\partial v_i}{\partial x_j}+\frac{\partial v_j}{\partial x_i}\right) = \frac{1}{2}\tilde{\sigma}_{ij}\tilde{\sigma}_{ij}, \tag{1.70}$$

where the last step follows from the implied summations and the fact that $\Sigma\,\sigma_{ii} = 0$. We see that both terms on the right-hand side of Equation (1.68) are positive. Note that, depending on the velocity field, either term on the right-hand side of Equation (1.68) may vanish. Hence, for the dissipative term in Equation (1.66) always to be negative, both viscosity coefficients must be positive.

It is clear that the same dissipative term must appear on the right-side of the internal energy equation with a *positive* sign. Hence, the "viscous" counterpart of Equation (1.40) reads

$$\frac{\partial e}{\partial t}+\frac{\partial}{\partial x_j}(e v_j) = -P\frac{\partial v_j}{\partial x_j}+\sigma_{jk}\frac{\partial v_j}{\partial x_k}. \tag{1.71}$$

Finally, adding Equation (1.71) and Equation (1.66), we obtain the "viscous" counterpart of the total energy equation (1.36)

$$\frac{\partial \mathcal{E}}{\partial t}+\frac{\partial}{\partial x_j}[(\mathcal{E}+P)v_j-\sigma_{jk}v_k] = \rho v_j F_j, \tag{1.72}$$

in which the dissipative terms on the right-hand side cancel. We are now ready to write the complete set of Eulerian equations describing the evolution of viscous fluids

$$\begin{aligned}\frac{\partial \rho}{\partial t} + \frac{\partial}{\partial x_j}(\rho v_j) &= 0 \\ \frac{\partial}{\partial t}(\rho v_i) + \frac{\partial}{\partial x_j}(\rho v_i v_j - \sigma_{ij}) &= -\frac{\partial P}{\partial x_i} + \rho F_i, \quad (i = 1, 2, 3) \\ \frac{\partial e}{\partial t} + \frac{\partial}{\partial x_j}(e v_j) &= -P\frac{\partial v_j}{\partial x_j} + \sigma_{jk}\frac{\partial v_j}{\partial x_k}\end{aligned} \tag{1.73}$$

which are known as the *Navier–Stokes equations*. In these equations, the index i refers to the velocity component and summation is implied over both indices j and k. Comparing the Navier–Stokes equations to their Eulerian counterparts (Equation 1.42), we see that the presence of friction only modifies the expressions for momentum and energy fluxes, while the general structure of all equations (time-derivative + divergence of flux = sources and sinks) remains unchanged. For the purpose of numerical calculations, the total energy equation (1.72) can be used in place of the third line of Equation (1.73).

The Navier–Stokes momentum equation is often seen in the form appropriate for an incompressible fluid (e.g., Landau and Lifshitz, 1959); however, this form is usually not suitable for astrophysical problems. In this case, $\frac{\partial v_j}{\partial x_j} = \text{div}\,\mathbf{v} = 0$ and the second of Equations (1.73) can be written in the following (vector) form

$$\rho\left[\frac{\partial \mathbf{v}}{\partial t} + (\mathbf{v}\cdot\text{grad})\mathbf{v}\right] = -\mathbf{grad}\ P + \rho\mathbf{F} + \eta\nabla^2\mathbf{v} \tag{1.74}$$

and the stress tensor is simply

$$\sigma_{ij} = \eta\left(\frac{\partial v_i}{\partial x_j} + \frac{\partial v_j}{\partial x_i}\right). \tag{1.75}$$

To highlight the effect of viscosity on the dynamics of the fluid, assume that the terms $\frac{\partial}{\partial t}(\rho v_i)$ and $\frac{\partial}{\partial x_j}\sigma_{ij}$ dominate the left-hand and the right-hand side of Equation (1.65), respectively. For simplicity, consider an incompressible fluid. We have

$$\frac{\partial}{\partial t}v_i \approx \frac{1}{\rho}\frac{\partial}{\partial x_j}\sigma_{ij}, \quad \text{or} \quad \frac{|\mathbf{v}|}{t_\nu} \sim \nu\frac{1}{l_{ch}}\frac{|\mathbf{v}|}{l_{ch}},$$

where l_{ch} and t_ν are characteristic length and time scales and

$$\nu \equiv \frac{\eta}{\rho} \tag{1.76}$$

is the *kinematic viscosity coefficient*. We see that viscosity causes the velocity field to evolve on a time scale $t_\nu = l_{ch}^2/\nu$. It is clear that viscous coupling between fluid elements causes faster elements to decelerate and slower ones to accelerate. As a result, the velocity field becomes more nearly uniform. This type of evolution is often called *momentum diffusion*.

We shall now perform a few simple estimates concerning the magnitude of viscous effects. As indicated above, a natural source of viscosity is random (peculiar) motions of fluid particles with respect to the bulk motion of fluid elements. Such motions produce the so-called *molecular viscosity* with a kinematic viscosity coefficient

$$\nu_{\rm mol} = \lambda \tilde{u}_{ch}, \tag{1.77}$$

where λ is the mean free path of fluid particles and $\tilde{u}_{ch}$ is a characteristic velocity of peculiar motions. Thus,

$$t_\nu = \frac{l_{ch}}{\lambda} \frac{l_{ch}}{\tilde{u}_{ch}}.$$

To estimate t_ν, $\tilde{u}_{ch}$ may be set to the velocity of sound given by Equation (2.65) in Chapter 2. Alternatively, either mean or mean-square particle velocity may be used (subtle differences between the three velocities are unimportant for this purpose). For a typical HI cloud with $l_{ch} \sim 10^{19}$ cm, $\lambda \sim 10^{14}$ cm, and $\tilde{u}_{ch} \sim 10^4$ cm s^{-1}, we obtain $t_\nu \sim 3 \times 10^{12}$ yr, which is more than two orders of magnitude longer than the age of the universe. Evidently, molecular viscosity is entirely unimportant in this case.

As a further example, consider a dense protostellar disk ($n \sim 10^{14}$ cm^{-3}) with $l_{ch} \sim 10^{14}$ cm, $\lambda \sim 10^{10}$ cm, and $\tilde{u}_{ch} \sim 10^5$ cm s^{-1}. In this case, t_ν is even longer ($\sim 3 \times 10^{14}$ yr). However, ample observational evidence indicates that disks evolve on a time scale of $\sim 10^7$ yr, and it is commonly adopted that the driving agent of their evolution is viscosity. A viscosity large enough to account for the observed time scales may be associated with turbulent flows within the disk. Turbulence may be roughly approximated by random motions of eddies with characteristic size l_t and characteristic velocity v_t. Such eddies generate a *turbulent* or *eddy* viscosity with a kinematic viscosity coefficient

$$\nu_{\rm t} = l_t v_t, \tag{1.78}$$

which may be many orders of magnitude larger than $\nu_{\rm mol}$. In the case of a protoplanetary disk (or, more generally, an accretion disk), a natural limit for l_t is set by its thickness, h (also called "height"). It is also reasonable to demand that v_t be smaller than the velocity of sound, c_s, because at $v_t > c_s$ shock fronts would be generated (see Section 6.1), resulting in a very high dissipation rate of the energy of turbulent motions. As a result, turbulence would be very strongly damped until v_t became smaller than c_s. Thus, we may write

$$\nu_{\rm t} = \alpha c_s h, \tag{1.79}$$

where $\alpha < 1$ is a free parameter, representing the ratio of mean turbulent speed to sound speed. This prescription was introduced in the early 1970s, and it is known in the astrophysical literature as the "α-*viscosity*" (Shakura and Sunyaev, 1973). A likely source of disk turbulence and associated viscosity has been pointed out by Balbus and Hawley (1991) and Hawley and Balbus (1991). It is the *magnetorotational instability* (or MRI; also frequently referred to as *Balbus–Hawley instability*), and it is further discussed in Chapter 8, Section 8.5.4. However, rigorous treatment of MRI is extremely demanding from the computational point of view, and the α-viscosity is still widely used for accretion disk modeling.

1.6 RADIATION TRANSFER

As we mentioned at the end of Section 1.2, the internal energy may be transported independently of the flow of the medium. One such process is transfer by photons, which are generated by atomic processes, such as transitions by free electrons or bound electrons from a higher to a lower energy state. In the interiors of stars, the matter is dense enough so that a photon is absorbed, again by interactions with atoms that result in a transition of an electron from a lower state to a higher state after traveling only a very short distance. The photons can also be scattered, again through interaction with matter, resulting in a change in direction of the photon with or without a change in frequency. Under conditions of very close coupling between radiation and matter, an equilibrium is reached, in which the "kinetic" temperature of the matter arising from its random (peculiar) motions is about the same as the temperature of the radiation, as measured by the distribution of radiation intensity as a function of frequency. However, in interstellar space a photon radiated, say, from the surface of a star may travel an enormous distance before being absorbed or scattered. Under these conditions, the "radiation" temperature can be quite different from the "kinetic" temperature of the medium through which it passes. In this section, we consider basic definitions of radiation quantities, derive the fundamental equation of radiation transport from the Boltzmann equation, and then consider some special cases and limiting forms of that equation. Radiative transfer calculations are a key element in converting the results of hydrodynamic calculations into astronomically observable quantities.

If dE_ν is the energy of photons that pass through an area dA (with normal $\hat{\mathbf{n}}$) at the point $\mathbf{x}$ from a direction $\hat{\mathbf{s}}$, within the element of solid angle $d\Omega$, in the frequency interval $(\nu, \nu + d\nu)$ in time dt, then the *radiation intensity*, I_ν, is defined from the following relation

$$dE_\nu = I_\nu(\mathbf{x}, \hat{\mathbf{s}}, \nu, t)\, \hat{\mathbf{n}} \cdot \hat{\mathbf{s}}\, dA\, d\Omega\, d\nu\, dt. \tag{1.80}$$

Here, $\hat{\mathbf{n}} \cdot \hat{\mathbf{s}}\, dA$ can also be written as $dA \cos\theta$, where θ is the angle between the direction of propagation and the normal to the surface dA. Some of the quantities that appear in Equation (1.80) are illustrated in Figure 1.4. In general, I_ν is a function of seven variables: three spatial, two directional, frequency, and time. Numerically, one can see that in order to store the intensity as a function of all these variables in a computer with reasonable numerical resolution (about 100 grid points for each variable), one needs to store 10^{12} numbers per time step. Obviously this number is too large to be handled by present-day computers, as is the computation time required to solve for all of these variables. Simplifications have to be made. For the time-independent case in 1-D (slab or spherical symmetry), 2-D (axial symmetry), or 3-D (no symmetry), the radiation intensity is a function of three, five, or six variables, respectively

$$I_\nu = I_\nu^{1d}\,(r, \theta', \nu) \quad I_\nu = I_\nu^{2d}\,(r, \theta, \theta', \phi', \nu) \quad I_\nu = I_\nu^{3d}\,(r, \theta, \phi, \theta', \phi', \nu), \tag{1.81}$$

where the unprimed variables denote positions in physical space in spherical coordinates, and the primed variables denote the directions for the radiation.

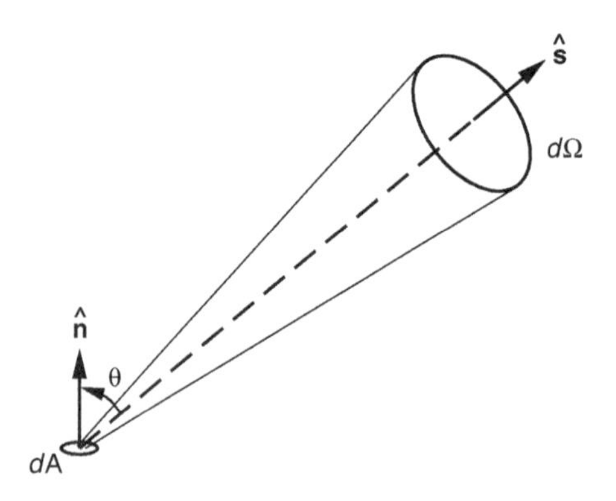

FIGURE 1.4 Illustration of the definition of specific intensity. Radiation is propagating in the direction $\hat{\mathbf{s}}$, inclined at an angle θ from the normal ($\hat{\mathbf{n}}$) to an element of area dA, into a very narrow cone, corresponding to solid angle $d\Omega$, about the direction $\hat{\mathbf{s}}$.

The photon distribution function, f_R, can be defined in exact analogy to the particle distribution function f introduced in Section 1.1. The quantity $f_R(\mathbf{x}, \mathbf{q}, t)\,\mathbf{dx}\,\mathbf{dq}$ is the number of photons at time t at positions between $\mathbf{x}$ and $\mathbf{x} + \mathbf{dx}$ with momenta in the range $\mathbf{q}$ to $\mathbf{q} + \mathbf{dq}$. Photons propagate with a velocity c, and each has energy $h\nu$ and a vector momentum $(h\nu/c)\,\hat{\mathbf{s}}$, where $\hat{\mathbf{s}}$ is the direction of propagation. We can now relate f_R to the specific intensity (Shu, 1992). We note that the total radiation energy in the phase space volume element is:

$$dE_\nu = h\nu f_R(\mathbf{x}, \mathbf{q}, t)\mathbf{dx}\,\mathbf{dq}. \tag{1.82}$$

We consider a beam of photons in the direction $\hat{\mathbf{s}}$ passing through an element of area dA with normal $\hat{\mathbf{n}}$, into a small element of solid angle $d\Omega$ about the direction $\hat{\mathbf{s}}$. Over a small time element dt the element of physical volume swept out by the beam is $\mathbf{dx} = cdt(\hat{\mathbf{n}} \cdot \hat{\mathbf{s}})dA$. If the magnitude of the momentum is q, and we consider the momentum range q to $q + dq$, then the corresponding volume element in momentum space is $\mathbf{dq} = q^2 dq d\Omega$. (If we considered all directions, this volume element would be $4\pi q^2 dq$.) Thus, $\mathbf{dq} = (h\nu/c)^2(h/c)d\nu d\Omega$. Substituting into Equation (1.82), we obtain the energy in the phase space volume element in the beam

$$dE_\nu = (h\nu/c)^2(h^2\nu/c)d\nu d\Omega \cdot c \cdot dt(\hat{\mathbf{n}} \cdot \hat{\mathbf{s}})\,dA\;f_R. \tag{1.83}$$

Comparing this expression with the definition of the specific intensity (Equation 1.80), we see that

$$I_\nu = (h\nu/c)^2(h^2\nu)f_R = \frac{h^4\nu^3}{c^2}f_R = C_1 f_R. \tag{1.84}$$

Thus, a Boltzmann equation can be written for photons at a given frequency ν

$$\frac{1}{C_1}\left[\frac{\partial I_\nu}{\partial t} + c(\hat{\mathbf{s}} \cdot \nabla)I_\nu\right] = \frac{1}{C_1}\left[\frac{\partial I_\nu}{\partial t}\right]_{\text{coll}} \tag{1.85}$$

since, in the absence of general relativistic effects, $\mathbf{F} = 0$ for photons. The collision term here is provided by interactions of photons with matter through absorption, emission, and scattering, as described in the next section.

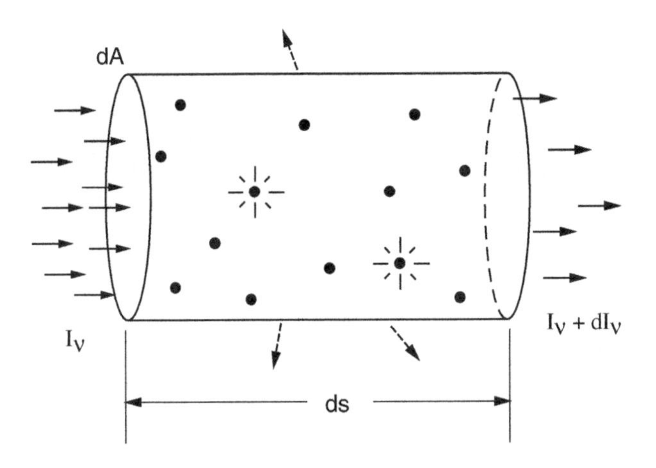

FIGURE 1.5 A beam of light with intensity I_ν is incident on a cylinder with length ds and cross section dA. The atoms or molecules in the cylinder absorb some of the photons (*flashlight symbols*) and scatter others (*dashed arrows*), resulting (assuming no emission) in a reduced intensity $I_\nu + dI_\nu$, where dI_ν is negative, emerging from the cylinder.

1.6.1 Absorption, Emission, and Scattering

The *mass* absorption coefficient κ_ν is defined so that if ΔE_ν^{abs} is the energy absorbed in time dt and in frequency interval ν to $\nu + d\nu$ by a volume element with mass density ρ, of cross-sectional area dA, and length ds, from a beam with intensity I_ν, incident normal to dA and contained within an element of solid angle $d\Omega$ (Figure 1.5), then

$$\Delta E_\nu^{abs} = \kappa_\nu \rho I_\nu \, d\Omega \, d\nu \, dA \, dt \, ds, \tag{1.86}$$

where the units of $\kappa_\nu \rho$ are 1/[length]. Similarly, the scattering coefficient σ_ν is defined by the energy scattered out of the same beam in the same volume

$$\Delta E_\nu^{scatt} = \sigma_\nu \rho I_\nu \, d\Omega \, d\nu \, dA \, dt \, ds. \tag{1.87}$$

The extinction ϵ_ν is then defined simply as the sum of the absorption and scattering

$$\epsilon_\nu = \kappa_\nu + \sigma_\nu. \tag{1.88}$$

The volume element also emits energy ΔE_ν^{em} into the direction of the beam within an element of solid angle $d\Omega$. The mass emission coefficient j_ν is defined by

$$\Delta E_\nu^{em} = j_\nu \rho d\Omega \, d\nu \, dA \, dt \, ds. \tag{1.89}$$

The emission coefficient is assumed here to include both the thermal emission from the volume element and the contribution from the energy scattered into the beam from all other directions.

The equation of transfer can then easily be derived by considering the energy added to and subtracted from a beam, or ray, as it travels through a volume element in direction $\hat{\mathbf{s}}$. The collision term in the Boltzmann equation for photons is the rate of change in intensity caused by absorption and emission. From the definitions of the absorption coefficient (Equation 1.86), the emission coefficient (Equation 1.89), and the specific intensity (Equation 1.80), we see that the change in intensity caused by absorption and scattering is $dI_\nu = -\epsilon_\nu \rho I_\nu ds$ and the change caused by emission is $dI_\nu = +j_\nu \rho ds$. Taking the time derivative, we see that the Boltzmann equation (multiplied by C_1) can be written as an equation for radiation transfer

$$\frac{1}{c}\frac{\partial I_\nu}{\partial t} + (\hat{\mathbf{s}} \cdot \nabla) I_\nu = -\epsilon_\nu \rho I_\nu + j_\nu \rho. \tag{1.90}$$

This fundamental equation, in its simplest form, can be seen to define the specific intensity as a function of time, if expressions for ϵ_ν and j_ν are given. In its more complicated form, the equation must be solved self-consistently, taking into account the effect of a change in I_ν on the absorption, scattering, and emission properties of the material (Shu, 1992). In most of the applications discussed in this book, however, it will be assumed that κ_ν and j_ν can be explicitly calculated as a function of temperature, density, chemical composition, and frequency, and that they do not depend on I_ν, and that σ_ν is negligible. In a few cases, the effects of σ_ν, which do depend on the radiation field, will be included.

We now mention a few specific cases of this equation, which are often used in practice. If we consider the time-independent case and define the source function $S_\nu = j_\nu/\epsilon_\nu$, then the equation of transfer along the direction of the beam $\hat{\mathbf{s}}$ takes the form

$$\frac{\partial I_\nu}{\partial s} = -\epsilon_\nu \rho (I_\nu - S_\nu). \tag{1.91}$$

In the case of a plane-parallel atmosphere with θ being the angle between the vertical (downward) direction (z) and the direction of propagation, so that $dz = -ds \cos\theta$, Equation (1.91) simplifies to

$$\cos\theta \frac{dI_\nu}{d\tau_\nu} = I_\nu - S_\nu, \tag{1.92}$$

where the optical depth τ_ν is defined by

$$d\tau_\nu = \epsilon_\nu \rho dz. \tag{1.93}$$

Going back to the time-dependent case, if we define $\mu = \cos\theta$, then in spherical symmetry Equation (1.90) reduces to

$$\frac{1}{c}\frac{\partial I_\nu}{\partial t} + \mu \frac{\partial I_\nu}{\partial r} + \frac{1-\mu^2}{r}\frac{\partial I_\nu}{\partial \mu} = -\epsilon_\nu \rho (I_\nu - S_\nu). \tag{1.94}$$

1.6.2 Moments of the Boltzmann Equation for Photons

The moments of the Boltzmann equation define the dynamical equations for the radiation field. We first define the moments, with respect to angle, of the radiation field itself. We define the direction cosines s_i by noting that, in an orthonormal coordinate system, the operator d/ds for an arbitrary direction can be written

$$\frac{d}{ds} = \hat{\mathbf{s}} \cdot \nabla = s_1 \frac{\partial}{\partial x_1} + s_2 \frac{\partial}{\partial x_2} + s_3 \frac{\partial}{\partial x_3}, \tag{1.95}$$

where $\hat{s}$ is the unit vector in the direction considered and s_i is the cosine of the angle between $\hat{s}$ and the ith coordinate axis. The *mean intensity* J_ν is defined to be:

$$J_\nu = \frac{1}{4\pi}\int_{4\pi} I_\nu d\Omega, \tag{1.96}$$

where $\int_{4\pi}(X)d\Omega$ means the integration of (X) over all directions, and the element of solid angle $d\Omega = \sin\theta d\theta d\phi$ in spherical coordinates. The first moment, $H_\nu^i = F_\nu^i/(4\pi)$ where F_ν^i is the net flux, is given by

$$H_\nu^i = \frac{1}{4\pi}\int_{4\pi} I_\nu s_i d\Omega. \tag{1.97}$$

The flux is another fundamental quantity in radiation transfer theory. From the unit of specific intensity, which is erg s^{-1} cm^{-2} Hz^{-1} steradian^{-1}, it is clear that the vector flux is the net energy transported by radiation per second per unit area per unit frequency interval over a given surface, taking into account radiation arriving from all directions. For example, one often encounters the radial component of the flux in a spherical coordinate system, in which case

$$F_\nu^r = \int_0^{2\pi}\int_0^{\pi} I_\nu \cos\theta \sin\theta d\theta d\phi, \tag{1.98}$$

where θ is the angle between the radial direction and the direction of the beam. The second moment is:

$$K_\nu^{ij} = \frac{1}{4\pi}\int_{4\pi} I_\nu s_i s_j d\Omega. \tag{1.99}$$

In the tensor for the second moment, $K_\nu^{ij} = K_\nu^{ji}$ and $\sum K_\nu^{ii} = J_\nu$ because $\sum s_i s_i = 1$. Thus, in general, there is one zeroth-moment, three first-moment, and five second-moment components.

We now consider the moment equations. In an orthogonal coordinate system, we integrate Equation (1.90) over all directions and obtain

$$\frac{1}{c}\frac{\partial J_\nu}{\partial t} + \nabla\cdot\mathbf{H}_\nu + \frac{1}{4\pi}\int_{4\pi}\epsilon_\nu\rho(I_\nu - S_\nu)d\Omega = 0. \tag{1.100}$$

If we multiply Equation (1.90) by s_i and integrate, we obtain the first-order moment equation with respect to the ith coordinate axis

$$\frac{1}{c}\frac{\partial H_\nu^i}{\partial t} + \sum_j \frac{\partial K_\nu^{ij}}{\partial x_j} + \frac{1}{4\pi}\int_{4\pi} s_i\epsilon_\nu\rho(I_\nu - S_\nu)d\Omega = 0. \tag{1.101}$$

If we consider only isotropic scattering and absorption and emission of thermal continuum radiation, we can write

$$S_\nu = \frac{\kappa_\nu B_\nu + \sigma_\nu J_\nu}{\kappa_\nu + \sigma_\nu}, \tag{1.102}$$

where $B_\nu(T)$ is the (isotropic) Planck function

$$B_\nu(T) = \frac{2h\nu^3}{c^2}\frac{1}{\exp(h\nu/kT) - 1}. \tag{1.103}$$

(See Gray 1992, Chapter 5.) Then the moment equations simplify to

$$\frac{1}{c}\frac{\partial J_\nu}{\partial t} + \nabla \cdot \mathbf{H}_\nu + \kappa_\nu \rho (J_\nu - B_\nu) = 0. \tag{1.104}$$

$$\frac{1}{c}\frac{\partial H_\nu^i}{\partial t} + \sum_j \frac{\partial K_\nu^{ij}}{\partial x_j} + \epsilon_\nu \rho H_\nu^i = 0. \tag{1.105}$$

Equation (1.104) is the radiation energy equation, as can be seen more clearly by use of the radiation energy density u_ν (radiation energy per unit volume), which is given by Mihalas (1978, Chapter 1, Section 1.2)

$$u_\nu = \frac{1}{c}\int_{4\pi} I_\nu d\Omega = \frac{4\pi}{c} J_\nu. \tag{1.106}$$

The energy equation then becomes

$$\frac{\partial u_\nu}{\partial t} + \nabla \cdot \mathbf{F}_\nu - 4\pi\kappa_\nu \rho B_\nu + c\kappa_\nu \rho u_\nu = 0. \tag{1.107}$$

In this form the radiation energy equation can be combined with the hydrodynamic energy equation to form part of the set of *radiation hydrodynamic* equations, which are discussed in Chapter 9. The radiation momentum equation (1.105) is another member of that set.

Equation (1.104) and Equation (1.105), which combined provide a total of four equations, are insufficient to solve for the nine unknowns: J_ν, the three components of $\mathbf{H}_\nu$, and the five independent components of K_ν^{ij}. Five additional equations, independent of the moment equations, as well as some boundary conditions, are necessary to have a well-posed problem. The five additional equations are called the closure relations and are given by

$$K_\nu^{ij} = f_\nu^{ij} J_\nu, \tag{1.108}$$

where $f_\nu^{ij} = f_\nu^{ji}$ and $\sum_i f_\nu^{ii} = 1$. This equation may be regarded as a definition of the generalized Eddington factors f_ν^{ij}. Unfortunately, these equations do not give any new information, because one still needs a procedure for calculating the f_ν^{ij} independently from the moment equations. Many approximate methods for solving radiative transfer problems have schemes for estimating f_ν^{ij}, which are valid for a limited number of problems. The so-called Eddington approximation

$$f_\nu^{ij} = 0 \quad \text{for [r m] } i \neq j, \quad \text{and} \quad f_\nu^{ij} = \frac{1}{3} \quad \text{for [r m] } i = j \tag{1.109}$$

is valid for the case of optically thick radiation transfer in the interior of a star, for example, but is not valid for the case of an extended low-density envelope. In the case of spherical symmetry, the moment equations then become

$$\frac{1}{c}\frac{\partial J_\nu}{\partial t} + \frac{1}{r^2}\frac{\partial}{\partial r}(r^2 H_\nu) + \kappa_\nu \rho (J_\nu - B_\nu) = 0. \tag{1.110}$$

$$\frac{1}{c}\frac{\partial H_\nu}{\partial t} + \frac{\partial}{\partial r}(f_\nu J_\nu) + \frac{3f_\nu - 1}{r} J_\nu + \epsilon_\nu \rho H_\nu = 0. \tag{1.111}$$

1.6.3 Optically Thick and Optically Thin Limits

In the interior of a star, conditions occur that are very close to thermodynamic equilibrium, usually referred to as *local thermodynamic equilibrium*. The absorption coefficient and density are high enough so that the mean free path of a photon (≈ 1 to 10 mm) is so small compared with the radius of the star (typically 10^9 m) that photons undergo enormous numbers of absorptions and re-emissions on their way from the interior to the surface. Since the temperature at which a photon is absorbed is practically the same as that at which it is emitted, the conditions appropriate to thermodynamic equilibrium hold

$$I_\nu \approx B_\nu(T) \quad \text{and} \quad j_\nu = \kappa_\nu B_\nu(T), \tag{1.112}$$

where $B_\nu(T)$ is the Planck function. The Planck function itself is isotropic; however, in the star there will be a slight degree of anisotropy in the radiation intensity because of the existence of a temperature gradient. In calculating the flux (Equation 1.97), one must take this small deviation into account. The implicit assumption made here is that the radiation temperature and kinetic temperature of the matter are practically the same. Under such conditions, as shown by Shu (1992, Chapter 2), the term $\frac{1}{c}\frac{\partial H_\nu^i}{\partial t}$ is smaller than the term $\epsilon_\nu \rho H_\nu$ by about 28 orders of magnitude for the case of the solar interior, mainly as a consequence of the 10-billion-year lifetime of the Sun. Also, as is evident from their definitions, under conditions of almost complete isotropy $K_\nu = \frac{1}{3}J_\nu$, where K_ν is defined by $K_\nu^{ij} = K_\nu I^{ij}$, where I^{ij} is the unit tensor with components (in a Cartesian coordinate system) of $I^{ii} = 1$, and $I^{ij} = 0$ for $i \neq j$. Under conditions of local thermodynamic equilibrium, as defined above, absorption processes must dominate over scattering processes so that the matter is tightly coupled to the radiation field. Thus, $\epsilon_\nu \approx \kappa_\nu$. Equation (1.105) then becomes

$$H_\nu = -\frac{1}{\kappa_\nu \rho}\nabla\cdot\mathbf{K}_\nu = -\frac{1}{3\kappa_\nu \rho}\frac{\partial B_\nu}{\partial T}\nabla T, \tag{1.113}$$

since J_ν is also very close to the Planck function.

We then define $F_\nu \equiv 4\pi H_\nu$ and $F \equiv \int_0^\infty F_\nu d\nu$, and integrate over all frequencies to obtain

$$\mathbf{F} = -\frac{4\pi}{3\kappa_R \rho}\nabla T \int_0^\infty \frac{\partial B_\nu}{\partial T} d\nu, \tag{1.114}$$

where the Rosseland mean opacity κ_R is defined by*

$$\frac{1}{\kappa_R} \equiv \frac{\int_0^\infty \frac{1}{\kappa_\nu}\frac{dB_\nu}{dT} d\nu}{\int_0^\infty \frac{dB_\nu}{dT} d\nu}. \tag{1.115}$$

But,

$$\int_0^\infty \frac{dB_\nu}{dT} d\nu = \frac{d}{dT}\int_0^\infty B_\nu(T) d\nu = \frac{d}{dT}\frac{\sigma_B T^4}{\pi}, \tag{1.116}$$

* More generally, if induced emission and scattering are taken into account, the Rosseland mean is modified. (See Clayton, 1968, Chapter 3 and Equation 5.96.)

where σ_B is the Stefan–Boltzmann constant. Thus, the radiative flux can be expressed as

$$\mathbf{F} = -\frac{c}{3\kappa_R\rho}\nabla(aT^4), \tag{1.117}$$

where a is the radiation density constant $a = 4\sigma_B/c$. This equation is often referred to as the "diffusion approximation," and it is generally used, in this frequency-averaged form, for calculations of stellar evolution (see Chapter 5). With regard to the opacity, note again that (1) Equation (1.115) was derived for pure absorption; if scattering is also present then the formula has to be corrected, and (2) the Rosseland mean is a function only of local density, temperature, and chemical composition as well as atomic constants, so that it can be computed separately, independently of the radiation field of a particular problem.

At the opposite extreme, consider an observer in a very diffuse medium observing the radiation from a central point source. If the medium is optically thin, the photons stream freely in the radial direction from the source without absorptions and scatterings. In the limit of extreme optical thinness and if the observer is at a large distance from the source, the intensity $I_\nu(\cos\theta)$ becomes essentially a delta function, peaked in the radial direction $\cos\theta = \mu = 1$. Then the definitions of the moments J, H, and K from Equation (1.96), Equation (1.97), and Equation (1.99) show that $J_\nu = H_\nu = K_\nu$, so the Eddington factor is 1. Furthermore, since the radiation pressure $P_{rad,\nu} = (4\pi/c)K_\nu$ and the radiation energy density $u_\nu = (4\pi/c)J_\nu$, then in this limit

$$P_{rad,\nu} = u_\nu = \frac{1}{c}F_\nu. \tag{1.118}$$

Also in this limit and for a time-independent source, Equation (1.110) shows that in the limit $\kappa_\nu \to 0$, $\partial/\partial r(r^2 H_\nu) = 0$, so the flux drops off as $1/r^2$. Similarly, Equation (1.111) shows that in the limit $\epsilon_\nu \to 0$, and with $f_\nu \to 1$, that $J_\nu \propto 1/r^2$ and $K_\nu \propto 1/r^2$.

If we define $u = \int_0^\infty u_\nu d\nu$ and integrate over all frequencies, Equation (1.118) becomes

$$u = \frac{1}{c}|\mathbf{F}|. \tag{1.119}$$

1.6.4 Flux-Limited Diffusion

A single, approximate expression for radiation transfer can be written that reduces to Equation (1.117) in the optically thick limit and to Equation (1.119) in the optically thin limit. Equation (1.117) can be written in a somewhat revised form

$$\mathbf{F} = -\frac{c\lambda}{\kappa_R\rho}\nabla u, \tag{1.120}$$

where λ is the so-called *flux limiter* and, in the limit of black-body radiation, $u = \frac{4\pi}{c}\int_0^\infty B_\nu(T)d\nu = aT^4$, where a is the radiation density constant. Note that $1/(\kappa_R\rho)$ is the mean free path for a photon, i.e., the characteristic distance it travels before being absorbed. In the limit that the mean free path is short, the diffusion limit applies, so in order for equation (1.120) to reduce to equation (1.117), λ must equal 1/3. However, in the limit $\kappa_R \rightarrow 0$, Equation (1.117) shows that the flux becomes unphysically large, so λ must be adjusted accordingly. The usual procedure (Levermore and Pomraning, 1981) is to define a dimensionless quantity R, which we can write in the form

$$R = \frac{|\nabla u|}{\kappa_R \rho u}, \tag{1.121}$$

which physically is the ratio of the mean free path of a photon to the scale height of the integrated energy density. Therefore, very small values of R correspond to the optically thick limit. There are various ways to express λ in terms of R, one of which is (Levermore and Pomraning, 1981)

$$\lambda = \frac{2 + R}{6 + 3R + R^2}. \tag{1.122}$$

Note that when $\kappa_R \rho \rightarrow \infty$ (short mean free path), $R \rightarrow 0$ and $\lambda \rightarrow 1/3$ as expected for the diffusion limit. But as $\kappa_R \rho \rightarrow 0$ (long mean free path), $R \rightarrow \infty$ and $\lambda \rightarrow 1/R$. Thus, it can be seen that the flux is limited to the value cu, as required physically. However, flux-limited diffusion is not the correct solution for radiation transfer in optically thin regions; it is simply an interpolation between the physically correct optically thick and optically thin limits. The main uncertainty is in the choice of λ. Different versions correspond to different choices in the assumed distribution of specific intensity with angle. It is not always clear which form of the flux limiter is appropriate for a given problem. Another form is (Minerbo, 1978)

$$\begin{aligned} \lambda(R) &= \frac{2}{3 + (9 + 12R^2)^{0.5}} \quad \text{if } 0 \le R \le 3/2 \\ \lambda(R) &= [1 + R + (1 + 2R)^{0.5}]^{-1} \quad \text{if } 3/2 \le R \le \infty. \end{aligned} \tag{1.123}$$

This expression also goes to the limit $\lambda = 1/3$ when $R \rightarrow 0$, and to the limit $F = cu$ when $R \rightarrow \infty$. In between the limits, however, the values of λ can be quite different, as shown in Figure 1.6. (For further discussion of the flux limiter, see Turner and Stone [2001].)

1.6.5 Energy Equation in the Optically Thick Limit

In material that is very optically thick, such as the interior of a star, an internal energy equation can be written that includes the effects of radiation. The radiative diffusion approximation (Equation 1.117) can be combined with the first law of thermodynamics

$$dQ = d\epsilon + PdV, \tag{1.124}$$

where Q is the heat energy per unit mass, ϵ is the internal energy per unit mass, and V is the specific volume $1/\rho$. This equation is simply the generalization of the Lagrangian adiabatic equation (1.60). In the case of spherical symmetry, we can convert the radiation flux obtained from Equation (1.117) into the luminosity L, the total energy per second crossing a spherical surface at a distance r from the center

$$L = -\frac{16\pi r^2 ac}{3\kappa_R \rho} T^3 \frac{dT}{dr}, \tag{1.125}$$

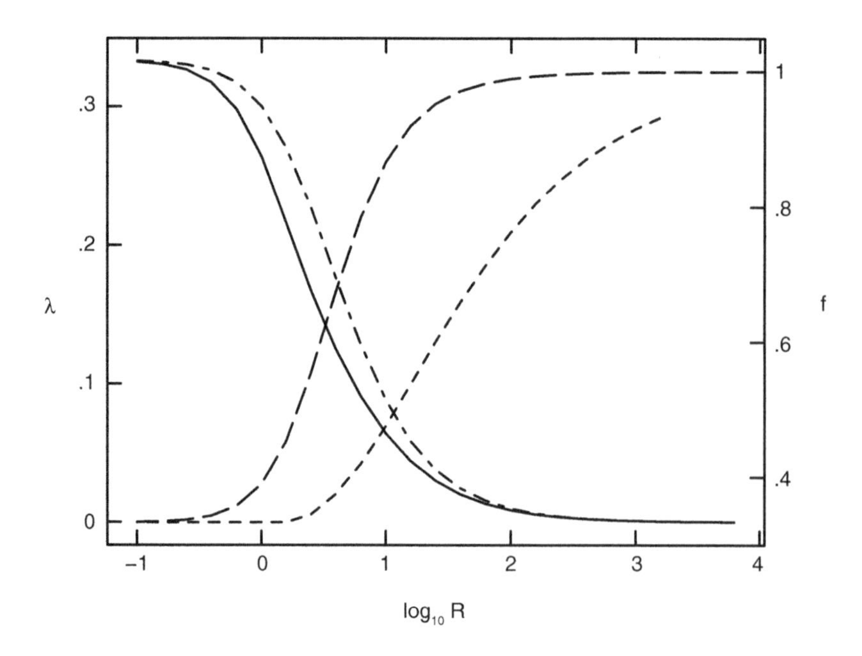

FIGURE 1.6 The flux-limiter λ (left-hand scale) and the Eddington factor f (right-hand scale) are plotted as a function of the optical thickness parameter R. (*Dash-dot curve and long-dashed curve*): Levermore–Pomraning values based on Equation (1.122). (*Solid curve and short-dashed curve*): Minerbo values based on Equation (1.123). Note that in the limit of large optical depth both λ and $f \rightarrow 1/3$. (Adapted from Turner, N.J. and Stone, J.M. (2001) *Astrophys. J. Suppl.* 135:95.)

where a is the radiation density constant defined earlier. The change per second in the heat energy in a layer of thickness Δr and mass $\Delta M = 4\pi\rho r^2 \Delta r$ as a result of radiation transport is $L(r + \Delta r) - L(r)$, so that we can write

$$dQ = dt\left(E_{\rm nuc} - \frac{1}{4\pi r^2 \rho}\frac{dL}{dr}\right), \tag{1.126}$$

where $E_{\rm nuc}$ represents the heating of the layer, in energy per unit mass per second, by other sources, such as nuclear reactions. Let us consider the simple case where the internal energy is only a function of temperature, $\epsilon = c_V T$, where c_V, the specific heat at constant volume, is taken to be a constant. Eliminating dQ, we obtain the Lagrangian equation

$$c_V \frac{dT}{dt} = \frac{4ac}{3\rho}\frac{1}{r^2}\frac{d}{dr}\left(\frac{r^2 T^3}{\kappa_R \rho}\frac{dT}{dr}\right) + E_{\rm nuc} - P\frac{dV}{dt}. \tag{1.127}$$

This equation shows that the change in internal energy (per unit mass) with time is a result of three effects: radiation transport, the compression term PdV, and the nuclear energy term. If the entire equation is multiplied by ρ, so that the left-hand side represents the change in energy per unit volume, then the radiative transport term, here given for spherical symmetry, is seen to be simply the negative of the divergence of the radiative flux **F**. The equivalent Eulerian equation for the internal energy per unit volume is (without the nuclear energy term)

$$\frac{\partial e}{\partial t} + \frac{\partial}{\partial x_j}(e v_j) = -P\frac{\partial v_j}{\partial x_j} - \nabla \cdot \mathbf{F}. \tag{1.128}$$

This equation is the simplified form of the energy equation in radiation hydrodynamics in which the internal energies of gas and radiation are combined. Thus, if e is expressed as $\rho c_V T$, where T is the temperature of both radiation and matter, the c_V must include the effects of radiation.

1.7 CONDUCTING AND MAGNETIZED MEDIA

An ionized medium (hereafter referred to as a *plasma*) is capable of conducting electric currents and, as such, it may interact with electromagnetic fields. When it is permeated by a magnetic field, a complex feedback originates between its motions and the field, severely influencing (or even dominating) its dynamics. In this section, we shall derive and discuss equations describing the evolution of magnetized plasmas in the simplest possible approximation, known as the *magnetohydrodynamic* or MHD limit. For this approximation to be applicable, the following conditions must be satisfied:

1. The medium fulfills all requirements listed in Section 1.3, i.e., it can be treated as a continuous fluid and the Navier–Stokes equations (1.73) can be used to describe it when the field is not present.
2. Positive and negative charges are globally and locally balanced at all times, i.e., all fluid elements are always and everywhere neutral.
3. Electrons are in statistical equilibrium with ions, i.e., they have the same temperature as the ions.
4. Interparticle collisions occur frequently enough for all effects of magnetic forces to be instantaneously transferred from electrons to ions and neutral particles if the latter are present.

For currents to flow, electrons and ions cannot move at exactly the same velocity. However, as we shall demonstrate below, their relative drift velocity is usually minute and it can be safely neglected.

1.7.1 MAXWELL'S EQUATIONS

In the MHD limit, we are allowed to combine the Navier–Stokes equations of hydrodynamics and Maxwell's equations of electrodynamics into a single set governing the evolution of conducting fluids. Unfortunately, when it comes to electrodynamics, we may easily get confused by various systems of units used both in scientific literature and textbooks. The real problem is that in different systems not only are the values of physical quantities different, but the equations themselves look different (Priest, 1987). Officially, the International System of Units (SI) is approved and recommended. However, it is rarely used by astrophysicists, who traditionally and stubbornly stick to Gaussian cgs (centimeter gram-second) units. It seems that the best we can do is to follow this tradition.

In Gaussian units, the basic version of Maxwell's equations takes the form

$$\begin{aligned} \nabla \cdot \mathbf{D} &= 4\pi q \\ \nabla \cdot \mathbf{B} &= 0 \\ \nabla \times \mathbf{E} &= -\frac{1}{c}\frac{\partial \mathbf{B}}{\partial t} \\ \nabla \times \mathbf{H} &= \frac{4\pi}{c}\mathbf{j} + \frac{1}{c}\frac{\partial \mathbf{D}}{\partial t}, \end{aligned} \tag{1.129}$$

where **D**, **E**, **B**, **H**, **j**, are, respectively, vectors of electric displacement, electric field, magnetic induction, magnetic field, and current density; q is charge density and c is the speed of light. Vectors **D** and **H** may be eliminated from Equation (1.129) with the help of constitutive relations

$$\begin{aligned} \mathbf{H} &= \frac{\mathbf{B}}{\mu_m} \\ \mathbf{D} &= \varepsilon_d \mathbf{E}, \end{aligned} \tag{1.130}$$

where μ_m (the *magnetic permeability*) and ε_d (the *dielectric constant*) are dimensionless parameters characterizing the medium in which electric and magnetic fields are generated.

In this approach a distinction is made between "free" and "bound" charge and current densities, where "bound" charge and "bound" current density, which are accounted for by auxiliary fields **D** and **H**, originate from polarization and magnetization of the medium. Such a distinction is generally impossible in a plasma, where all charges are free to move and all currents are free to flow. Consequently, while dealing with the plasma, it is better to use the vacuum version of Maxwell's equations, in which q and **j** include all the charges and currents and no reference is made to auxiliary fields. In Gaussian units, both μ_m and ε_d of the vacuum are equal to 1, so Maxwell's equations reduce to

$$\nabla \cdot \mathbf{E} = 4\pi q \tag{1.131}$$

$$\nabla \cdot \mathbf{B} = 0 \tag{1.132}$$

$$\nabla \times \mathbf{E} = -\frac{1}{c}\frac{\partial \mathbf{B}}{\partial t} \tag{1.133}$$

$$\nabla \times \mathbf{B} = \frac{4\pi}{c}\mathbf{j} + \frac{1}{c}\frac{\partial \mathbf{E}}{\partial t}. \tag{1.134}$$

Let (t_{ch}, l_{ch}) be, as before, the characteristic time and length scales of the problem, and (E_{ch}, B_{ch}) be typical values of fields **E** and **B**. According to Equation (1.133), spatial variations of the electric field, E_{ch}/l_{ch}, are of the same order as typical temporal variations of the magnetic field, B_{ch}/t_{ch} divided by c. Given l_{ch} and t_{ch}, the characteristic speed of the plasma, v_{ch}, is l_{ch}/t_{ch}. We assume that the plasma is nonrelativistic, i.e.,

$v_{ch} \ll c$. Then, the last term in Equation (1.134) has an approximate magnitude

$$\frac{E_{ch}}{ct_{ch}} \approx \frac{B_{ch}l_{ch}}{c^2 t_{ch}^2} \approx \frac{v_{ch}^2}{c^2}|\nabla \times \mathbf{B}| \ll |\nabla \times \mathbf{B}|, \tag{1.135}$$

and we see that it can be neglected, reducing Equation (1.134) to

$$\nabla \times \mathbf{B} = \frac{4\pi}{c}\mathbf{j}. \tag{1.136}$$

From the above equation, the drift velocity of electrons with respect to ions may be estimated (Shu, 1992). Consider a globally neutral hydrogen plasma in which the bulk velocity of electrons, $\mathbf{v}_e$, is slightly different from the bulk velocity of protons, $\mathbf{v}_p$. The difference in velocity causes a current $\mathbf{j}$ such that

$$\mathbf{j} = q_e(n_e\mathbf{v}_e - n_p\mathbf{v}_p) = q_e n_e(\mathbf{v}_e - \mathbf{v}_p) = q_e n_e \mathbf{v}_{\text{drift}}, \tag{1.137}$$

where n_e is the number density of free electrons (i.e., the number of free electrons per cm^{-3}), n_p is the number density of free protons, and $\mathbf{v}_{\text{drift}} \equiv (\mathbf{v}_e - \mathbf{v}_p)$ is the drift velocity.

As an example (Shu, 1992, Chapter 21), take the solar magnetic field, believed to originate at the bottom of the convection zone, where $n_e \sim 10^{23}$ cm^{-3}, $l_{ch} \sim 10^{10}$ cm, and the characteristic value of B_{ch} is $\sim 10^3$ Gauss. Approximating as before $|\nabla \times \mathbf{B}| \approx B_{ch}/l_{ch}$, then from Equation (1.136) we obtain $v_{\text{drift}} \approx 10^{-12}$ cm s^{-1} — a value which is indeed negligible. We may safely forget differences between various charge carriers, referring to solar plasma simply as conducting fluid. Even under interstellar conditions, for example l_{ch} = 1 pc, $B_{ch} = 10^{-5}$ Gauss, and $n_e = 10^{-3}$ cm^{-3}, the drift velocity ($\sim 10^{-2}$ cm s^{-1}) is much less than typical thermal or turbulent velocities ($\sim$ 1 km s^{-1}).

Exercise

Prove that

$$\frac{\partial}{\partial t}(\nabla \cdot \mathbf{B}) = 0. \tag{1.138}$$

Equation (1.138) illustrates a very important property of magnetic fields. We see that if the field is divergence-free (i.e., sourceless) at $t = 0$, it will always preserve that state; no magnetic sources or sinks can be created as a result of electromagnetic interactions. On the other hand, we know that $\mathbf{B}$ must be divergence-free initially because no "natural monopoles" (i.e., particles carrying "magnetic charges") exist except in yet unconfirmed theories unifying all elementary interactions. Thus, we arrive at the basic requirement that all numerical methods employed to solve MHD problems must satisfy — they may not generate $\nabla \cdot \mathbf{B} \neq 0$.

Exercise

Based on Equation (1.131) and Equation (1.134), prove that

$$\frac{\partial q}{\partial t} + (\nabla \cdot \mathbf{j}) = 0. \tag{1.139}$$

Just as the continuity equation (1.24) expresses the law of mass conservation, Equation (1.139) expresses conservation of electric charge. The local charge density changes due to currents. According to Equation (1.136), the currents create magnetic fields and, according to Equation (1.133), variable magnetic fields influence electric fields. The latter in turn modify the flow of charge, thus, closing the electromagnetic feedback loop.

1.7.2 Equations of Magnetohydrodynamics

According to Ohm's law, the current in the plasma at rest is proportional to the electric field

$$\mathbf{j} = \sigma_e \mathbf{E}, \tag{1.140}$$

where $\sigma_e = n_e e^2/(m_e \nu_c)$ is the *electric conductivity* (m_e is the electron mass and ν_c is the collision frequency between ions and electrons). In the plasma moving at a nonrelativistic velocity, an additional component of the $\mathbf{E}$-field is generated and Ohm's law takes the form

$$\mathbf{j} = \sigma_e \left(\mathbf{E} + \frac{\mathbf{v}}{c} \times \mathbf{B} \right). \tag{1.141}$$

A current flowing through a medium with nonzero σ_e converts electromagnetic energy into internal energy of the medium at a rate

$$\frac{\partial e}{\partial t} = \frac{1}{\sigma_e} |\mathbf{j}|^2, \tag{1.142}$$

which, for finite σ_e, has to be included as a source term in the Eulerian energy equation (1.36) or Equation (1.40). The process of current decay due to finite conductivity is commonly referred to as *Joule* or *Ohmic dissipation.*

We shall now derive the fundamental equation of magnetohydrodynamics, known as the *induction equation*. As a first step, eliminate $\mathbf{j}$ from Equation (1.141) with the help of Equation (1.136). The result is:

$$\mathbf{E} = \frac{c}{4\pi\sigma_e} \nabla \times \mathbf{B} - \frac{\mathbf{v}}{c} \times \mathbf{B}. \tag{1.143}$$

The induction equation may now be obtained by substituting Equation (1.143) into Equation (1.133). We get

$$\frac{\partial \mathbf{B}}{\partial t} = \nabla \times (\mathbf{v} \times \mathbf{B}) - \nabla \times (\eta_e \nabla \times \mathbf{B}), \tag{1.144}$$

where

$$\eta_e \equiv \frac{c^2}{4\pi\sigma_e} \tag{1.145}$$

is the *electric resistivity*. In the plasma at rest ($\mathbf{v} = 0$), the induction equation reduces to

$$\frac{\partial \mathbf{B}}{\partial t} = -\nabla \times (\eta_e \nabla \times \mathbf{B}), \quad \text{or} \quad \frac{B}{t_{ch}} \sim \left(-\eta_e \frac{B}{l_{ch}^2}\right),$$

where t_{ch} and l_{ch} are, as before, characteristic time and length scales. We see that finite resistivity causes B to decay on a characteristic time scale $t_{ch} \sim l_{ch}^2/\eta_e$. (For that reason η_e is also referred to as the *magnetic diffusivity*.)

The induction equation describes the reaction of the magnetic field to the motions of the plasma. On the other hand, the field causes the plasma to move in response to the Lorentz force

$$\begin{aligned} \mathbf{F}_L &= \frac{1}{c}\mathbf{j} \times \mathbf{B} = \frac{1}{4\pi}(\nabla \times \mathbf{B}) \times \mathbf{B} \\ &= \frac{1}{4\pi}(\mathbf{B} \cdot \nabla)\mathbf{B} - \frac{1}{8\pi}\nabla(B^2), \end{aligned} \tag{1.146}$$

where B is the length of the vector $\mathbf{B}$ (hereafter referred to as "magnetic field intensity" or just "magnetic field") and $\mathbf{F}_L$ is the force per unit volume. The first component of the Lorentz force is related to *magnetic tension* caused by the curvature of magnetic field lines. (Note that it vanishes when $\mathbf{B}$ is constant on straight lines.) The second term is the gradient of the isotropic *magnetic pressure* defined by

$$P_m \equiv \frac{1}{8\pi}B^2. \tag{1.147}$$

We are now ready to write the full set of Eulerian MHD equations without viscous effects. To save space, we shall use the vector notation

$$\begin{aligned} \frac{\partial \rho}{\partial t} &= -\nabla \cdot (\rho \mathbf{v}) \\ \rho \frac{\partial \mathbf{v}}{\partial t} &= -\rho(\mathbf{v} \cdot \nabla)\mathbf{v} - \boldsymbol{\nabla} P + \frac{1}{4\pi}(\mathbf{B} \cdot \nabla)\mathbf{B} - \frac{1}{8\pi}\nabla(B^2) + \rho \mathbf{F} \\ \frac{\partial \mathcal{E}}{\partial t} &= -\nabla \cdot [(\mathcal{E} + P)\mathbf{v}] + \rho \mathbf{v} \cdot \mathbf{F} + \frac{4\pi\eta_e}{c^2}|\mathbf{j}|^2 \\ \frac{\partial \mathbf{B}}{\partial t} &= -\nabla \times (\eta_e \nabla \times \mathbf{B}) + \nabla \times (\mathbf{v} \times \mathbf{B}), \end{aligned} \tag{1.148}$$

where $\mathbf{j}$ is given by Equation (1.141), $\mathbf{F}$ is the external force, such as gravity, per unit mass, and $\mathcal{E}$ is the sum of the internal energy density per unit volume of the gas and kinetic energy per unit volume.

1.7.3 Limits of the MHD Approximation

In MHD we assume that the drift velocity of electrons with respect to ions is very low. This is true as long as the electromagnetic forces due to large-scale fields are balanced by frictional forces due to interparticle collisions. However, with decreasing density of the plasma, the frequency of collisions eventually becomes so low that the balance cannot be maintained and the electrons effectively "decouple" from the ions. The critical density n_e^{crit} at which this happens can be estimated by equating the Larmor (cyclotron) frequency

$$\nu_e^L = \frac{q_e B}{2\pi m_e c} = 2.8 \times 10^6 \, B \, [\text{s}^{-1}]. \tag{1.149}$$

with which electrons gyrate in a uniform magnetic field, to the frequency

$$\nu_e^{\text{coll}} \approx 3.7 \ln \Lambda \, T^{-3/2} \, n_e \, [\text{s}^{-1}], \tag{1.150}$$

with which collisions between free electrons occur. In the above formulae, q_e is the electron charge, the unit of B is Gauss, T is measured in Kelvins (K), and $\ln \Lambda$ is a slowly varying function of temperature and electron density, known as the *Coulomb logarithm*. For a pure hydrogen plasma with $10^2 \leq T \leq 10^8$ and $1 \leq n_e \leq 10^{24}$, one obtains $\sim 5 \leq \ln \Lambda \leq\sim 35$, and a value $\ln \Lambda = 20$ may be used for rough estimates (Spitzer, 1962).

In a typical star-forming interstellar cloud, composed mainly of molecular hydrogen, we have $T \sim 10$ K, $B \sim 10^{-5}$ Gauss, and $n_e^{\text{crit}} \approx 10 \text{ cm}^{-3}$. This is rather high, especially if one remembers that the ionization degree is very low. Formally, the cloud approaches the *collisionless regime*, in which more sophisticated physics must be employed to adequately describe the plasma. In practice, however, the simple MHD formalism applied to evidently collisionless cases yields surprisingly good results. Probably the best example of such a situation is the flow of the solar wind past Earth's magnetosphere. Here, the Larmor frequency is several orders of magnitude higher than the collision frequency, and the mean free path of wind particles reaches several astronomical units (AU). Yet, a shock front (see Section 6.1) forms on the dayside of the Earth, which is only a fraction of the Earth's radius wide, and many features of the wind-magnetosphere interaction are accurately reproduced by MHD models.

Why then does MHD work where it should formally fail? Very loosely speaking, highly nonlinear phenomena (which are not accounted for by the MHD model) cause the collisionless plasma to behave *as if it were a continuous medium*. However, this does not mean that MHD can be regarded as a universal tool. On the contrary, the further we venture into parameter space away from the safe area defined by the conditions listed at the beginning of Section 1.7, the more cautious we should be, the more rigorously the assumptions underlying our MHD models should be discussed, and the more critically the results should be evaluated.

1.7.4 Field Freezing

In a medium with vanishing resistivity (or, equivalently, practically infinite conductivity), the Ohm's Equation (1.141) reduces to

$$\mathbf{E} + \frac{\mathbf{v}}{c} \times \mathbf{B} = 0 \tag{1.151}$$

(because the current must stay finite when we increase the conductivity while keeping **E** and **B** unchanged). Assume that Equation (1.151) holds exactly, and define the magnetic flux through a surface S as

$$\Phi_m \equiv \int_S \mathbf{B} \cdot \hat{\mathbf{n}}\, dA. \tag{1.152}$$

(Note that Φ_m may be thought of as a bunch of field lines intersecting S.) Consider a *co-moving* surface, defined at all times by the same particles. The time derivative of magnetic flux through such a surface is:

$$\frac{\mathrm{d}\Phi_m}{\mathrm{d}t} = \int_S \frac{\partial \mathbf{B}}{\partial t} \cdot \hat{\mathbf{n}}\, dA + \int_C \mathbf{B} \cdot (\mathbf{v} \times d\hat{\mathbf{l}}), \tag{1.153}$$

where the curve C is the boundary of S, $d\hat{\mathbf{l}}$ is the unit vector tangent to C, and the symbol $\frac{\mathrm{d}}{\mathrm{d}t}$ is used to accentuate the "Lagrangian character" of S. The first and second terms of the derivative are related to time-variation of the field over S and motion of S across the field, respectively. (Note that $\mathbf{v} \times d\hat{\mathbf{l}}$ is the area swept by $d\hat{\mathbf{l}}$ per unit time.) Elements of the triple vector product may be permuted, changing the second term into

$$-\int_C (\mathbf{v} \times \mathbf{B}) \cdot d\hat{\mathbf{l}}.$$

Using Equation (1.133) and the Stokes theorem, we obtain

$$\begin{aligned}\frac{\mathrm{d}\Phi_m}{\mathrm{d}t} &= -c\int_S (\nabla \times \mathbf{E}) \cdot \hat{\mathbf{n}}\, dA - \int_S \nabla \times (\mathbf{v} \times \mathbf{B}) \cdot \hat{\mathbf{n}}\, dA \\ &= -c\int_S \left[\nabla \times \left(\mathbf{E} + \frac{\mathbf{v}}{c} \times \mathbf{B}\right)\right] \cdot \hat{\mathbf{n}}\, dA = 0.\end{aligned} \tag{1.154}$$

The above result tells us that, in the limit of very high conductivity, the flux is conserved; the field lines originally intersecting S intersect it at all times, i.e., they are inseparable from the particles defining S. This situation is commonly referred to as *field freezing*. In fact, it is often useful to visualize a magnetic field as a collection of highly elastic bands that either drag plasma particles or are dragged by them, depending on which term dominates the right-hand side of the MHD momentum equation. However, one should not think of particles being entirely frozen, or glued, to field lines. While they cannot separate from the lines they occupied initially, their freedom of motion *along the lines* is preserved. An analogy of beads sliding along rubber bands is much more appropriate.

In Gaussian units the resistivity of a fully ionized hydrogen plasma is given by

$$\eta_e = 6.5 \times 10^{12} \frac{\ln \Lambda}{T^{3/2}}\ [\mathrm{cm}^2\mathrm{s}^{-1}] \tag{1.155}$$

(Spitzer, 1962). For a typical photo-ionized HII region with $T \sim 10^4$, $n_e = 10-100$, and $\ln \Lambda \approx 20$, this formula yields $\eta_e \sim 10^8$. If the extent of the ionized region is ~ 1 pc, the field diffuses on a time scale $l_{ch}^2/\eta_e \sim 10^{21}$ years — more than 10 orders of magnitude larger than the age of the universe. The field freezing approximation is certainly very good in this case, and similar estimates show that it may be safely applied in many astrophysical problems. It breaks down, however, when the field becomes very tangled or when oppositely directed fields come in close contact (both situations corresponding to small l_{ch}). On the other hand, magnetic diffusivity may become very large (and the corresponding decay time very short) when the plasma is strongly turbulent. As before, turbulence may be approximated by random motions of eddies with characteristic size l_t and characteristic velocity v_t. Such motions generate an *eddy* or *turbulent* magnetic diffusivity, for which $\eta_e^t \equiv v_t l_t$. (Note that $\eta_e^t = \nu_t$.)

When the plasma is not fully ionized, Equation (1.155) loses its validity. In the limit of very low ionization degree ($n_e \ll n_n$, where n_n is the number density of neutral particles), the expression

$$\eta_e = 3 \times 10^2 \frac{n_n}{n_e} \sqrt{T} \ [\mathrm{cm^2 s^{-1}}] \tag{1.156}$$

should be used instead, valid up to $\sim 10^4$ K (Gammie and Menou, 1998). However, the field freezing approximation may still work well. As an example, consider a protoplanetary disk with height-to-radius ratio of 0.1 and $T \sim 10^3$ K, for simplicity neglecting any effects of turbulence. Taking $l_{ch} = 1.5 \times 10^{12}$ cm (i.e., the height of the disk at Earth's orbit) and $t_{ch} \sim 10^7$ yr (i.e., the standard lifetime of such a disk), we obtain $n_e/n_n \sim 10^{-6}$. For the time scale of field decay to become shorter than the lifetime of the disk, the ionization degree would have to be still lower than that, and only then would the field freezing approximation break down.

1.7.5 Summary

The inclusion of magnetic fields in astrophysical flows introduces numerous complications, which, to treat numerically, require ingenious techniques and careful testing. Many solutions that have been accomplished to date have assumed strictly ideal MHD, in which Equations (1.148) are solved without the terms involving the resistivity η_e. In this case the energy equation can be replaced by Equation (1.40). This approximation can be justified in many astrophysical situations because the relevant length scales are so long and the resistivity small enough so that the time scale for magnetic decay is much longer than the typical flow time scales. Chapter 8 discusses numerical techniques in this approximation. However, it should be noted that in many significant problems, for example, magnetic field generation in the interior of the Earth (Glatzmaier and Roberts, 1996), the inclusion of finite resistivity is required.

REFERENCES

Balbus, S. A. and Hawley, J. F. (1991) *Astrophys. J.* **376**: 214.

Clayton, D. (1968) *Principles of Stellar Evolution and Nucleosynthesis* (New York: McGraw-Hill).

Gammie, C. F. and Menou, K. (1998) *Astrophys. J.* **492**: L75.

Glatzmaier, G. A. and Roberts, P. H. (1996) *Science* **274**: 1887.

Gray, D. F. (1992) *The Observation and Analysis of Stellar Photospheres,* 2nd ed. (Cambridge: Cambridge University Press).
Hawley, J. F. and Balbus, S. A. (1991) *Astrophys. J.* **376**: 223.
Huang, K. (1987) *Statistical Mechanics* (New York: John Wiley & Sons).
Landau, L. D. and Lifshitz, E. M. (1959) *Fluid Mechanics* (London: Pergamon Press).
Levermore, C. D. and Pomraning, G. C. (1981) *Astrophys. J.* **248**: 321.
Lifshitz, E. M. and Pitaevskii, L. P. (1980) *Statistical Physics,* 3rd ed. (Oxford: Pergamon Press).
Mihalas, D. (1978) *Stellar Atmospheres,* 2nd ed. (San Francisco: W H Freeman).
Minerbo, G. N. (1978) *J. Q. S. R. T.* **31**: 149.
Press, W. H., Teukolsky, S. A., Vetterling, W. T., and Flannery, B. P. (1992) *Numerical Recipes in Fortran,* 2nd ed. (Cambridge: Cambridge University Press).
Priest, E. R. (1987) *Solar Magnetohydrodynamics* (Dordrecht: D. Reidel).
Shakura, N. I. and Sunyaev, R. A. (1973) *Astron. Astrophys.* **24**: 337.
Shu, F. H. (1991) *The Physics of Astrophysics,* vol. I (Mill Valley, CA: University Science Books).
Shu, F. H. (1992) *The Physics of Astrophysics,* vol. II (Mill Valley, CA: University Science Books).
Spitzer, L., Jr, (1962) *Physics of Fully Ionized Gases*, 2nd ed. (New York: Interscience).
Turner, N. J. and Stone, J. M. (2001) *Astrophys. J. Suppl.* **135**: 95.

2 Numerical Approximations to Partial Differential Equations

The emphasis in this chapter is on solving physical problems by numerical methods on a fixed Eulerian grid, using finite difference techniques. Examples of the equations to be treated in more detail in later sections, including the wave equation and the diffusion equation, are introduced in Section 2.1. The detailed examples given in this chapter are generally limited to problems in one space dimension and time. The basic procedure for representing the derivative of a function as a finite difference is discussed in Section 2.2, and a method for extending it to higher-order derivatives is presented. The following section goes into more detail on how functions are represented on a grid, on the accuracy of various finite difference approximations, and how time derivatives, as opposed to space derivatives, are represented. An important aspect of numerical computation, comparison with a known analytic solution, is illustrated for two special cases. Section 2.4 and Section 2.5 introduce the concept of numerical stability. Certain numerical procedures lead to spurious numerical oscillations that grow in amplitude and eventually mask the real physical solution. A technique for detailed analysis of stability is discussed, and critical limits on the time step, required for stability, are derived for the hydrodynamic equations and for the diffusion equation. Discussed briefly in Section 2.6 are some further difficulties that must be overcome in numerical computations: diffusion and dispersion. The final section mentions briefly several other numerical techniques, besides finite differences on a grid, that are commonly employed.

2.1 NUMERICAL MODELING WITH FINITE-DIFFERENCE EQUATIONS

Partial differential equations (PDEs) are usually divided into three general categories: elliptic, parabolic, or hyperbolic. However, in practical applications of astrophysical interest this classification is rarely important. For our purposes it is sufficient to know that the three types of PDEs correspond to different problems, for which different solution strategies must be applied, and the reader interested in precise mathematical definitions is referred to standard textbooks (e.g., Wendt, 1992; Ames, 1992, Section 1–3).

In the realm of PDEs, the term *modeling* is essentially equivalent to solving *initial value* and/or *boundary value problems*. In the first case, the evolution of a system described by PDEs is followed in time. In the second case, one or more functions describing the system at a given instant of time are found.

Elliptic equations are associated with boundary value problems in which at every point inside the domain of interest the solution depends on the data provided on the boundary of the domain. A change in the data at just one boundary point changes the solution everywhere inside the domain, but the time does not appear explicitly as the independent variable. An example of an elliptic PDE is the Poisson equation for gravitational potential

$$\frac{\partial^2 u}{\partial x^2} + \frac{\partial^2 u}{\partial y^2} + \frac{\partial^2 u}{\partial z^2} = 4\pi G\rho(x, y, z), \tag{2.1}$$

where G is the gravitational constant, ρ is the mass density of the medium, u is the gravitational potential itself, and the equation is expressed in a Cartesian coordinate system (x, y, z).

In hyperbolic equations, one of the independent variables is the time. To solve such an equation one has to specify the values of the dependent variables at some initial time (e.g., locate an expanding fireball in a uniform medium). Additional conditions may be specified on the boundary of the domain (e.g., free outflow or no flow through the boundary). In more complicated problems, the ambient (surrounding) medium may influence the interior of the domain (e.g., by irradiation, long-range forces, or inflow of matter), but once these conditions are specified, the solution only depends on the initial data, i.e., we deal with an initial value problem. The simplest example of a hyperbolic PDE is the one-dimensional wave equation

$$\frac{\partial^2 u}{\partial t^2} = v^2 \frac{\partial^2 u}{\partial x^2}, \tag{2.2}$$

where u is the density perturbation of the fluid owing to the passage of the wave, and v is the propagation velocity of the wave.

Parabolic PDEs are also related to initial value problems. However, while hyperbolic equations may be solved with just the initial condition provided (as in the fireball example above), parabolic equations *require* additional conditions to be specified on the boundary of the domain at all times. The simplest parabolic PDE is the one-dimensional diffusion equation

$$\frac{\partial u}{\partial t} = \frac{\partial}{\partial x}\left(\nu_d \frac{\partial u}{\partial x}\right), \tag{2.3}$$

where u represents a physical variable (e.g., the temperature T), and $\nu_d > 0$ is the diffusion coefficient. When applied to the problem of radiation transfer in an optically thick medium between x_1 and x_2, it demands that both the initial distribution $T(x)$ be specified for $x_1 \le x \le x_2$ and the boundary values $T(x_1, t)$, $T(x_2, t)$ be provided at every instant of time t.

It must be remembered that in hydrodynamics or magnetohydrodynamics a single flow may be described by equations of mixed type. For example, in one part of the domain, where the fluid velocity is supersonic, the equations are hyperbolic, while in the other part, where the fluid velocity is subsonic, the equations are elliptic. By purely mathematical methods, these problems are very difficult or just impossible to solve. However, provided that viscous effects are negligible, unsteady flows (which we most frequently deal with in astrophysics), are governed by the hyperbolic equations independently of how fast the fluid moves. Such equations can be solved with the help of reliable numerical methods. Similar methods can also be applied to more complicated cases in which complex physical processes operate within the fluid.

To solve a differential equation on a computer one has to *discretize* it, i.e., to transform it into an algebraic *finite difference equation* or, if necessary, into a system of algebraic equations. Such transformation (*discretization*) is the basis of every procedure of numerical modeling. Discretization is performed with the help of discrete grid points, also called "mesh points," suitably chosen in the interior and on the borders of the domain of interest (also called the *computational domain*). All grid points constitute the *grid* or the *mesh*. When the original equation contains time derivatives, a similar grid of discrete values must be set up in time. As a result of discretization, the original derivatives are replaced with corresponding difference quotients.

In general, the grid points may be distributed nonuniformly in space and their positions may vary in time. The total number of space grid points may also vary, depending on how the solution evolves; as long as it is smooth, only a few points may be needed, but when small-scale structures begin to appear, additional points may have to be introduced. The time intervals are also not necessarily uniform, as discussed below.

2.2 DIFFERENCE QUOTIENT

Let $y = f(x)$ be a function of a single variable x. Then the derivative of the function at a given point x is defined by

$$\frac{df}{dx} \equiv \lim_{h\to 0} \frac{f(x+h) - f(x)}{h}. \tag{2.4}$$

The representation of this derivative by a difference quotient simply means that the interval h (often referred to as Δx) is taken to be finite, but small.

In general, let $f(x_1, \ldots, x_N)$ be a function of N independent variables. The partial derivative of f with respect to x_i

$$\frac{\partial f}{\partial x_i} \equiv \lim_{h\to 0} \frac{f(x_1, \ldots, x_i + h, \ldots, x_N) - f(x_1, \ldots, x_i, \ldots, x_N)}{h} \tag{2.5}$$

can be approximated with the right-hand (or *forward*) difference quotient

$$\frac{f(x_1, \ldots, x_i + h, .., x_N) - f(x_1, \ldots, x_i, \ldots, x_N)}{h}, \tag{2.6}$$

where $h > 0$ has a finite value. By analogy, the left-hand (or *backward*) difference quotient is defined as

$$\frac{f(x_1, \ldots, x_i, \ldots, x_N) - f(x_1, \ldots, x_i - h, \ldots, x_N)}{h}. \tag{2.7}$$

For simplicity, in the following the general list of arguments, $x_1, \ldots, x_N$ will be replaced by just one space variable, x, and the time variable, t. One should remember, however, that in hydrodynamical applications one can deal with up to three space variables, and additional independent variables may appear (see, for example, the radiative transfer equation 1.90).

Another approximation of $\frac{\partial f}{\partial x}$ is given by the centered difference quotient

$$\frac{f(x+h,t) - f(x-h,t)}{2h} . \tag{2.8}$$

Note, however, that the limit

$$\lim_{h \to 0} \frac{f(x+h,t) - f(x-h,t)}{2h} \tag{2.9}$$

can exist without f having a derivative at x. As an example, the function $f(x,t) = |x|$ can be taken, for which the limit (Equation 2.9) exists at $x = 0$ (it is equal to zero), but the derivative does not. In general, $f(x,t)$ need not be differentiable or even continuous at all times everywhere in the domain of interest. Many numerical schemes include techniques that enable the handling of discontinuities (see Chapter 6, Section 6.1 and Section 6.3.2), and centered difference quotients along with the right-hand and left-hand ones are used to approximate the derivatives.

Consider for the moment a function of a single variable $f(x)$ and its derivatives denoted by $f'(x) = df/dx$, $f''(x) = d^2f/dx^2, \ldots f^{(n)}(x)$. Approximations for arbitrary derivatives $f^{(n)}(x)$, which are accurate to a desired order, can be obtained by using the Taylor series. Consider for the moment an equidistant grid $\ldots, x-2h, x-h, x, x+h, x+2h, \ldots$ for which the values of a function $f(x+jh)$, where $j = \ldots, -2, -1, 0, 1, 2, \ldots$ are known, and consider the n linearly independent equations:

$$f(x+jh) - f(x) = f'(x)(jh) + f''(x)(jh)^2/2! + \cdots + f^{(n)}(jh)^n/n! + O(h^{n+1}), \tag{2.10}$$

where j takes on the n values $j = j_{\min}, j_{\min}+1, \ldots, j_{\max}$ ($j = 0$ is excluded). The left-hand sides of these equations and the coefficients of the derivatives $f', f'', \ldots, f^{(n)}$ are known, whereas the derivatives themselves, defined at the point x, are unknown. By ignoring for the moment the error $O(h^{n+1})$, the n linear equations can be solved for the n unknown derivatives using standard techniques. By including the error $O(h^{n+1})$ in the derivation, it can be shown that the resulting formula for $f^{(k)}(x)$ is accurate to $O(h^{n-k+1})$ for $k = 1, 2, \ldots, n$.

As a specific example, let us derive a formula for $f'(x)$, accurate to $O(h^3)$ from the known values $f(x-h)$, $f(x)$, $f(x+h)$, and $f(x+2h)$. Here $j_{\min} = -1$, $j_{\max} = 2$, $n = 3$, and $k = 1$. The three linear equations for the three unknowns, f', f'', and f''' are:

$$\begin{aligned} f(x-h) - f(x) &= (-h)f' + (h^2/2)f'' + (-h^3/6)f''' \\ f(x+h) - f(x) &= (h)f' + (h^2/2)f'' + (h^3/6)f''' \\ f(x+2h) - f(x) &= (2h)f' + (2h^2)f'' + (4h^3/3)f''' \end{aligned} \tag{2.11}$$

from which we derive the formula

$$f' = \frac{-2f(x-h) - 3f(x) + 6f(x+h) - f(x+2h)}{6h} + O(h^3). \tag{2.12}$$

(Note that this procedure can be generalized to non-equidistant grids.)

Exercise

Calculate the second derivative to order h^2 and the third derivative to order h from Equations (2.11).

2.3 DISCRETE REPRESENTATION OF VARIABLES, FUNCTIONS, AND DERIVATIVES

Let $u(x, t)$ be defined as a function of a spatial variable x, which varies between x_{min} and x_{max}, and a time variable t, which varies between 0 and t_{max}. At a given instant t, we can discretize u with the help of a space grid $x_j = x_{min} + j\Delta x$, where Δx is a small space interval and the integer j runs from 0 to some maximum value J. Similarly, discretization in time can be performed on a grid of discrete values $t^n = n\Delta t$, where the integer n runs from 0 to a maximum value N and Δt is a small interval of time. Thus, in effect, we set up a two-dimensional grid, with coordinates (x, t), as illustrated in Figure 2.1. The time interval Δt is often called the *time step*, and the space interval between neighboring grid points is commonly referred to as a *grid cell* or *zone*. In our example, Δx and Δt are constant (i.e., both the space grid and the time grid are uniform), but one should remember that in general both space and time intervals can be functions of x and t.

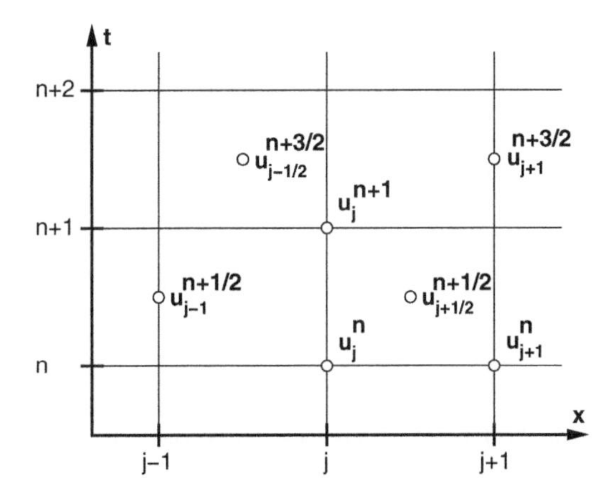

FIGURE 2.1 A simple space–time grid. The values of the discretized function $u(x, t)$ can be defined either at cell corners (e.g., u_j^n), cell edges (e.g., $u_{j-1}^{n+1/2}$), or cell centers (e.g., $u_{j+1/2}^{n+1/2}$).

The function u can be defined at the actual grid points as

$$u(x_j, t^n) = u_j^n \tag{2.13}$$

or halfway between grid points as

$$u(x_j + 0.5\Delta x, t^n) = u_{j+1/2}^n. \tag{2.14}$$

In Figure 2.1, one sees that the former representation defines a variable at a cell corner, while the latter defines it on a cell edge. A function can also be defined at a cell center. Note that in a difference representation of a differential equation, different functions can be defined at different locations, in which case the grid is called *staggered*.

The space derivative of the function u at the grid point x_j can be calculated from the Taylor series about x_j

$$u_{j+1}^n = u_j^n + \left.\frac{\partial u}{\partial x}\right|_j \Delta x + \frac{1}{2}\left.\frac{\partial^2 u}{\partial x^2}\right|_j (\Delta x)^2 + O(\Delta x)^3, \tag{2.15}$$

where derivatives are evaluated at x_j, and the last term indicates that the error introduced by terminating the Taylor series after only two terms is of order $(\Delta x)^3$. Solving for $\frac{\partial u}{\partial x}|_j$, we obtain

$$\left.\frac{\partial u}{\partial x}\right|_j = \frac{u_{j+1}^n - u_j^n}{\Delta x} - \frac{1}{2}\left.\frac{\partial u^2}{\partial x^2}\right|_j \Delta x + O(\Delta x)^2, \tag{2.16}$$

where the first term on the right-hand side is the forward difference quotient. Equation (2.16) shows that the commonly used "forward-difference" approximation to the derivative

$$\left.\frac{\partial u}{\partial x}\right|_j = \frac{u_{j+1}^n - u_j^n}{\Delta x} \tag{2.17}$$

is only accurate to first order in Δx, that is, the dominant error term, the second term on the right in Equation (2.16), is of order Δx. Similarly, the "backward-difference" approximation

$$\left.\frac{\partial u}{\partial x}\right|_j = \frac{u_j^n - u_{j-1}^n}{\Delta x} \tag{2.18}$$

also is only first-order accurate.

To improve the accuracy, we can use the "centered" difference quotient. We first define $\Delta x_{j+1/2} = x_{j+1} - x_j$ and $\Delta x_{j-1/2} = x_j - x_{j-1}$. Up to this point, we have assumed that these two quantities are the same and equal to Δx, but in general they can be different. If we now let

$$\Delta x_j = 0.5(\Delta x_{j+1/2} + \Delta x_{j-1/2}) \tag{2.19}$$

and expand u_j^n at x_j into both forward and backward series, we obtain, after subtracting the two expressions,

$$\left.\frac{\partial u}{\partial x}\right|_j = \frac{u_{j+1} - u_{j-1}}{2\Delta x_j} - \frac{1}{2}\frac{\partial^2 u}{\partial x^2}\frac{(\Delta x_{j+1/2})^2 - (\Delta x_{j-1/2})^2}{2\Delta x_j} + O(\Delta x)^2, \tag{2.20}$$

TABLE 2.1
Difference Approximations to $d/dx(e^x)$ at $x = 0$

Δx	Forward	Backward	Centered
1.0	1.718	0.632	1.175
0.1	1.05	0.95	1.00167
0.01	1.005	0.995	1.0000167
0.001	1.0005	0.9995	1.000000167

where the first term on the right-hand side is the desired centered difference quotient. On a uniform grid (but not otherwise) this approximation is second-order accurate

$$\left.\frac{\partial u}{\partial x}\right|_j = \frac{u_{j+1} - u_{j-1}}{2\Delta x} + O(\Delta x)^2. \tag{2.21}$$

As an example, consider the evaluation of the derivative of the function $u(x) = e^x$ at $x = 0$. Table 2.1 shows the numerically evaluated derivative for different values of Δx and for the three difference schemes (Scheme 2.17), (Scheme 2.18), and (Scheme 2.21). Comparing these derivatives with the actual value of 1, one sees that the errors in the forward or backward schemes decrease only linearly with Δx, while those in the centered scheme decrease quadratically.

Exercise

If the function u_j^n is defined at cell corners and its spatial derivative is defined at cell edges, show by Taylor expansion that the derivative

$$\left.\frac{\partial u}{\partial x}\right|_{j+1/2}^{n} = \frac{u_{j+1}^n - u_j^n}{\Delta x_{j+1/2}}$$

is accurate to second order even on a nonuniform grid.

The above exercise shows that the forward difference quotient is a good approximation for $\partial u/\partial x$ at $x + \Delta x/2$; similarly the backward difference quotient is a good approximation for the first derivative at $x - \Delta x/2$. Combining these two expressions, we obtain a centered difference approximation for the second derivative at x

$$\frac{\partial^2 u}{\partial x^2} \approx \frac{\frac{\partial u}{\partial x}(x + \Delta x/2) - \frac{\partial u}{\partial x}(x - \Delta x/2)}{\Delta x} \tag{2.22}$$

or

$$\frac{\partial^2 u}{\partial x^2} \approx \frac{u(x + \Delta x) - 2u(x) + u(x - \Delta x)}{(\Delta x)^2}. \tag{2.23}$$

Exercise

Expand the function $u(x + \Delta x)$ in a Taylor series, including the term with the third derivative. Do the same for $u(x - \Delta x)$. Add the two expressions to show that Equation (2.23) is accurate to order $(\Delta x)^2$.

We now consider the numerical evaluation of time derivatives. To illustrate the various possibilities, we shall use a simple equation

$$\frac{\partial u(x,t)}{\partial t} = h(u,x,t). \tag{2.24}$$

In a single time step the numerical scheme advances the solution from the time t^n to the time $t^{n+1} = t^n + \Delta t$. The advanced solution can be determined by any of the following expressions

$$u_j^{n+1} = u_j^n + h^n \Delta t \qquad \text{(forward time differencing)} \tag{2.25}$$

$$u_j^{n+1} = u_j^n + h^{n+1} \Delta t \qquad \text{(backward time differencing)} \tag{2.26}$$

$$u_j^{n+1} = u_j^n + \frac{h^{n+1} + h^n}{2} \Delta t \qquad \text{(centered time differencing).} \tag{2.27}$$

The first of these expressions is an *explicit* scheme, involving only the known values of the function at time t^n. The second and third are *implicit* schemes, where the solution involves the initially unknown function h^{n+1}. In general, the h_j^{n+1} must be obtained together with u_j^{n+1} by the solution of a set of linear equations or by an iterative technique. Note that although time derivatives are defined similarly to space derivatives, in practice there is not a full analogy between the two cases. Since we know u_j^n at every j, we also know all its derivatives, whereas neither u_j^{n+1} nor its derivatives are known before the solution is advanced to t^{n+1}.

Scheme (2.25) and Scheme (2.26) are *first-order accurate*, whereas Scheme (2.27) is *second-order accurate*. Define the error ϵ_{nm} of a numerical solution $u(x,t)$ as

$$\epsilon_{nm}(t) = max_x |u(x,t) - w(x,t)|, \tag{2.28}$$

where $w(x,t)$ is the analytical solution of the original differential equation. In the following, we shall demonstrate that the error grows more slowly when higher-order accuracy is used. To that end, we shall integrate the simplest possible version of Equation (2.24), in which the solution does not depend on x

$$\frac{du(t)}{dt} = -u(t). \tag{2.29}$$

The results for the three different schemes are shown in Figure 2.2. We can clearly see that higher-order schemes are indeed more accurate, i.e., the formal accuracy of a scheme translates directly into its practical accuracy.

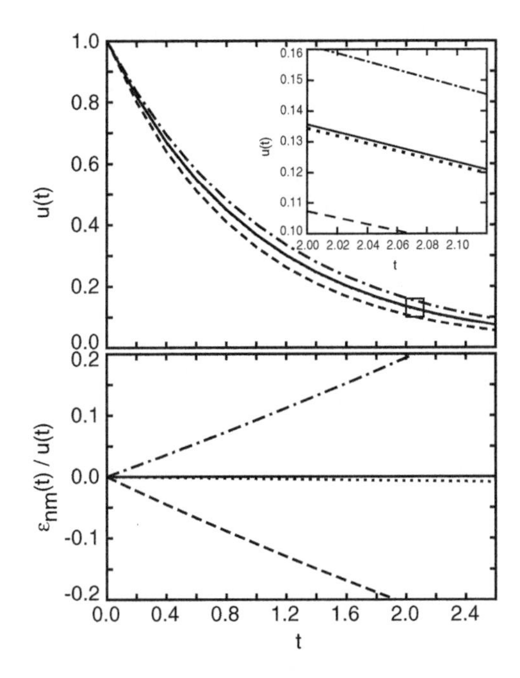

FIGURE 2.2 The upper panel shows the ordinary differential equation $du/dt = -u(t)$ and initial condition $u(0) = 1$ integrated with forward differencing according to Equation (2.25) (*dashed line*), backward differencing according to Equation (2.26) (*dash-dotted line*), and the time-centered second-order scheme according to Equation (2.27) (*dotted line*, indistinguishable from the solid line in the main figure). The time step $\Delta t = 0.2$. The solutions, starting from the point $u = 1$, are compared with the exact solution (*solid line*). The small box near $t = 2$ is shown on an expanded scale in the inset. The relative error of the solution $\epsilon_{nm}(t)/u(t)$, shown in the lower panel with lines diverging from point (0,0), increases much more slowly for the second-order scheme. The errors, of course, decrease with smaller Δt.

The next example involves both time and space differencing. We shall consider a simplified version of the one-dimensional diffusion equation (2.3), in which the viscosity coefficient ν_d is constant in space and time

$$\frac{\partial u}{\partial t} = \nu_d \frac{\partial^2 u}{\partial x^2}. \tag{2.30}$$

We shall follow the diffusion of an infinitely thin and infinitely high spike placed at $t = 0$ in the middle of the integration domain $0 \leq x \leq 1$. With the initial condition $u(x, 0) = \delta_D(x - 0.5)$ (where δ_D is the Dirac δ function*), and the boundary condition $u(0, t) = u(1, t) = 0$ at all times, the analytical solution to this problem is given by

$$w(x, t) = 2 \sum_{n=1}^{\infty} \sin\left(\frac{\pi n}{2}\right) \sin(\pi n x) \exp\left(-\pi^2 n^2 \nu_d t\right). \tag{2.31}$$

*The function $\delta_D(x)$ is everywhere zero except at $x = 0$, where it $\to \infty$ under the condition that $\int \delta_D(x) dx = 1$.

(see Morse and Feshbach, 1953). With first-order forward time differencing (Equation 2.25) applied to the time derivative, and second-order centered space differencing (Equation 2.23) applied to the space derivative, Equation (2.30) yields a simple explicit scheme

$$u_j^{n+1} = u_j^n + \frac{\nu_d \Delta t}{(\Delta x)^2} \left(u_{j+1}^n + u_{j-1}^n - 2u_j^n \right). \tag{2.32}$$

Because of the singular initial condition at $t = 0$, the solution must be advanced analytically to a smooth state at some $t > 0$ from which it can be followed numerically. The results of numerical integrations starting from the advanced state are shown in Figure 2.3. One can clearly see that the relative error $\epsilon_{nm}(t)/w(t)$ decreases with decreasing Δx or Δt. Note, however, that it is impossible to obtain arbitrarily high accuracy of integrations while keeping Δt constant and decreasing Δx (or keeping Δx constant and decreasing Δt). For our numerical solution to converge to the analytical one, both Δt *and* Δx must decrease *simultaneously*.

Exercise

Repeat the integration illustrated in Figure 2.3. Choose Δt and, keeping it constant, vary Δx until ϵ_{nm} stops decreasing. Repeat the same with constant Δx and varying Δt.

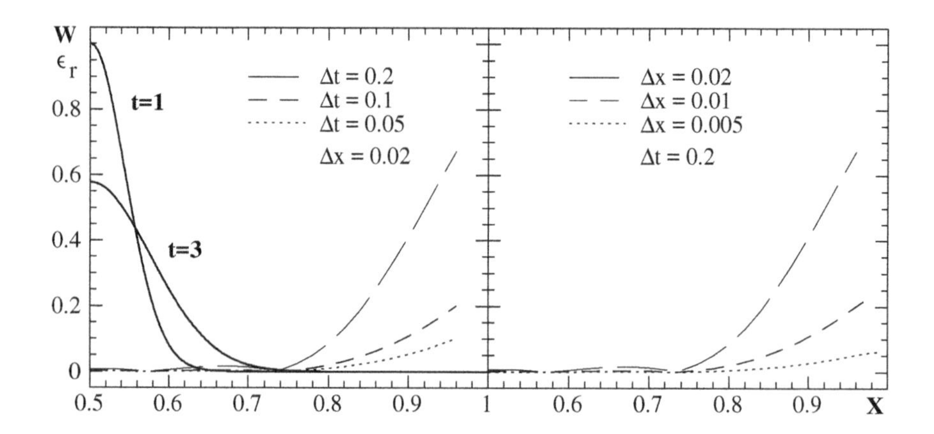

FIGURE 2.3 The left panel shows Equation (2.30) integrated with Scheme (2.32). The initial condition for integrations (*solid line* labeled "$t = 1$") is the analytical curve $w(x, 1)$ obtained from Equation (2.31) for $t = 1$. The integrations from $t = 1$ to $t = 3$ are performed for indicated values of Δt and Δx. For $t = 3$, at the resolution of the figure, they are indistinguishable from each other and from the exact solution shown with the solid line labeled "$t = 3$". However, the relative errors $\epsilon_r = |u(x, 3) - w(x, 3)|/w(x, 3))$, shown with *dashed* and *dotted* lines, are already substantial. The vertical scale is the same for all curves, and the diffusion coefficient ν_d is in all cases equal to 0.001. The right panel shows only the error curves for three different values of Δx.

2.4 STABILITY OF FINITE-DIFFERENCE METHODS

Formal accuracy, which we briefly discussed in Section 2.3, is a desirable property for a difference equation, but even more important is its *stability*. A difference scheme for the original equation is considered to be stable if the error (Equation 2.28) introduced at any stage of the computation does not grow.

A scheme is said to be *unconditionally stable* if errors decrease with time and ultimately vanish. A scheme is *conditionally stable* if the errors decrease with time, provided that the time step Δt is limited to be smaller than some critical value. On the other hand, a scheme is *unconditionally unstable* if arbitrarily small errors grow without bound and mask the real physical solution. We can check by trial-and-error that Scheme (2.26) and Scheme (2.27) are unconditionally stable, Scheme (2.25) is conditionally stable (the condition being $\Delta t < 2$ for the particular case $\frac{du}{dt} = -u$), whereas the apparently flawless scheme

$$\frac{u_j^{n+1} - u_j^{n-1}}{2\Delta t} = h^n \tag{2.33}$$

turns out to be unconditionally unstable (Figure 2.4).

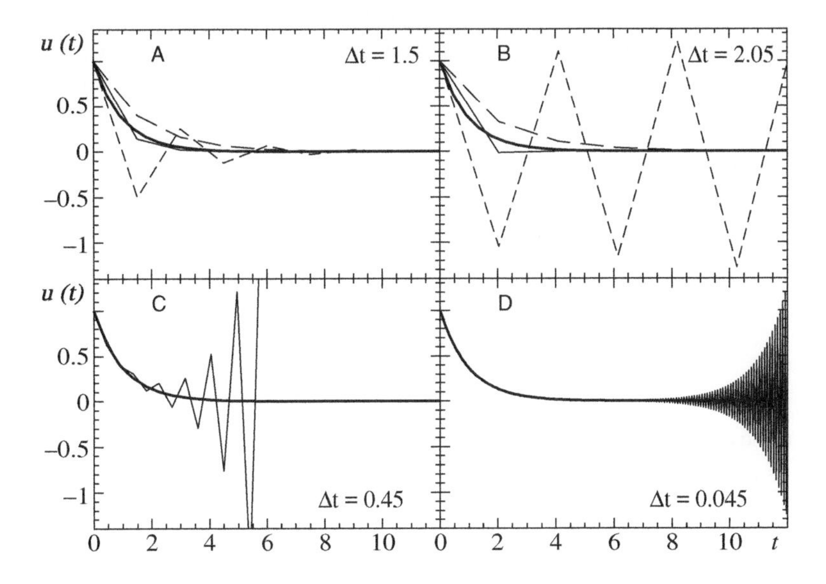

FIGURE 2.4 Numerical solutions of the equation $\frac{du}{dt} = -u(t)$ with the initial condition $u(0) = 1$. (A,B): *Long-dashed* and *thin solid lines*: implicit Scheme (2.26) and Scheme (2.27), respectively; *short-dashed line*: explicit Scheme (2.25). (C,D): Implicit Scheme (2.33). In the latter case the error ϵ_{nm} grows without limits no matter how small Δt is, i.e., the scheme is unconditionally unstable. In the case of the explicit Scheme (2.25), ϵ_{nm} grows without limits only when $\Delta t > 2$ is used, i.e., the scheme is conditionally stable. In the remaining two cases ϵ_{nm} does not grow even for very large Δt, i.e., Scheme (2.26) and Scheme (2.27) are unconditionally stable. For comparison, the exact analytical solution is shown with the heavy solid line in all panels.

In practice, the stability of a numerical scheme is, of course, analyzed by much more sophisticated methods. The most commonly used among them is the von Neumann stability analysis. It is relatively easy to perform, but it is limited only to linear initial value problems with constant or slowly varying coefficients. To apply it to a nonlinear system, one must linearize the equations first, and what one obtains is only a "necessary," but not "sufficient" condition for stability.

To perform an illustrative, but not too complicated example of the von Neumann analysis, consider once more the simplified version of the one-dimensional diffusion equation (2.30) with a general initial condition $u(x,0) = u_0(x)$. The scheme to be analyzed

$$\frac{u_j^{n+1} - u_j^{n-1}}{2\Delta t} = \nu_d \frac{u_{j+1}^n + u_{j-1}^n - 2u_j^n}{(\Delta x)^2}, \tag{2.34}$$

is second-order accurate in both space and time. Let

$$d = \frac{2\nu_d \Delta t}{(\Delta x)^2} > 0. \tag{2.35}$$

Then

$$u_j^{n+1} = u_j^{n-1} + d\left(u_{j+1}^n + u_{j-1}^n - 2u_j^n\right). \tag{2.36}$$

The reader who wishes to bypass the detailed analysis can skip to the final result, Equation (2.52), which shows that the scheme is unconditionally unstable.

Let the function $u(x,t)$ be defined on the interval [0,X] and let it be periodic in x so that $u(x + nX) = u(x)$, $n = 1, 2, \ldots$. Then, if Δx is constant, the function u can be represented as a Fourier series

$$u_j = \frac{1}{L} \sum_{k=1}^{L} \hat{u}_k e^{-2\pi i k \frac{j}{L}} \tag{2.37}$$

where

$$\hat{u}_k = \sum_{j=1}^{L} u_j e^{2\pi i k \frac{j}{L}}, \tag{2.38}$$

and i is the imaginary unit $i = \sqrt{-1}$. The amplitudes $\hat{u}_k$ will evolve in time. At $t = t^n$ and $x = x_j$, we have

$$u_j^n = \frac{1}{L} \sum_{k=1}^{L} \hat{u}_k^n e^{-ij\cdot\frac{2\pi k \Delta x}{X}} = \frac{1}{L} \sum_{k=1}^{L} \hat{u}_k^n e^{-ij\phi_k}, \tag{2.39}$$

where ϕ_k is known as the phase angle. The difference equation (2.34) is linear and can be decomposed into its separate Fourier modes. Dividing the decomposed Equation (2.36) by $e^{-ij\phi_k}$, we obtain for each k

$$\hat{u}_k^{n+1} = \hat{u}_k^{n-1} + d\left(\hat{u}_k^n e^{-i\phi_k} + \hat{u}_k^n e^{i\phi_k} - 2\hat{u}_k^n\right) \tag{2.40}$$

or

$$\hat{u}_k^{n+1} = \hat{u}_k^{n-1} + 2d\hat{u}_k^n(\cos\phi_k - 1). \tag{2.41}$$

With the k index dropped for brevity, this expression can be written as

$$\begin{bmatrix} \hat{u}^{n+1} \\ \hat{u}^{n} \end{bmatrix} = \mathbf{G} \cdot \begin{bmatrix} \hat{u}^{n} \\ \hat{u}^{n-1} \end{bmatrix} \tag{2.42}$$

or, to further simplify the notation,

$$\mathrm{F}^{n} = \mathbf{G} \cdot \mathrm{F}^{n-1}, \tag{2.43}$$

where $\mathbf{G}$, the so-called *amplification matrix*, does not depend on time and is given by

$$\mathbf{G} = \begin{bmatrix} 2d(\cos\phi - 1) & 1 \\ 1 & 0 \end{bmatrix} \tag{2.44}$$

Assuming we know F^0, then n repeated multiplications by $\mathbf{G}$ give

$$\mathrm{F}^{n} = \underbrace{\mathbf{G} \cdot \ldots \cdot \mathbf{G}}_{n} \cdot \mathrm{F}^{0}. \tag{2.45}$$

The stability condition may be stated as follows: if the amplitudes of the Fourier modes are finite at $t = 0$, then they must remain finite for all time steps.

Let λ_1 and λ_2 be the eigenvalues of $\mathbf{G}$, and G_1 and G_2 the eigenvectors. Then

$$\mathrm{F}^{0} = \sum_i g_i^0 G_i \qquad i = 1, 2 \tag{2.46}$$

and

$$\mathbf{G} \cdot \mathrm{F}^{0} = \mathbf{G} \cdot \sum_i g_i^0 G_i = \sum_i g_i^0 \mathbf{G} \cdot G_i = \sum_i g_i^0 \lambda_i G_i \tag{2.47}$$

Similarly,

$$\mathrm{F}^{n} = \sum_i g_i^0 (\lambda_i)^{n'}. \tag{2.48}$$

The stability requirement will be fulfilled when

$$|\lambda_i|^{n'} < M, \tag{2.49}$$

where M is a positive finite real number, or

$$|\lambda_i| < M^{1/n} \quad \text{or} \quad |\lambda_i| < 1 \tag{2.50}$$

as $M^{1/n} \to 1$ when the number of time steps is large. In our example, for the difference scheme to be stable the condition

$$|\lambda_{1,2}| \equiv |\lambda_{\pm}| = |d(\cos\phi - 1) \pm \sqrt{d^2(\cos\phi - 1)^2 + 1}| \leq 1 \tag{2.51}$$

must be satisfied for all ϕ. But, for example,

$$\lambda_{-}(\phi = \pi/2) = d + \sqrt{1 + d^2} > 1 \tag{2.52}$$

for all $d > 0$, so the scheme is intrinsically (i.e., unconditionally) unstable, even for very short time steps.

It must be remembered that a seemingly insignificant change in the difference representation may severely influence the stability of the resulting scheme. For example, when in the unstable scheme (2.34) u_j^n is replaced by its time average: $u_j^n \rightarrow 0.5(u_j^{n+1} + u_j^{n-1})$, then the resulting difference formula

$$\frac{u_j^{n+1} - u_j^{n-1}}{2\Delta t} = \nu_d \frac{u_{j+1}^n + u_{j-1}^n - u_j^{n+1} - u_j^{n-1}}{(\Delta x)^2}, \tag{2.53}$$

(known as the *DuFort–Frankel* scheme) is unconditionally stable with no limit on Δt.

Exercise

Show that the amplification matrix for the DuFort–Frankel scheme (2.53) is:

$$\mathbf{G} = \begin{bmatrix} \frac{2d\cos\phi}{1+d} & \frac{1-d}{1+d} \\ 1 & 0 \end{bmatrix}$$

with

$$\lambda_{\pm} = \frac{d\cos\phi}{1+d} \pm \sqrt{\left(\frac{d\cos\phi}{1+d}\right)^2 + \frac{1-d}{1+d}}.$$

We are now ready to perform the stability analysis of Scheme (2.32), which we used to integrate the diffusion equation in Section 2.3. If we define for this case $d = \nu_d \Delta t/(\Delta x)^2$, then the amplification factor is just a 1×1 matrix

$$\mathbf{G} = 1 - 2d(1 - \cos\phi). \tag{2.54}$$

As the maximum of $|1 - \cos\phi| = 2$, the stability requirement is fulfilled when $d < 0.5$ or

$$\Delta t < \frac{1}{2}\frac{(\Delta x)^2}{\nu_d}. \tag{2.55}$$

It is important to know that similar limits apply to Δt for all explicit schemes.

Exercise

Applying Scheme (2.25), Scheme (2.26), Scheme (2.27), and Scheme (2.33) to Equation (2.29) prove that indeed

- Scheme (2.26) and Scheme (2.27) are unconditionally stable.
- Scheme (2.25) is conditionally stable, the condition being $\Delta t < 2$.
- Scheme (2.33) is unconditionally unstable.

2.5 PHYSICAL MEANING OF STABILITY CRITERION

The limits on Δt for explicit schemes have a clear physical meaning, which we will illustrate in the next example, the equation of continuity in one dimension

$$\frac{\partial \rho}{\partial t} + v \cdot \frac{\partial \rho}{\partial x} = 0 \tag{2.56}$$

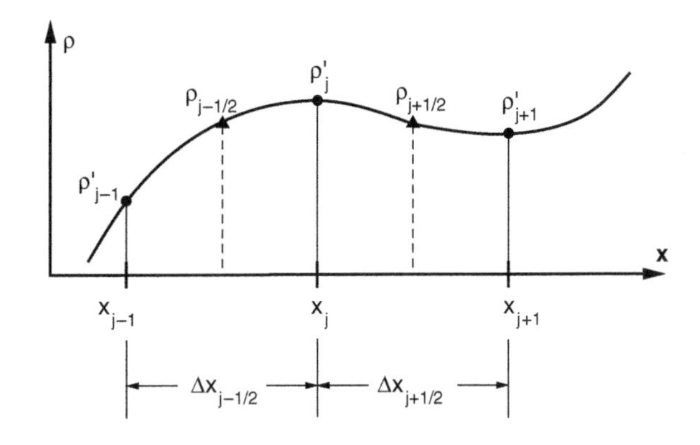

FIGURE 2.5 The one-dimensional staggered grid used to integrate the continuity equation (2.56) with ρ defined halfway between grid points (triangles) and its derivative $\rho' \equiv \frac{\partial \rho}{\partial x}$ defined at the grid points (circles).

where ρ is the mass density and v is the velocity, which we here assume to be a positive constant. The general solution to this equation is a wave propagating in the positive x-direction, $\rho = f(x - vt)$ where f is an arbitrary function.

A straightforward scheme that is accurate to first order in time is:

$$\rho_{j+1/2}^{n+1} = -d\left(\rho_{j+1/2}^{n} - \rho_{j-1/2}^{n}\right) + \rho_{j+1/2}^{n}, \tag{2.57}$$

where $d = \frac{v\Delta t}{\Delta x}$. Note that a *staggered* grid is used here, with ρ defined halfway between the grid points x_j and $\frac{\partial \rho}{\partial x}$ defined at the grid points themselves (Figure 2.5). In this case, the amplification factor is complex

$$G = 1 - d + d\cos\phi - id\sin\phi. \tag{2.58}$$

From the requirement that $GG^* < 1$, we get $0 < d \leq 1$ or $\Delta t \leq \Delta x/v$, which simply states that in one time step we cannot move a fluid element by more than one cell width. In other words, in one time step only fluid from its nearest neighbors is allowed to flow into a given cell or, in still other words, information may not propagate with a velocity larger than the characteristic velocity of the fluid.

Exercise

For a certain value of Δt satisfying the stability criterion, the numerical solution obtained from Equation (2.57) is 100% accurate and there is no numerical diffusion. What is this value of Δt and why can it not be used in practice?

Similarly, the time step limit (Equation 2.55) for the diffusion equation is closely related to the characteristic diffusion time $\tau_d = (\Delta x)^2/\nu_d$, i.e., the time for broadening an initially very thin spike in, say, density to a width of Δx (Figure 2.6). The limit on Δt simply requires the computed diffusion to be slower than the physical diffusion across one zone.

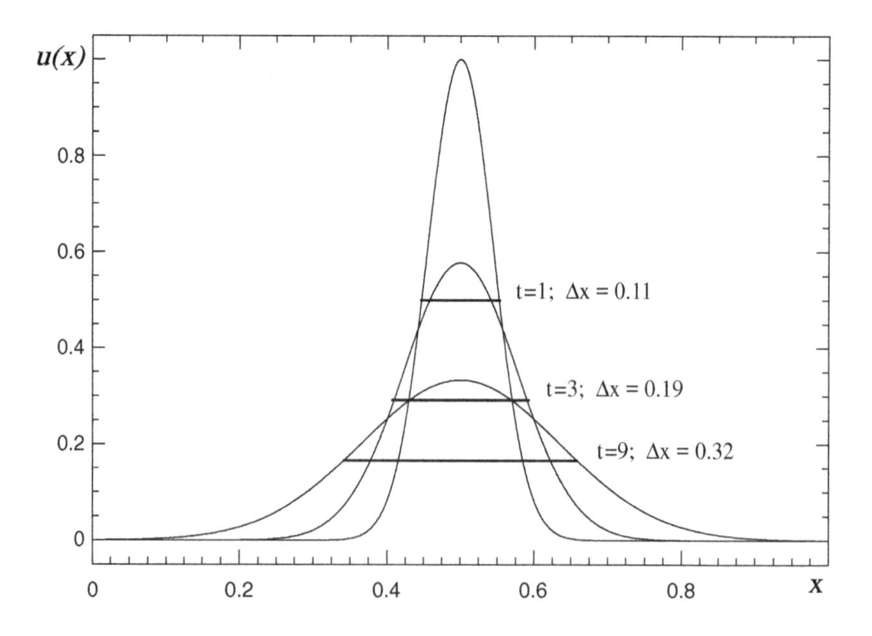

FIGURE 2.6 Numerical solution of the diffusion equation (2.30) obtained in the same way as in Figure 2.3 for $\nu_d = 0.001$, $\Delta t = 0.2$, and $\Delta x = 0.02$. The full width Δx (at half maximum) of an initially very thin spike increases proportionally to $\sqrt{t}$.

Our next example is the coupled equations of continuity and of momentum describing the flow of a fluid in one dimension, where we do not assume that the velocity is constant

$$\begin{aligned} \frac{\partial \rho}{\partial t} + \frac{\partial(\rho v)}{\partial x} &= 0 \\ \frac{\partial v}{\partial t} + v \cdot \frac{\partial v}{\partial x} &= -\frac{1}{\rho} \cdot \frac{\partial P}{\partial x}, \end{aligned} \tag{2.59}$$

where P is the pressure. This simple system has the unpleasant property of being nonlinear. In order to apply the von Neumann stability analysis, we first must linearize it. Let us assume that the medium is initially uniform and static, with $P(x, 0) = P_0$, $\rho(x, 0) = \rho_0$, and $v(x, 0) = 0$. We shall further assume that the gas is isothermal both in space and in time, so that the sound speed $c_s^2 \equiv P/\rho$ is constant. This medium is then perturbed in such a way that

$$\rho = \rho_0 + \rho_1, \quad P = P_0 + P_1, \quad v = v_1, \tag{2.60}$$

where (ρ_1, P_1) are small compared with the unperturbed values and v_1^2 is small compared with P/ρ. Then, neglecting quantities of second order

$$\begin{aligned} \frac{\partial v_1}{\partial t} &= -\frac{1}{\rho_0} \cdot \frac{\partial P_1}{\partial x} \\ \frac{\partial \rho_1}{\partial t} &= -\rho_0 \cdot \frac{\partial v_1}{\partial x}. \end{aligned} \tag{2.61}$$

Let $\phi = P_1$ and $\psi = \rho_0 c_s v_1$. Then

$$\frac{\partial\psi}{\partial t} = \rho_0 c_s \frac{\partial v_1}{\partial t} = -c_s \frac{\partial P_1}{\partial x} = -c_s \cdot \frac{\partial\phi}{\partial x}$$
$$\frac{\partial\psi}{\partial x} = \rho_0 c_s \frac{\partial v_1}{\partial x} = -c_s \cdot \frac{\partial\rho_1}{\partial t} = -\frac{1}{c_s}\frac{\partial P_1}{\partial t} = -\frac{1}{c_s} \cdot \frac{\partial\phi}{\partial t}, \tag{2.62}$$

i.e.,

$$\frac{\partial\psi}{\partial t} = -c_s \cdot \frac{\partial\phi}{\partial x}$$
$$\frac{\partial\phi}{\partial t} = -c_s \cdot \frac{\partial\psi}{\partial x}. \tag{2.63}$$

One can check that Equations (2.63) are equivalent to the wave equation (2.2) by differentiating the first of them with respect to x, the second with respect to t, and using $\partial^2\psi/\partial x\partial t = \partial^2\psi/\partial t\partial x$. The result is:

$$\frac{\partial^2\phi}{\partial t^2} = c_s^2 \frac{\partial^2\phi}{\partial x^2}, \tag{2.64}$$

where

$$c_s = \sqrt{\frac{P}{\rho}} \tag{2.65}$$

is the *isothermal velocity of sound.*

To solve Equation (2.63), we shall use a staggered grid, with ϕ defined at ${}^{n}_{j+1/2}$ and ψ at ${}^{n+1/2}_{j}$ (Figure 2.7). The difference scheme is then straightforward

$$\frac{1}{\Delta t}\left(\psi_j^{n+1/2} - \psi_j^{n-1/2}\right) = -\frac{c_s}{\Delta x}\left(\phi_{j+1/2}^n - \phi_{j-1/2}^n\right)$$
$$\frac{1}{\Delta t}\left(\phi_{j+1/2}^{n+1} - \phi_{j+1/2}^n\right) = -\frac{c_s}{\Delta x}\left(\psi_{j+1}^{n+1/2} - \psi_j^{n+1/2}\right), \tag{2.66}$$

so that, starting with ϕ at t^n, we get ψ at $t^{n+1/2}$ and then ϕ at t^{n+1}. Note that the scheme is accurate to second order in both space and time.

To begin the stability analysis, we first perform the Fourier expansions

$$\psi_j^{n+1/2} = \sum_{k=1}^{L} \hat{\psi}_k^{n+1/2} e^{-ij\theta_k},$$
$$\phi_{j+1/2}^{n+1} = \sum_{k=1}^{L} \hat{\phi}_k^{n+1} e^{-i(j+1/2)\theta_k}, \tag{2.67}$$

where $\theta_k = 2\pi k\Delta x/X$ and where, again, the functions are defined on the interval [0,X].

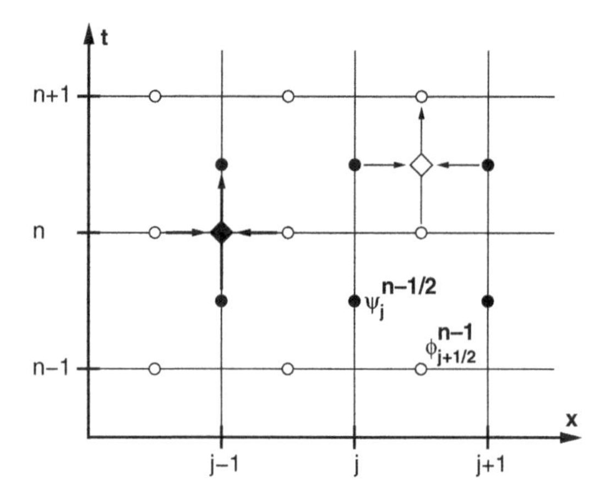

FIGURE 2.7 The staggered grid used to integrate the set of equations (Equation 2.63). The functions and their derivatives are defined at locations indicated by circles and diamonds, respectively. *Open circles:* $\phi(x, t)$; *filled circles*: $\psi(x, t)$; *filled diamond:* the location of the space derivative of ϕ and the time derivative of ψ; *open diamond:* the location of the space derivative of ψ and the time derivative of ϕ. Based on this grid, a scheme that is second-order accurate both in space and time naturally emerges. The horizontal arrows indicate how the space derivatives are calculated; the vertical arrows, how the functions are advanced in time.

Exercise

Show that when the Fourier expansions (2.67) are substituted into Equations (2.66) the result for a given Fourier mode k is:

$$\hat{\psi}_k^{n+1/2} - \hat{\psi}_k^{n-1/2} = s\hat{\phi}_k^n \cdot 2i \sin\frac{\theta_k}{2}$$

$$\hat{\phi}_k^{n+1} - \hat{\phi}_k^n = s\hat{\psi}_k^{n+1/2} \cdot 2i \sin\frac{\theta_k}{2}, \tag{2.68}$$

where s is defined to be $c_s \Delta t/\Delta x$.

From the result of the exercise, it is easy to show that the $\hat{\psi}_k$ and the $\hat{\phi}_k$ can be advanced in time through the relations

$$\begin{bmatrix} \hat{\psi}_k^{n+1/2} \\ \hat{\phi}_k^{n+1} \end{bmatrix} = G \begin{bmatrix} \hat{\psi}_k^{n-1/2} \\ \hat{\phi}_k^{n} \end{bmatrix}, \tag{2.69}$$

where

$$G = \begin{bmatrix} 1 & 2s \cdot i \sin\frac{\theta_k}{2} \\ 2s \cdot i \sin\frac{\theta_k}{2} & 1 - 4s^2 \sin^2\frac{\theta_k}{2} \end{bmatrix}. \tag{2.70}$$

The equation for the eigenvalues λ is:

$$\lambda^2 + \left(4s^2 \sin^2\frac{\theta_k}{2} - 2\right)\lambda + 1 = 0, \tag{2.71}$$

with $|\lambda_{\pm}| \leq 1$ when $s \leq 1$; thus, the stability criterion is:

$$\Delta t \leq \Delta x / c_s. \tag{2.72}$$

Exercise

Work out Equation (2.71).

This limit on Δt has the same physical meaning as the limit $\Delta t \leq \Delta x / v$ for the continuity equation. Again, information may not spread through the grid at a speed greater than the characteristic speed defined by the problem, which in this case is the sound speed. This requirement, which applies to explicit hydrodynamical schemes, is known as the *Courant–Friedrichs–Lewy* or CFL condition (Courant et al., 1928), and it is the most commonly encountered limit in hydrodynamical simulations.

To see what happens when the CFL limit is violated, let us integrate the linearized equations (2.63) with $s = 1$ and $s > 1$. The results are shown in Figure 2.8. We see that for s only slightly greater than 1, numerical oscillations with a short wavelength and large amplitude are excited. It should be remembered that despite the fact that $s = 1$ marginally works in this case, it would not work for the original nonlinear equation. As stated in Section 2.4, $s \leq 1$ is only a necessary condition for stability, not a sufficient condition. For this reason, when a hydrodynamic problem is solved with an explicit technique subject to the CFL condition, the actual time step is taken to be C_0 times the CFL limit, where $0 < C_0 < 1$. The "safety factor" C_0 is known as the *Courant number*, and it is typically set to be 0.5.

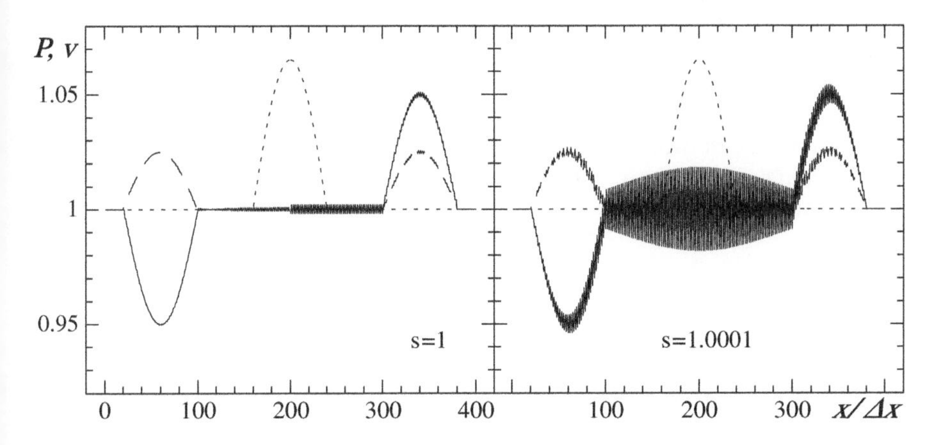

FIGURE 2.8 Propagation of a small-amplitude sound wave followed with the help of Equation (2.63) and Scheme (2.66). A numerically stable solution can only be obtained when the CFL limit is not exceeded. *Dotted and dashed lines*: initial and final pressure distribution, respectively (normalized to the unperturbed initial pressure). *Solid line*: final velocity distribution (arbitrarily normalized). The quantity is $s = c_s \Delta t / \Delta x$, where c_s is the sound speed. The integration from $t = 0$ to $t = 140$ is performed with $\Delta x = c_s = 1$, so that $s = \Delta t = 1$ and 1.0001 in the left and right frames, respectively. The spatial grid is uniform.

Implicit schemes have less stringent stability requirements than explicit ones; in fact, they are generally unconditionally stable. But, note the following **caution**: The fact that a scheme is formally stable without any limits on Δt does not mean that Δt can be arbitrarily large in practice. Practical limits on Δt are set in this case by the fact that usually we have to iterate to advance the solution from t^n to t^{n+1}, and the interations do not converge if Δt is too large. Even if we do not have to iterate, the size of Δt in implicit schemes is limited by the error growth rate. To keep the error low, the change in a physical quantity over one time step must be less than, say 5%, so the time step in fact is limited, by reasons of accuracy, to the physical time scale on which quantities change. However, if the physical time scale to be followed is much longer than the time for a sound wave to cross the grid, then the use of an implicit method may be required.

2.6 A USEFUL IMPLICIT SCHEME

In this section, we shall illustrate by a particular example that an implicit scheme can give an accurate and stable solution with a time step that is considerably larger than the diffusion limit given by Equation (2.55). At the same time, we describe a solution method for implicit equations that is often used in practice. A general difference representation of the diffusion equation (2.30) is given by

$$u_j^{n+1} = u_j^n + \alpha b\left(u_{j+1}^n - 2u_j^n + u_{j-1}^n\right) + (1-\alpha)b\left(u_{j+1}^{n+1} - 2u_j^{n+1} + u_{j-1}^{n+1}\right), \quad (2.73)$$

where $0 \leq \alpha \leq 1$, $b \equiv \Delta t/(\Delta x)^2$, and it is assumed for simplicity that $\nu_d = 1$. For $\alpha = 1$, we recover the explicit scheme (2.32), whereas for any $0 \leq \alpha < 1$, the scheme becomes implicit. For the particular case $\alpha = 0.5$, the difference equations are centered and second-order accurate in time, and the method is known as the *Crank–Nicolson* scheme.

Consider a case with boundary conditions $u(0,t) = 1$, $u(1,t) = 0$, and initial condition $u(x,0) = 0$ for $x > 0$. The analytical solution is given by the sum of the special solution

$$w_s(x,t) = 1 - x, \quad (2.74)$$

and the series

$$w(x,t) = -\frac{2}{\pi}\sum_{n=1}^{\infty}\frac{\sin(\pi n x)}{n}\exp(-\pi^2 n^2 t). \quad (2.75)$$

(see Morse and Feshbach, 1953). The correct initial conditions are recovered for most values of x if the series includes on the order of 200 terms; however, many more terms must be added if x is close to 0.

As long as $\alpha \neq 1$, the method can be expressed in terms of a matrix equation

$$[s]\mathbf{u^{n+1}} = -\mathbf{E}, \quad (2.76)$$

where the tridiagonal matrix $[s]$ and the vector $\mathbf{E}$ are known, and the unknowns u_j^{n+1} are to be solved for, given u_j^n. In more detail

$$\begin{bmatrix} B_1 & C_1 & 0 & 0 & \cdots & \cdots & \cdots \\ A_2 & B_2 & C_2 & 0 & 0 & \cdots & \cdots \\ 0 & A_3 & B_3 & C_3 & 0 & \cdots & \cdots \\ \cdots & \cdots & \cdots & \cdots & \cdots & \cdots & \\ 0 & 0 & 0 & 0 & 0 & A_J & B_J \end{bmatrix} \cdot \begin{bmatrix} u_1^{n+1} \\ u_2^{n+1} \\ u_3^{n+1} \\ \cdots \\ u_J^{n+1} \end{bmatrix} = \begin{bmatrix} -E_1 \\ -E_2 \\ -E_3 \\ \cdots \\ -E_J \end{bmatrix}. \tag{2.77}$$

Suppressing the superscript $(n+1)$ for the moment, the inner boundary zone corresponds to the equation

$$B_1 u_1 + C_1 u_2 = -E_1, \tag{2.78}$$

while a general point on the grid gives

$$A_j u_{j-1} + B_j u_j + C_j u_{j+1} = -E_j, \tag{2.79}$$

and the outer boundary gives

$$A_J u_{J-1} + B_J u_J = -E_J. \tag{2.80}$$

Exercise

Show that for the general point with the Crank–Nicolson scheme $B_j = 1+b$. Calculate A_j, C_j, and E_j.

We then assume that the solution follows the linear relation

$$u_{j+1} = H_j u_j + Y_j. \tag{2.81}$$

At the outer boundary, using Equation (2.80) we obtain $H_{J-1} = -A_J/B_J$ and $Y_{J-1} = -E_J/B_J$. In fact, H and Y at the outer boundary are determined from the boundary conditions. For example, if $u_J = 0$, then $H_{J-1} = Y_{J-1} = 0$. Or, if u_J has a fixed value determined from an independent boundary calculation, of say U_B, then $H_{J-1} = 0$ and $Y_{J-1} = U_B$. For the general point, using Equation (2.79)

$$A_j u_{j-1} + (B_j + C_j H_j) u_j = -E_j - C_j Y_j \tag{2.82}$$

or

$$u_j = -\frac{A_j}{(B_j + C_j H_j)} u_{j-1} - \frac{E_j + C_j Y_j}{(B_j + C_j H_j)} \tag{2.83}$$

Thus, we can identify

$$H_{j-1} = -\frac{A_j}{B_j + C_j H_j} \tag{2.84}$$

$$Y_{j-1} = -\frac{E_j + C_j Y_j}{B_j + C_j H_j}. \tag{2.85}$$

Starting at the outer boundary, H_{j-1} can be calculated recursively from H_j, and Y_{j-1} can be calculated recursively from Y_j and H_j. Then, at the inner boundary where $u_2 = H_1 u_1 + Y_1$, we solve two simultaneous equations to obtain

$$u_1 = -\frac{E_1 + C_1 Y_1}{B_1 + C_1 H_1}. \tag{2.86}$$

Here the inner boundary condition is applied, so, for example, if $u_1 = 0$, then $E_1 = C_1 = 0$. If, on the other hand, the derivative of u is zero at the inner boundary, a situation that often occurs, one obtains, from $u_1 = u_2$, $Y_1 = 0$ and $H_1 = 1$. Thus, for example, the value of the function itself can be specified at the outer edge, while its derivative can be specified at the inner edge.

Once the inner boundary has been reached, all the values of Y_j and H_j are available, so that u_2, u_3, etc. can be calculated recursively from Equation (2.83), working back from the inner edge toward the outer edge. The procedure is quick and efficient and has many applications. In particular, it is used to solve the diffusion approximation (Equation 1.117) to the radiation transfer equation (see Chapter 6, Section 6.6.2).

The analytical solution and the numerical solutions obtained with $\Delta t = 0.3$ and $\Delta t = 0.03$ are compared in Figure 2.9 for $t = 0.3$, at which time $u(x, t)$ is already very

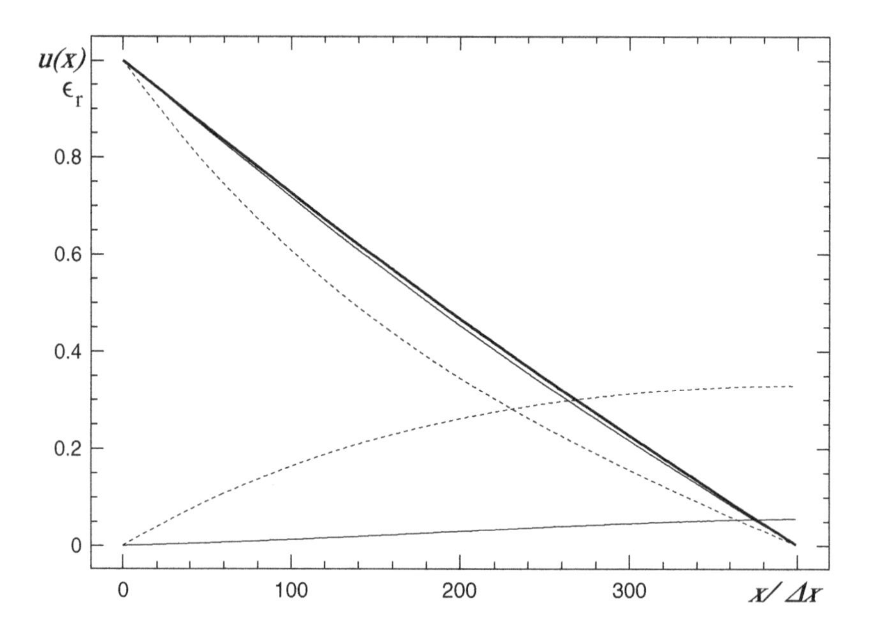

FIGURE 2.9 Numerical solution of the diffusion equation (2.30) obtained with Scheme (2.73) for $\alpha = 0$ (i.e., fully implicit). *Initial condition*: $u(x, 0) = 0$ for $x > 0$. *Boundary conditions*: $u(0, t) = 1$, $u(1, t) = 0$. The curves originating at the point (0, 1) are the results of integrations from $t = 0$ to $t = 0.3$ with $\Delta t = 0.3$ (*dots*) and $\Delta t = 0.03$ (*thin solid line*). They are compared with the analytical solution (Equation 2.74 and Equation 2.75) calculated for $t = 0.3$ (*heavy solid line*). The curves originating at the point (0, 0) give the relative errors of the integrations. The integrations are performed with $\Delta x = 0.0025$ and $\nu_d = 1$. The vertical scale is the same for all curves.

close to a steady state $u(x,t) = 1 - x$. The accuracy of the integration deteriorates significantly when the larger of the two time steps is used. Note the large number (400) of zones in x required to get the match to be as good as shown in the figure. However, also note that in both integrations $\Delta t \gg (\Delta x)^2$, i.e., that the limit on Δt set for explicit schemes by the stability condition (Equation 2.55), is exceeded by a large factor.

To summarize and extend slightly the important practical results of Section 2.4, Section 2.5, and Section 2.6:

- In the solution of a one-dimensional diffusion equation by an explicit technique, the time step is limited, by considerations of stability, to

$$\Delta t < 0.5\frac{(\Delta x)^2}{\nu_d} \tag{2.87}$$

  where Δx is the smallest space interval on the grid and ν_d is the diffusion coefficient.
- If the diffusion equation is to be solved in N-space dimensions, the time step limit becomes

$$\Delta t < \frac{0.5}{N}\frac{(\Delta x)^2}{\nu_d}. \tag{2.88}$$

  This restriction may make the explicit technique inappropriate for many problems. In contrast, the implicit Crank–Nicolson scheme and other implicit schemes are numerically stable for all Δt, and the time step is limited only by the physical time scale of the problem.
- In the solution of the one-dimensional hydrodynamic equations by explicit techniques, the continuity equation leads to a time step limit of

$$\Delta t < \min\frac{\Delta x}{|v|}, \tag{2.89}$$

  where v is the (in general, variable) velocity. The minimum is taken across the entire grid. Furthermore, the linearized stability analysis of a simple explicit hydrodynamic system gives the time step limit

$$\Delta t < \min\frac{\Delta x}{c_s}, \tag{2.90}$$

  where c_s is the sound speed and the minimum is taken across the entire grid.
- In a general Eulerian hydrodynamic system, where the velocity is not necessarily small compared with the sound speed, the two criteria are usually combined according to the equation

$$\Delta t < \min\frac{\Delta x}{\left(c_s^2 + v^2\right)^{1/2}} \tag{2.91}$$

  or

$$\Delta t < \min\frac{\Delta x}{(c_s + |v|)}. \tag{2.92}$$

In spite of this restriction, many current hydrodynamic codes, particularly those in two or three space dimensions, use explicit Eulerian techniques (see Chapter 6). Note that a switch to a Lagrangian explicit code changes the time step limitation somewhat (Richtmyer, 1957) in that Equation (2.90) can be used. In hydrodynamic systems, however, the actual physical time scale for significant change in the variables is not much longer than that given by Equation (2.91). Thus, considerations of accuracy require a short time step in any case, and going to an implicit scheme offers not much advantage in terms of time step and a disadvantage in terms of programming complexity.

- The limiting time steps given in Equation (2.88) and Equation (2.91) are in common practice multiplied by a factor of approximately 0.5, the Courant number.

2.7 DIFFUSION, DISPERSION, AND GRID RESOLUTION LIMIT

Stability and accuracy are not the only factors to be taken into account in the evaluation of a numerical scheme. If we expand the DuFort–Frankel Scheme (2.53) in Taylor series centered at point u_j^n and neglect third-order terms, we obtain

$$\frac{\partial u}{\partial t} = \nu \frac{\partial^2 u}{\partial x^2} - \nu \frac{\partial^2 u}{\partial t^2} \cdot \frac{(\Delta t)^2}{(\Delta x)^2} \tag{2.93}$$

which is an equation *qualitatively* different from the original one: It has the properties of a wave equation instead of a diffusion equation.

Unfortunately, this is a very general property of numerical schemes. Discretization introduces terms that are not present in the original equation. Schemes in which the leading error term is composed of second-order spatial derivatives suffer loss of accuracy from *numerical diffusion* (also referred to as *numerical viscosity* or *internal viscosity*). The problem, of course, is reduced with the use of smaller space and time intervals. In the case of a propagating wave, the diffusion corresponds to damping of the amplitude of the wave. This effect may or may not be important, depending on the difference scheme. Higher-order schemes result in improvement.

Schemes in which the leading error term is composed of third-order spatial derivatives introduce spurious *numerical dispersion*, that is, the speed with which waves are propagated across the grid becomes wavelength-dependent. The origin of the problem is that the calculated phase angle of the wave does not equal the analytical value of 2π after the wave has propagated one wavelength, and this effect depends on the grid resolution.

In addition, we have to require that the difference scheme be *consistent*, i.e., the original differential equation be recovered in the limit $\Delta t, \Delta x \rightarrow 0$. In the case of Equation (2.93), it is clear that if the ratio $\Delta t/\Delta x$ remains finite as both go to zero, the difference equation does not reduce to the differential equation. In general, this requirement can only be checked by recalculating the same problem on a finer grid and showing that the solution converges as Δt, Δx become arbitrarily small.

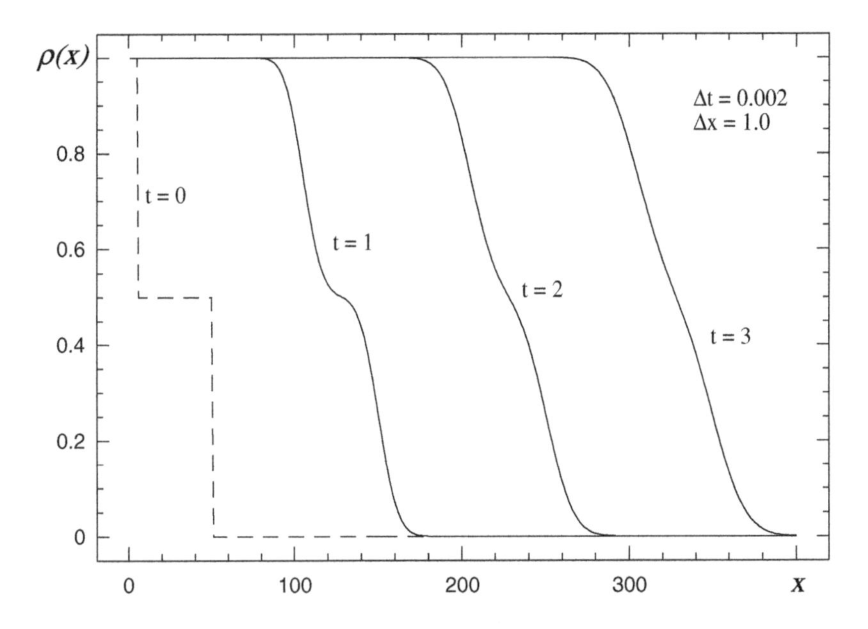

FIGURE 2.10 Solution of the continuity equation (2.56), with $v = 100$, obtained with the help of Scheme (2.57). The large internal viscosity of the scheme quickly spreads initial discontinuities in the density distribution into broad "ramps," even though the time step is only 1/5 that at the stability limit $\Delta t < \Delta x / v$.

.

The effects of numerical diffusion are best illustrated with the help of the continuity equation. Also, in this case, our difference approximation (Equation 2.57) is not an exact representation of the equation itself. To see this, let us expand Equation (2.56) in a Taylor series centered at $\rho^n_{j+1/2}$. We obtain

$$\frac{\partial \rho}{\partial t} + v \cdot \frac{\partial \rho}{\partial x} = \frac{1}{2} \cdot v \cdot \Delta x \cdot \frac{\partial^2 \rho}{\partial x^2} - \frac{1}{2} \cdot \Delta t \cdot \frac{\partial^2 \rho}{\partial t^2} + O(\Delta x)^2, \tag{2.94}$$

and we see that the first term on the right-hand side of this equation is a diffusion term (actually Equation (2.94) is a wave equation, but the diffusion term usually dominates). In fact, the spurious viscosity coefficient $\nu = \frac{1}{2} \cdot v \cdot \Delta x$ may become so large that small-scale details of the $\rho(x, 0)$ distribution are completely washed out (Figure 2.10) after a few hundred time steps.

Exercise

The dire situation shown in Figure 2.10 can be improved upon, at the cost of additional programming effort and computing time. One might think that centering the space derivative would help

$$\rho^{n+1}_{j+1/2} = -0.5d\left(\rho^n_{j+3/2} - \rho^n_{j-1/2}\right) + \rho^n_{j+1/2}, \tag{2.95}$$

where $d = v\Delta t/\Delta x$. However, it has been shown (Bowers and Wilson, 1991, Section 2.2) that this scheme is unconditionally unstable.

Try an implicit scheme to see if that helps

$$\rho^{n+1}_{j+1/2} = -0.5d\left(\alpha\left[\rho^{n}_{j+3/2} - \rho^{n}_{j-1/2}\right] + [1.-\alpha]\left[\rho^{n+1}_{j+3/2} - \rho^{n+1}_{j-1/2}\right]\right) + \rho^{n}_{j+1/2}, \quad (2.96)$$

where α is an arbitrary weighting between 0 and 1. Program this scheme, using the parameters shown in Figure 2.10, to determine which values of α give a reasonable solution. The numerical scheme will require an iteration at each time step to determine ρ^{n+1} at all points.

Numerical dispersion becomes particularly troublesome when we approach the grid resolution limit. Before small-scale structures are washed out by diffusion, a dispersive scheme may drown them entirely in numerical noise. To demonstrate this, let us employ Equations (2.63) to follow the propagation of sound waves with progressively shorter wavelengths. To see how the wavelength enters the problem, consider the simple one-dimensional wave

$$u(x,t) = \sin\left[2\pi\left(\frac{v}{\lambda}t - \frac{1}{\lambda}x\right)\right], \quad (2.97)$$

where λ is the wavelength and v is the propagation velocity, which in this case is c_s. Differentiation shows that this function is a solution to the wave equation (2.2). The wave frequency is given by $\nu = c_s/\lambda$.

The results for two different wavelengths are shown in Figure 2.11, and the conclusion is that short waves cannot be reliably propagated across the grid: dispersive

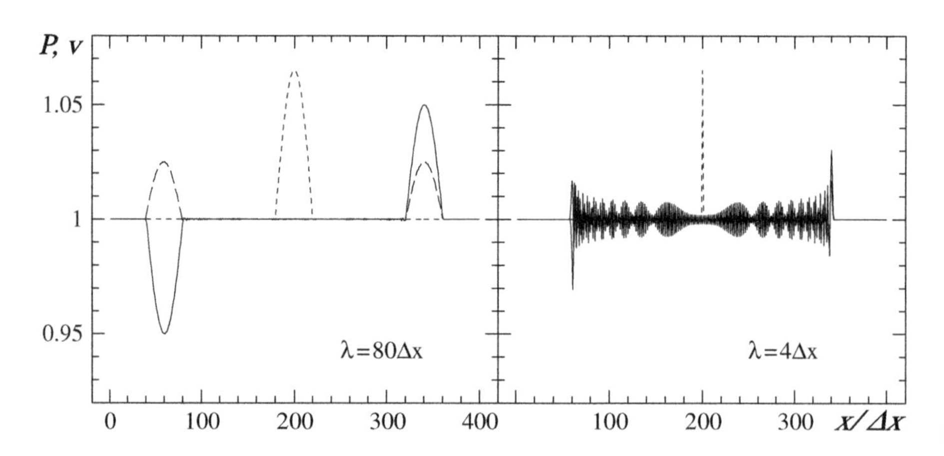

FIGURE 2.11 Propagation of small-amplitude sound waves followed with the help of Equation (2.63) and Scheme (2.66). An attempt to propagate waves with λ shorter than $\sim 10\Delta x$ results in generation of numerical noise. *Dotted and dashed lines*: initial and final pressure distribution, respectively (normalized to the unperturbed initial pressure). *Solid line*: final velocity distribution (arbitrarily normalized). Integration from $t = 0$ to $t = 140$; performed with $\Delta t = 0.99$ and $\Delta x = c_s = 1$.

effects cause the whole solution to oscillate whenever small-scale structures appear in the flow. The important practical meaning of this observation is that the grid cannot reliably resolve details smaller than about $10\Delta x$, which are much larger than the formal resolution limit that is set by the *Nyquist wavelength* λ_{Ny}.

To estimate λ_{Ny}, consider the simple wave from Equation (2.97). To uniquely represent such a wave on a grid, one has to sample it at least at the crest (where $u = 1$) and the nearest trough (where $u = -1$). This means that the distance between the crest and the nearest trough cannot be smaller than Δx. On the other hand, at all times the distance between the crest and the nearest trough is $\lambda/2$. Therefore,

$$\lambda \geq 2\Delta x \equiv \lambda_{Ny}. \tag{2.98}$$

The corresponding frequency limit, $\nu_{Ny} = v/\lambda_{Ny}$, is called the *Nyquist frequency*. As noted above, the practical limiting wavelength for a wave that can reliably propagate across the grid is significantly larger than λ_{Ny} (obviously, the limiting frequency is significantly lower than ν_{Ny}).

We conclude this section with a word of caution. A scheme that is accurate, stable, and consistent may prove useless because it is not *efficient* in terms of CPU time and/or the programmer's time for developing and debugging the numerical code, which constitutes its practical implementation. While the programmer's time is never easy to estimate before the task is accomplished, the efficiency of the code may be assessed with the help of a procedure outlined in Chapter 6, Section 6.3.5.

2.8 ALTERNATIVE METHODS

In finite difference schemes, differentials are replaced by finite differences, and the solution is computed on a grid of discrete points. An entirely different approach is employed in schemes based on finite element, Monte Carlo, and spectral methods. Another alternative to grid-based methods is provided by smoothed particle hydrodynamics (commonly referred to as SPH), which we will discuss in detail in Chapter 4.

In the finite element method, the entire solution domain is divided into smaller "finite elements" (Ames, 1992, Chap. 1). A finite element is an arbitrarily defined control volume, which could be a simple triangle in two space dimensions, an irregular tetrahedron in three space dimensions, or, in general, a volume bounded by curved surfaces. Within each element the solution is described by a simple function, such as a low-order polynomial. These functions may vary from element to element. The solution is found with the help of conditions defined at surfaces separating the elements. Additional conditions may occur at the boundary of the computational domain (see, e.g., Reddy and Gartling, 2001). The main advantage of finite element schemes is that they can deal with complex geometries and can arrange the elements into structures suited for various engineering applications (e.g., airflow over an airplane wing). Another advantage results from the fact that boundary conditions are easy to handle. For example, the elements can be adjusted to the shape of the wing, and the velocity component perpendicular to the surface of the wing can be set equal to zero. However, static structures — that is, fixed, special-purpose elements — generated by this method are generally of little interest in astrophysical problems, where nonstationary flows often occur. On the other hand, dynamical structures, in which the finite elements adjust themselves to the flow, would suffer from the same problems as do multidimensional Lagrangian grids (see the discussion at the end of Chapter 6, Section 6.2). For more details on this method see Griffiths (1986), Strang and Fix (1973), Wait and Mitchell (1985), Johnson (1987), and other textbooks on the subject.

Monte Carlo methods involve statistical sampling techniques in which the solution is gradually built up from simulations involving random numbers. For example, particles randomly drawn from a prescribed distribution are repeatedly launched into an absorbing and/or scattering medium, and their propagation accompanied by random interactions with the medium is followed until the emergent particle spectrum stabilizes. In astrophysics, Monte Carlo methods are particularly useful for treatment of complex radiation transfer problems, e.g., when the radiation and the medium through which photons propagate are far from a thermodynamical equilibrium, or when Comptonization must be accounted for.

A simplified possible procedure for a radiative transfer calculation runs as follows (Wolf et al., 1999). First, a volume in space is defined, e.g., a dusty cloud in space, with a given distribution of density and temperature. A large number of individual photons is introduced into the volume, with a range of frequencies appropriate to the source, each with a random initial position and direction. The source could be an imbedded star or an external interstellar radiation field. The second step involves propagation of each photon through the volume, taking into account its absorbing and scattering properties. The volume is subdivided into a large number of cells. Once the photon has propagated approximately one mean free path, it is absorbed or scattered, and reemitted in some other direction. The change in intensity is calculated according to Equation (1.90). The frequency of reemission of the photon, which depends in general on the local temperature, is generally different from its initial frequency. Its path is then followed through the volume in a straight line until the next scattering or absorption takes place. The trajectory of each photon is, thus, a random walk. The energy lost as a result of an absorption can be used to recalculate the temperature of the cell, so that, at the end, the radiation field and temperature distribution are consistent. The temperature in the cell determines the properties of the reemitted spectrum. The third step involves the analysis of the photons that leave the volume. Their spectrum, their spatial distribution of intensity at various frequencies, and other properties can be determined. The results will depend on the direction from which the cloud is observed. Once a sufficient number of photons has been introduced, the results should reach steady state. In general, Monte Carlo schemes are relatively simple to implement, but they may require very long CPU times (see Liu, 2002).

The key idea of spectral methods is to substitute the sought-for function $u(x, t)$ with its expansion into a series

$$u = \sum_{k=0}^{K} \hat{u}_k(t)\varphi_k(x), \tag{2.99}$$

where the functions $\varphi_k(x)$ form a complete basis, and expansion coefficients vary in time. Unlike the finite element method, the same expansion functions apply over the entire solution domain. Most frequently applied are Fourier expansions and/or expansions into the orthogonal polynomials of Chebyshev or Legendre. Such a substitution transforms any linear operation on u (e.g., partial derivative) into matrix multiplication

$$\frac{\partial}{\partial x}u = \sum_{k=0}^{K} \hat{u}_k \frac{\partial \varphi_k}{\partial x} = \sum_{k=0}^{K} \hat{u}_k \sum_{l=0}^{K} a_{kl}\varphi_l = \sum_{l=0}^{K}\left(\sum_{k=0}^{K} a_{kl}\hat{u}_k\right)\varphi_l, \tag{2.100}$$

where $\sum_{l=0}^{K} a_{kl}\varphi_l$ is the expansion of $\frac{\partial u}{\partial x}$. Thus, instead of the original PDE, a system of K ordinary differential equations for $\hat{u}_k(t)$ is solved. In many cases, the basis φ_k may be chosen such that its elements automatically satisfy the boundary conditions of the problem. Alternatively, the boundary conditions may be used to modify the expansion coefficients.

As an example, following Canuto et al. (1988, Chapter 3), consider the equation in one space dimension

$$\frac{\partial u}{\partial t} + G(u) + Lu = 0, \tag{2.101}$$

where $G(u)$ is a nonlinear operator, for example, $u(\partial u/\partial x)$, and L is a linear operator, for example, $\partial^2 u/\partial x^2$. We look for a solution that is periodic on the interval $[0, 2\pi]$ and call u^N the approximate solution to the problem. This function is expanded in a Fourier series

$$u^N(x,t) = \sum_{k=-N/2}^{N/2-1} \hat{u}_k(t)\exp(ikx), \tag{2.102}$$

where the $\hat{u}_k(t)$ are the unknowns to be found. The criterion to be met by the approximate solution is based on the minimization of the weighted residuals, where the weighting function, in this case, is chosen to be $\exp(-ikx)$ because of the orthonormality property

$$\int_0^{2\pi} \exp(ikx)\exp(-ilx)dx = 2\pi\delta_{kl}. \tag{2.103}$$

The weighted residual condition is:

$$\int_0^{2\pi}\left(\frac{\partial u^N}{\partial t} + G(u^N) + Lu^N\right)\exp(-ikx)dx = 0 \tag{2.104}$$

for $k = -N/2, \cdots, N/2-1$. This expression then reduces to

$$\frac{\partial \hat{u}_k}{\partial t} + \frac{1}{2\pi}\int_0^{2\pi} G(u^N)\exp(-ikx)dx + \frac{1}{2\pi}\int_0^{2\pi} Lu^N \exp(-ikx)dx = 0. \tag{2.105}$$

The problem reduces to N ordinary differential equations for the change in the $\hat{u}_k$ with time. (For technical reasons, it is customary to set $u_{-N/2} = 0$.) An initial condition for $u(x,t)$ must be specified, in which case the starting values for the $\hat{u}_k$ are:

$$\hat{u}_k(0) = \frac{1}{2\pi}\int_0^{2\pi} u(x,0)\exp(-ikx)dx. \tag{2.106}$$

This method is known as the *Fourier–Galerkin* technique. Although the function is represented spectrally in space, the time dependence is handled through standard finite difference techniques.

Exercise

Take G(u) = 0 and $Lu = -\frac{\partial^2 u}{\partial x^2}$. What are the resulting differential equations?

Another approach to the spectral method is to use Chebyshev polynomials. They are often convenient when the boundary conditions are not periodic. Suppose we consider Equation (2.101) on the interval $[-1, 1]$. Let u be expanded into a series of Chebyshev polynomials T_k

$$u = \sum_{k=0}^{N} \hat{u}_k(t) T_k(x) \tag{2.107}$$

where

$$T_k(x) = \cos(k \cos^{-1} x). \tag{2.108}$$

Because $T_k(\pm 1) = (\pm 1)^k$, we have

$$\begin{aligned} u(-1, t) &= \sum_{k=0}^{N} (-1)^k \hat{u}_k(t) \\ u(1, t) &= \sum_{k=0}^{N} \hat{u}_k(t). \end{aligned} \tag{2.109}$$

Suppose we assume a Dirichlet boundary condition

$$u(-1, t) = u(1, t) = 0. \tag{2.110}$$

We illustrate here the *collocation* method, in which the function u is determined on a finite set of grid points; the weighted residual method is not used, but instead it is simply required that the equation be satisfied at each of the grid points (note that Chebyshev polynomials can also be used in the Galerkin method and Fourier series in the collocation method). A popular set of grid points is:

$$x_j = \cos \frac{\pi j}{N} \tag{2.111}$$

for $j = 0, 1, \cdots, N$. Note that the collocation points are not distributed uniformly across the grid; they are closer together near the boundaries, giving higher accuracy there. The corresponding polynomials are:

$$T_k(x_j) = \cos \frac{\pi j k}{N}. \tag{2.112}$$

The equations to be solved are:

$$\frac{\partial u^N}{\partial t} + G(u^N) + Lu^N = 0 \tag{2.113}$$

for each j from 1 to $N-1$; at $j = 0$ and $j = N$ the boundary conditions are provided. The spatial derivatives involved in the operators are evaluated by series expansion as indicated schematically in Equation (2.100). For more details see Canuto et al. (1988, Section 2.4.2). In this case, the boundary conditions of the problem do not agree with the boundary conditions for the Chebyshev polynomials. The Dirichlet boundary condition can be satisfied simply by setting u^N to 0 at $x = -1$ and $+1$.

Another way to satisfy the boundary conditions is to modify the expansion coefficients, as mentioned above. As an example of the latter procedure consider a simple linear equation, e.g., the 1-D continuity equation (2.56), assuming that $-1 \leq x \leq +1$. Using again the Chebyshev collocation scheme, the time-integration scheme may be represented as a set of linear equations

$$L_{ij} u_j^{n+1} = w_i, \tag{2.114}$$

where i and j enumerate grid points; right-hand sides w_i are calculated from known values of u at previous time steps; and the coefficients L_{ij} characterize the scheme itself. Assume a boundary condition

$$u(-1, t) = g(t), \tag{2.115}$$

where g is a known function of time. Let $\tilde{u}^{n+1}$ be a solution of the system

$$L_{ij} \tilde{u}_j^{n+1} = z_i, \tag{2.116}$$

where all Chebyshev coefficients of z are equal to 0 except the last one

$$\hat{z}_k = (0, \ldots, 0, 1). \tag{2.117}$$

Then, the linear combination $U = u + \lambda \tilde{u}$, where

$$\lambda = \frac{g(t^{n+1}) - u(-1, t^{n+1})}{\tilde{u}(-1, t^{n+1})}, \tag{2.118}$$

satisfies the boundary condition at t^{n+1} by construction. Obviously, U does not exactly satisfy the original Equation (2.114). However, since only the last Chebyshev coefficient $\hat{u}_K$ was modified, U should not differ much from u.

As long as we deal with smooth solutions, spectral schemes (with a given number of polynomials) can be considerably more accurate than finite-difference schemes with the corresponding number of grid points. In fact, it can be shown (e.g., Marcus, 1986) that for analytical (i.e., infinitely differentiable) functions, the error decays exponentially with the number of expansion terms, while for a second-order finite difference scheme, the error decays only quadratically with the number of grid points. Another advantage of spectral schemes is their vanishing internal diffusivity. However, for nonsmooth solutions (which are standard in astrophysics), the error in the kth coefficient in the expansion decays like k^m, where m is the order of the first noncontinuous derivative of u. Thus, the method is not well adapted to the treatment of discontinuities. In fact, shocks can be followed with a spectral method, but they must be artificially broadened over a few grid points. Hydrodynamical applications of the spectral method are thoroughly discussed by Canuto et al. (1988). An interesting example is provided by Godon (1997), who does a two-dimensional hydrodynamic calculation of a disk, using a Chebyshev collocation method in the radial direction and a Fourier collocation method in the azimuthal direction.

In conclusion, we mention the *anelastic approximation* that is useful for certain problems. The normal equations of hydrodynamics allow for the propagation of pressure waves at the speed of sound, with an associated limit on the time step in an explicit procedure. If the sound waves can be eliminated and if the fluid velocities are very small compared to sound velocity, then, in principle, a considerable increase in the allowed time step is possible. The anelastic method filters out acoustic motions (Gough, 1969) and is formulated as follows:

1. All thermodynamic variables are expressed as the sum of an average value at a given point, plus a small, linearized perturbation. The approximation requires that the perturbation be small.
2. The hydrodynamic equations are expressed in the form that calculates the evolution of the perturbations, assuming that the average state, or *reference state*, does not change.
3. The perturbed velocities are not assumed to be small and they are not linearized.
4. The continuity equation is replaced by $\nabla \cdot \bar{\rho} \mathbf{v} = 0$ where $\bar{\rho}$ is the density of the reference state and $\mathbf{v}$ is the perturbed velocity vector.

It is this last approximation that is the key to the filtering out of the sound waves. The reference state is provided as an initial condition, as, for example, a stellar model. For this method to work the perturbations on the reference state must actually be small, $\approx 1\%$. If over time the cumulative perturbation becomes too large, the reference state itself must be evolved (Rogers and Glatzmaier, 2006). Since the approximation involves highly subsonic flow, a natural technique to use with it is the spectral method, but other numerical methods, such as finite differences, can be used as well.

REFERENCES

Ames, W. F. (1992) *Numerical Methods for Partial Differential Equations,* 3rd ed. (Boston: Academic Press).

Bowers, R. L. and Wilson, J. R. (1991) *Numerical Modeling in Applied Physics and Astrophysics* (Boston: Jones and Bartlett).

Canuto, C., Hussaini, M. Y., Quarteroni, A., and Zang, T. A. (1988) *Spectral Methods in Fluid Dynamics* (Berlin: Springer-Verlag).

Courant, R., Friedrichs, K. O. and Lewy, H. (1928) *Mathematische Annalen* **100**: 32.

Godon, P. (1997) *Astrophys. J.* **480**: 329.

Gough, D. O. (1969) *J. Atmos. Sci.* **26**: 448.

Griffiths, D. F. (1986) *Astrophysical Radiation Hydrodynamics*, K.-H. Winkler and M. L. Norman, Eds. (Dordrecht: Reidel), p. 327.

Johnson, C. (1987) *Numerical Solution of Partial Differential Equations by the Finite Element Method* (Cambridge: Cambridge University Press).

Liu, J. S. (2002) *Monte Carlo Strategies for Scientific Computing* (Berlin: Springer-Verlag).

Marcus, P. S. (1986) *Astrophysical Radiation Hydrodynamics*, K.-H. Winkler and M. L. Norman, Eds. (Dordrecht: Reidel) p. 359.
Morse, P. M. and Feshbach, H. (1953) *Methods of Theoretical Physics* (New York: McGraw-Hill).
Reddy, J. N. and Gartling, D. K. (2001) *The Finite Element Method in Heat Transfer and Fluid Dynamics* (Boca Raton, FL: CRC Press).
Richtmyer, R. D. (1957) *Difference Methods for Initial-Value Problems* (New York: Interscience).
Rogers, T. and Glatzmaier, G. (2006) *Astrophys. J.* (in press).
Strang, G. and Fix, G. J. (1973) *An Analysis of the Finite Element Method* (Englewood Cliffs, NJ: Prentice-Hall).
Wait, R. and Mitchell, A. R. (1985) *Finite Element Analysis and Applications* (New York: John Wiley & Sons).
Wendt, J. F. (1992) *Computational Fluid Dynamics* (Berlin: Springer-Verlag).
Wolf, S., Henning, Th., and Stecklum, B. (1999) *Astron. Astrophys.* **349**: 839.

3 N-Body Particle Methods

3.1 INTRODUCTION TO THE N-BODY PROBLEM

The classical astrophysical "N-body" problem is deceptively simple in appearance, and its formulation is a model of brevity: Each member within an aggregate of N ($i = 1, \ldots, N$) point mass bodies, having masses m_i, experiences an acceleration that arises from the gravitational attractions of all the other bodies in the system

$$\frac{d^2\mathbf{x}_i}{dt^2} = -\sum_{j=1;\, j\neq i}^{N} \frac{Gm_j(\mathbf{x}_i - \mathbf{x}_j)}{|\mathbf{x}_i - \mathbf{x}_j|^3}. \tag{3.1}$$

The description of the problem is completed by specifying the initial velocities $\mathbf{v}_i(t = 0)$ and positions $\mathbf{x}_i(t = 0)$ for the N particles.

Equation (3.1) was originally stated and used by Newton and has been intensively studied for more than 300 years. Solutions to this equation describe intricate and diverse phenomena ranging from the orbit of the Moon around the Earth to the structure of the Kirkwood gaps in the asteroid belt to the evolution of globular clusters to the spiral structure of galaxies. Indeed, one could assemble a brilliant and fully modern career in astronomy by focusing exclusively on aspects of Equation (3.1). The richness of Equation (3.1) derives from its nonlinearity. The positions, $\mathbf{x}_i$, which are doubly differentiated to yield accelerations, appear in a sum of terms involving $|\mathbf{x}_i - \mathbf{x}_j|^{-2}$. Because the rate of change of each particle's velocity arises from a sum of inverse-square dependences, a slight change in initial conditions can lead to a complete change of long-term behavior. The goal of an N-body numerical method is to capture, as accurately as possible, this endless complexity.

The N-body problem divides naturally into two basic parts: (1) calculating the net force on a given particle at a given time, and (2) determining the new position of the particle at a somewhat advanced time. In this chapter we consider five different methods for advancing the positions of the particles. The first, Runge–Kutta integration, is a standard technique for solving ordinary differential equations. The second, Bulirsch–Stoer, provides a highly accurate method, which in practice is usually limited to systems of only a few bodies, such as nine planets orbiting a central star. The core of this method is *Richardson extrapolation* or *Richardson's deferred approach to the limit*, in which the position of a particle is integrated up to a given time with several different time step lengths, then the results are extrapolated to what they would be for a time step of zero length. The third method is known as the *symplectic map*, which has its major application in the integration of the orbits in a planetary system for very long times, but within the context where close encounters between bodies do not occur. Its speed is derived by capitalizing on the near Keplerian motion of a system of planets. The fourth method is essentially a predictor–corrector method, as applied by Aarseth to systems with moderate to large numbers of particles when reasonable accuracy is still required. The fifth, a second-order leapfrog method, is used for extremely large numbers of particles when integrations over only a few dynamical times are required and high-order accuracy is not required. All of these methods may run into difficulties when two particles pass within short distances of each other. Such close encounters may be treated either by an inaccurate method, known as *softening*, or by an accurate method, known as *regularization*. Finally in this chapter, we give two examples of the approximate calculation of gravitational forces that are particularly applicable to large N systems: the hierarchical *tree* method and the method of Fourier analysis. A more detailed discussion of methods for calculating gravitational forces is given in Chapter 7.

3.2 EULER AND RUNGE–KUTTA METHODS

In attacking Equation (3.1) numerically (or, for that manner, any higher-order ordinary differential equation), one first dispenses with second derivatives by recasting the N second-order equations as a set of $2N$ coupled first-order equations. That is, any ordinary second-order differential equation of the generic form

$$A(x,t)\frac{d^2x}{dt^2} + B(x,t)\frac{dx}{dt} + C(x,t) = 0, \tag{3.2}$$

can be rewritten as a pair of coupled first-order differential equations

$$\frac{dx}{dt} = v(t), \tag{3.3}$$

$$\frac{dv}{dt} = -\frac{B(x,t)}{A(x,t)}v(t) - \frac{C(x,t)}{A(x,t)}. \tag{3.4}$$

In this manner, the N second-order equations (3.1) can be expressed as a coupled set of $2N$ first-order equations

$$\frac{d\mathbf{x}_i}{dt} = \mathbf{v}_i, \tag{3.5}$$

$$\frac{d\mathbf{v}_i}{dt} = -\sum_{j=1;\, j\neq i}^{N} \frac{Gm_j(\mathbf{x}_i - \mathbf{x}_j)}{|\mathbf{x}_i - \mathbf{x}_j|^3}. \tag{3.6}$$

If we write $\mathbf{w}_i \equiv [\mathbf{x}_i, \mathbf{v}_i] \equiv (w_{i1}, w_{i2}, w_{i3}, w_{i4}, w_{i5}, w_{i6})$ to represent the six-dimensional phase space coordinates of particle i, then we can describe the state of the entire N-body system by a single $6N$-dimensional vector

$$\mathbf{W} \equiv [\mathbf{w}_1, \ldots, \mathbf{w}_N]. \tag{3.7}$$

The components, W_l, of $\mathbf{W}$ run from $l = 1 \rightarrow 6N$. For example, the x-component of the velocity of the $j = 3$ particle corresponds to the term W_{16}. The evolution of the system as described by Equation (3.5) and Equation (3.6), thus, takes the form

$$\frac{dW_l}{dt} = g_l(\mathbf{W}), \tag{3.8}$$

where the $6N$ functions g_l are given by the right-hand sides of Equation (3.5) and Equation (3.6).

When the equations are written in the form given by Equation (3.8), the notation reinforces the point that a single integration routine can be used for many problems by simply respecifying the driving functions g_l. Note that while the gravitational force law (Equation 3.1) has no explicit time dependence, a general-purpose integrator has no problem advancing functions of the form $f_i(t, \mathbf{W})$ that do contain explicit time dependences. Certainly, when it seems that a specialized N-body algorithm is a hopelessly dense thicket of arcane techniques, it helps to keep in mind that at the end of the day, one is just integrating a fixed set of first-order ordinary differential equations.

Revisiting the discussion from the previous chapter, recall that an ordinary differential equation describes how a dependent variable, such as a particle's position or velocity, changes smoothly in response to variation of the independent variable (which in the N-body problem is the time). The differential equations are a recipe that describes a continuous sequence of changes, and in order to obtain the particles' trajectories, the specification of $6N$ boundary values is required. Since, in general, the positions and velocities of the particles are all known at a specified starting moment, the N-body problem is an "initial value problem."

The simplest solution algorithm for the gravitational N-body problem proceeds by constructing a basic finite difference representation of the differential equations (3.8) over the interval $h \equiv \Delta t \equiv t^{n+1} - t^n$

$$W_l^{n+1} = W_l^n + h\, g_l\left(W_1^n, \ldots, W_l^n, \ldots, W_{6N}^n\right) = W_l^n + h\, g_l(\mathbf{W}^n). \tag{3.9}$$

With this formula, the updated set of dependent variables, W_l^{n+1}, is computed using only information available at the initial time, t^n. (Note that the superscripts n and $n+1$ are time step indices, and do not indicate powers of n.) By repeatedly applying the finite difference approximation (Equation 3.9), particle trajectories are marched forward in intervals of length h, and the solution is built up as a set of explicit updates to the initial positions and velocities. This straightforward approach is known as *Euler's method*, and although it is not appropriate for production work, it is well worth coding up to form the simplest possible incarnation of an N-body program. If you are new to numerical methods, there is no substitute for such a hands-on introduction.

Any finite-difference scheme is necessarily an approximation to the continuous variations dictated by the actual differential equations, and for the N-body problem, we get an enormous advantage from the global conservation of energy and angular momentum. That is, $E_{\rm tot}$ and $\mathbf{L}_{\rm tot}$ both remain constant as the particles trace through their trajectories, where

$$E_{\rm tot} = \frac{1}{2} m_{\rm tot} v_{\rm com}^2 - \frac{1}{2} \sum_{i=1}^{N} \sum_{j=1;j\neq i}^{N} \frac{G m_i m_j}{r_{ij}} + \sum_{i=1}^{N} \frac{1}{2} m_i v_i^2, \tag{3.10}$$

and

$$\mathbf{L}_{\rm tot} = \sum_{i=1}^{N} m_i (\mathbf{x}_i \times \mathbf{v}_i), \tag{3.11}$$

where v_{com} is the velocity of the center of mass of the system, r_{ij} is the distance between particles i and j, and

$$m_{\mathrm{tot}} = \sum_{i=1}^{N} m_i \,. \tag{3.12}$$

The degree to which these quantities are conserved in a numerical solution scheme, for example,

$$\Delta E = \frac{E^{\mathrm{final}} - E^{\mathrm{initial}}}{E^{\mathrm{initial}}}, \tag{3.13}$$

is a measure of the fidelity to which the solution is being obtained. Good conservation of E_{tot} and $\mathbf{L}_{\mathrm{tot}}$ greatly bolsters confidence in the integrity of a numerical N-body integration. In general, however, for an arbitrary set of ODEs, convenient diagnostics like E_{tot} and $\mathbf{L}_{\mathrm{tot}}$ are not available. To stay on the right track, one needs to monitor accuracy by evaluating the fractional change in the solution obtained with varying step sizes, h.

Figure 3.1 shows the fractional change in energy of an Earth-mass planet started in a circular 1 AU orbit about a star of 1 $M_{\odot}$, as described by Equation (3.5) and Equation (3.6), numerically integrated with Euler's method for 100 years. The updates to $\mathbf{W}$ are performed with time steps of various lengths Δt. It is clear that at first, as

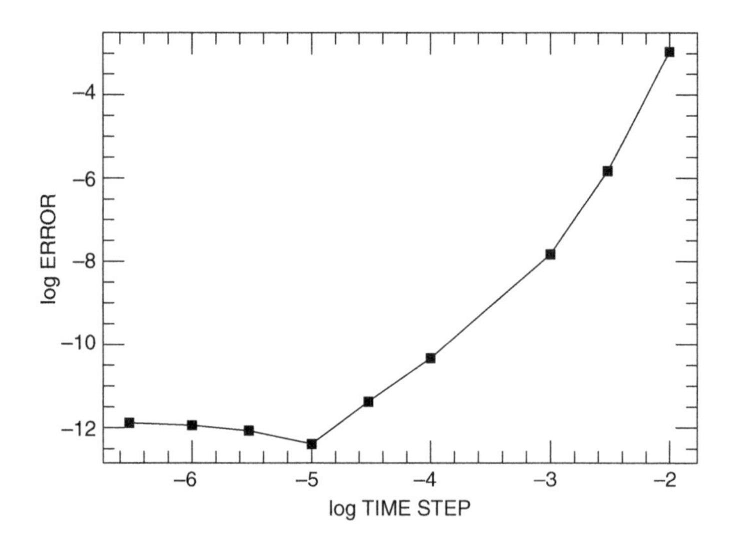

FIGURE 3.1 The log of the absolute value of the error ΔE is plotted as a function of the log of the time step (in years) for the integration by Euler's method of the orbit of a planet for 100 years at 1 AU from a star of 1 solar mass.

Δt decreases, the solution becomes increasingly accurate, albeit at an ever-increasing computational cost.

Curiously, however, Figure 3.1 also shows that the benefit obtained from using ever-smaller time step intervals saturates at a time step $\Delta t = 1 \times 10^{-5}$. This results from the computer's ability to represent numbers to only a finite number of decimal places. In the example shown, the variables are defaulted to "double precision," i.e., 14-digit precision. As the error ΔE approaches this limiting resolution, successive updates are given stochastic contributions resulting from roundoff error and the accuracy goes down. The combination of roundoff error and excessive computer time demands effectively shuts down the ability of Euler's method to follow the system for periods exceeding 10^5 years (which would, in this case, require 67 hours on a fast 2004-era desktop computer) and, more devastatingly, would entail an error buildup $\Delta E \sim 1 \times 10^{-1}$.

To trace the oppressive error buildup produced by Euler's method, we note that the exact value of the advanced dependent variable W_l^{n+1} (Equation 3.9) is given by a Taylor series composed of an infinite number of terms

$$W_l^{n+1} = W_l^n + \frac{h}{1!}\frac{dW_l}{dt} + \frac{h^2}{2!}\frac{d^2W_l}{dt^2} + \cdots + \frac{h^n}{n!}\frac{d^nW_l}{dt^n} + \cdots \tag{3.14}$$

where the derivatives are evaluated at time n. Note that the first two terms in this series comprise Euler's formula. When Euler's method is employed, the rest of the terms sum to give the total error in the finite difference estimate of W_l^{n+1}. The numerical time derivative, as determined by Euler's method, is said to be "first-order" accurate.

The fundamental unworkability of Euler's formula stems from its asymmetry; the increment $h dW_l/dt \equiv h g_l(t^n, \mathbf{W^n})$ to the dependent variable is based solely on the value of $\mathbf{W}$ evaluated at the beginning of the interval h. Any nonlinearity in g implies an inaccuracy in the updated value of the dependent variables. The key to improving Euler's method is the realization that the derivative function g_l, in general, can be computed for any trial values of $(t, \mathbf{W})$, where here we include the implicit time dependence in g_l. A higher order (and faster and more accurate) integration scheme, thus, can be designed in which the behavior of $g_l(t, \mathbf{W})$ in essence provides better guidance for the augmentation of W_l^n. In drawing together information regarding the topography of $g_l(t, \mathbf{W})$, a refined estimate for the slope $\overline{g_l(t, \mathbf{W})}$, needed to accurately update W_l over a particular interval h, can be obtained in terms of a weighted sum of values $g_l(t, \mathbf{W})$. Implementations of this strategy are called Runge–Kutta methods and the weightings applied to each of the k estimates of the slope can be tuned to provide cancellation of error terms in the Taylor series up to order $k + 1$.

The simplest Runge–Kutta integration scheme consists of first using the slopes $\mathbf{g}(t^n, \mathbf{W^n})$ and the Euler method to estimate the values $\mathbf{W}(t^n + h/2)$ at a point halfway through the desired interval h

$$\mathbf{W}\left(t^n + \frac{h}{2}\right) = \mathbf{W}_b = \mathbf{W}(t^n) + \frac{h}{2}\mathbf{g}(t^n, \mathbf{W}^n). \tag{3.15}$$

The values $\mathbf{W}_b$ better typify the true values of $\mathbf{W}$ within the interval $(t^n, t^n + h)$. A refined estimate $\mathbf{g}_b(t^n + h/2, \mathbf{W}_b)$, thus, provides a more accurate value for the slope required to augment $\mathbf{W}$ over the interval h.

A particular implementation of a Runge–Kutta integration formula is the heavily used classical fourth-order scheme, which consists of the following sequence of operations, involving four evaluations of the derivative functions $\mathbf{g}(t, \mathbf{W})$

$$\mathbf{f}_a = \mathbf{g}(t^n, \mathbf{W}^n) \tag{3.16}$$

$$\mathbf{W}_b = \mathbf{W}^n + \frac{h}{2}\mathbf{f}_a \tag{3.17}$$

$$\mathbf{f}_b = \mathbf{g}\left(t^n + \frac{h}{2}, \mathbf{W}_b\right) \tag{3.18}$$

$$\mathbf{W}_c = \mathbf{W}^n + \frac{h}{2}\mathbf{f}_b \tag{3.19}$$

$$\mathbf{f}_c = \mathbf{g}\left(t^n + \frac{h}{2}, \mathbf{W}_c\right) \tag{3.20}$$

$$\mathbf{W}_d = \mathbf{W}^n + h\mathbf{f}_c \tag{3.21}$$

$$\mathbf{f}_d = \mathbf{g}(t^n + h, \mathbf{W}_d) \tag{3.22}$$

$$\mathbf{W}^{n+1} = \mathbf{W}^n + \frac{1}{6}h\mathbf{f}_a + \frac{1}{3}h\mathbf{f}_b + \frac{1}{3}h\mathbf{f}_c + \frac{1}{6}h\mathbf{f}_d. \tag{3.23}$$

That is, a single integration step proceeds by evaluating (1) the initial slope, (2, 3) two successive slope estimates at the interval midpoint, and (4) an estimate of the slope at the end of the step interval. The four slopes are then combined, with weights $\frac{1}{6}, \frac{1}{3}, \frac{1}{3}, \frac{1}{6}$ to produce an average slope capable of accurately bridging the interval h with a single Euler-method step.

To see in detail how this works, consider the concrete example of a nonlinear, first-order ordinary differential equation

$$\frac{dy}{dx} = f(x, y) = \left(x + \frac{y}{4}\right)^{1/2}, \tag{3.24}$$

to be integrated from the initial conditions $x = 1$, $y(1) = 1$ to $x = 2$. Using Euler's method to step across the interval yields a value

$$y(2) \sim y(1) + \Delta x\left[x(1) + \frac{y(1)}{4}\right]^{1/2} = 2.118. \tag{3.25}$$

The second-order Runge–Kutta formula gives a refined intermediate value $y(1.5) \sim 1.559$ and a revised slope estimate $f(1.5, 1.559) = 1.375$, yielding a revised end-point solution $y(2) \sim y(1) + \Delta x f(1.5, 1.559) = 2.37468$. The fourth-order Runge–Kutta Scheme (3.23) produces a value $y(2) = 2.37523$, which is quite close to the true value.

Fourth-order Runga–Kutta integration is at its best when ease of programming is of greater concern than actual computer speed. For example, one might want to check whether two planets orbiting an oblate star are in resonance or, alternately, one might be interested in tracing the orbital behavior of stars in a particular realization of a galactic potential. In both of these cases, the functions $g_l(t, \mathbf{W})$ require modification from those in Equation (3.1), which might be time consuming to implement in a specialized preexisting code.

Often, in order to make a Runga–Kutta integration practical, one needs the ability to adaptively increase and decrease the time step as the integration proceeds. When bodies are close together or, more generally, when $g_l(t, \mathbf{W})$ is large, a small time step is required to maintain accuracy. If g_l becomes very small, a calculation can effectively grind to a halt if the time step is too small. *Step doubling* provides a simple way to implement adaptive time step control. One first advances the solution vector $\mathbf{W}^n$ by taking two Runge-Kutta time steps to cover an interval $2h$. Upon completion of the two steps, one has an estimate of the solution vector, which we call $\mathbf{W}_1$. One then makes a second estimate, $\mathbf{W}'_1$ by bridging the same interval with one large step of duration $(2h)$. Because the same starting derivative can be used in both cases, computing the time step of length $2h$ adds an overhead factor of $11/8 = 1.375$ to the computational cost.

Step doubling presents us with two estimates of the solution at $t = t^n + 2h$ and, thus, allows us to obtain a fractional estimate of the error

$$\delta_e = \left| \frac{(\mathbf{W}_1 - \mathbf{W}'_1)}{\mathbf{W}_1} \right| . \tag{3.26}$$

If $\delta_e \geq \delta_{max}$, where δ_{max} is a predefined relative time step accuracy criterion, then adequate convergence has not been achieved. In this event, the time step is decreased (repeatedly if necessary) and the estimates $\mathbf{W}_1$ and $\mathbf{W}'_1$ are recomputed until $\delta_e \leq \delta_{max}$. On the other hand, if $\delta_e \leq \delta_{min}$, where δ_{min} is another predefined criterion defined such that $\delta_{min} < \delta_{max}$, then an attempt can be made to integrate the next $2h$ interval with a larger time step. The simple adaptivity imparted by step doubling can dramatically increase the efficiency of an integration.

Often, however, in a standard N-body integration, both small and large time steps are required simultaneously as one advances the set of particles. In a globular cluster, the tight binary stars that provide a source of gravitational energy for the cluster center have periods measured in hours, while loosely bound stars on the distant fringes take literally millions of years to fall from one side of the cluster to the other. In such cases, simple adaptive methods like time step doubling will not be adequate, and more specialized techniques, such as following each particle with its own time step, must be adopted.

3.3 THE DESCRIPTION OF ORBITAL MOTION IN TERMS OF ORBITAL ELEMENTS

A common problem in N-body dynamics involves the motion of one or more low-mass objects in orbit around a central body of considerably higher mass. The next four sections are primarily devoted to this problem. The motion of two point masses with any relative position and velocity can be described exactly, and it is no exaggeration to say that this result is one of the high-water marks of astronomy. For a pair of particles,

Equation (3.1) reduces to a single expression that encapsulates the time evolution of the separation $\mathbf{r}$ of the two bodies

$$\frac{d^2\mathbf{r}}{dt^2} = \frac{-G(m_1+m_2)}{r^3}\mathbf{r}, \tag{3.27}$$

where $r = |\mathbf{r}|$. Conservation of angular momentum requires that the motion of the two bodies takes place in a plane, so we can switch to a polar coordinate system in which the origin is located at the position of mass m_1, and the angular position θ of mass m_2 is measured relative to a specified (but arbitrary) reference direction. It is then straightforward to show that the equation of relative motion for the bodies becomes

$$\frac{d^2r}{dt^2} - r\left(\frac{d\theta}{dt}\right)^2 = \frac{-G(m_1+m_2)}{r^2}. \tag{3.28}$$

Because the specific angular momentum $l = r^2\dot{\theta}$ is conserved, the above equation can be written

$$\ddot{r} = \frac{-G(m_1+m_2)}{r^2} + \frac{l^2}{r^3}, \tag{3.29}$$

indicating that the separation of the particles can be viewed as governed by the attractive force of gravity, as well as a repulsive restoring force provided by the angular momentum.

The time dependence in the equation of motion can be removed with the goal of obtaining a solution $r(\theta)$ that describes the entire locus of the relative positions of the particles in space. This is done by defining $u = 1/r$ and recasting Equation (3.28) as

$$\frac{d^2u}{d\theta^2} + u = \frac{G(m_1+m_2)}{l^2}. \tag{3.30}$$

The solution to this ordinary, linear, second-order differential equation is:

$$u = \frac{G(m_1+m_2)}{l^2}[1 + e\cos(\theta - \varpi)], \tag{3.31}$$

or, rewriting using r,

$$r = \frac{p}{1 + e\cos(\theta - \varpi)}. \tag{3.32}$$

The quantity $p = l^2/[G(m_1+m_2)]$ is (rather antiquatedly) known as the *semilatus rectum*. The quantities e and ϖ are the two required constants of integration. The full generic range of two-body motion is provided by varying the constant e. The phase ϖ simply describes the orientation of the motion relative to the arbitrarily chosen reference direction. Because $\cos(\theta) = -\cos(\theta + \pi)$, we lose no generality in restricting the constant e (called the orbital eccentricity) to values $e > 0$.

When $e = 0$, the motion is simply a circle of radius $r = p$ with the familiar expression of the circular orbital velocity $v = \sqrt{G(m_1+m_2)/r}$. Whenever $0 < e < 1$, the vector separating the bodies explores the full 2π range of angle, and is, therefore, bounded. The motion traces out an ellipse of semimajor axis $a = p/(1 - e^2)$, with

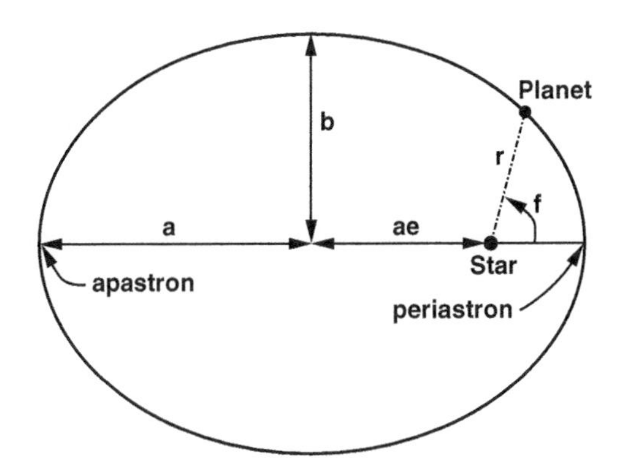

FIGURE 3.2 Diagram for an ellipse, illustrating the orbital elements a and e, and the semi-minor axis b.

the origin (the position of body m_1) located at one of the foci. For $e > 1$, the motion is hyperbolic, whereas $e = 1$ yields the separating case of a parabolic orbit.

The solution of the two-body problem indicates that, if the velocities and positions of the bodies are known at an initial time, then the subsequent motion is completely determined. The most analytically natural and intuitively pleasing description of the initial condition occurs when the orbit is described in terms of orbital elements (Figure 3.2 and Figure 3.3), which are a coordinate system basis that combines both positions and velocities. The full set of six orbital elements that describe a particle's position and velocity is given in the following list.

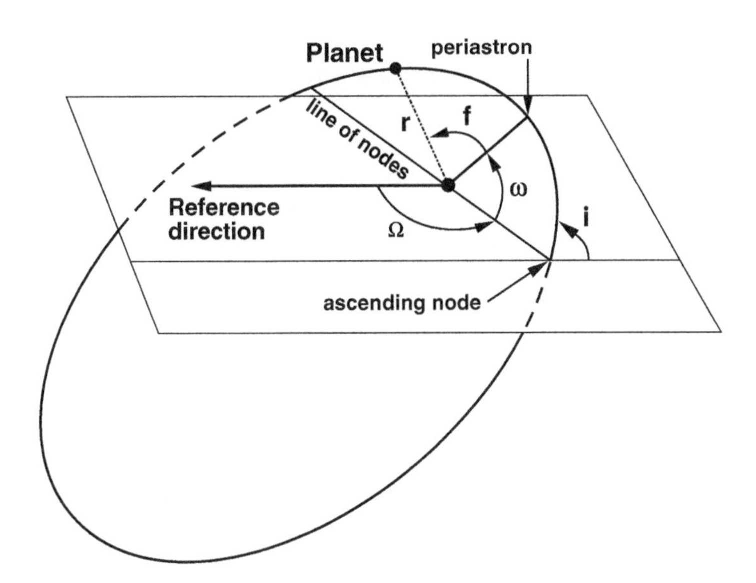

FIGURE 3.3 Diagram for an ellipse, illustrating the orbital elements Ω, ω, i, and f. The parallelogram is the reference plane inclined to the orbital plane by the angle i.

1. P, the period of the orbit, is related to the semimajor axis a via Kepler's Third Law

$$P^2 = \frac{4\pi^2}{G(m_1 + m_2)} a^3 \tag{3.33}$$

The quantity $n = 2\pi/P$, called the *mean motion*, is sometimes listed in lieu of P.

2. The eccentricity, e, determines the shape of the orbit. It is related to the fundamental constants of the system by

$$e = \left(1 + \frac{2El^2}{G^2(m_1 + m_2)^2}\right)^{1/2} \tag{3.34}$$

where E is the specific energy of the orbit

$$E = \frac{1}{2}v^2 - \frac{G(m_1 + m_2)}{r} = -\frac{G(m_1 + m_2)}{2a} \tag{3.35}$$

or, in terms of the semimajor axis a and the semiminor axis b,

$$e^2 = 1 - \frac{b^2}{a^2}. \tag{3.36}$$

3. M, the *mean anomaly*, is an angular measure that increases uniformly with time (a "clock hand") from 0 to 2π during the course of an orbit. M is related to the time through

$$M = n(t - T_{peri}), \tag{3.37}$$

where the *time of periastron passage*, T_{peri}, is the time at which m_1 and m_2 are at minimum separation, and n is the mean motion. From M and e, one can derive the angle f (*the true anomaly*), which describes the actual position of the planet in its orbit, relative to periastron, at time t.

4. The inclination, i, is the angle between an arbitrary reference plane and the plane of the orbit. In the convention shown in Figure 3.3, i is the angle between the orbital plane and the (x, y) plane of the (x, y, z) Cartesian coordinate system. In the solar system, the reference plane is usually taken to be that of the Earth's orbit.

5. The longitude of the ascending node, Ω, is defined using the *line of nodes*, which is the intersection line between the orbital plane and the (x, y) (reference) plane. Ω is then the angle between the line of nodes and an arbitrary reference direction in the reference plane, which in Figure 3.3 is the $-x$ direction.

6. The longitude of periastron, ϖ, is defined by

$$\varpi = \Omega + \omega \tag{3.38}$$

where ω, the *argument of periastron*, is the angle from the line of nodes to the periastron point. As such, ϖ is a dogleg angle, since it is composed as the sum of angles defined in two different planes.

The last three orbital elements fix the orbit's orientation in space. The transformations between the Cartesian system $(x, y, z, \dot{x}, \dot{y}, \dot{z})$ and the orbital element system $(P, M, e, i, \Omega, \varpi)$ are described in Murray and Dermott (1999). In general, if a system contains more than two bodies, then the concept of an elliptical orbit is only approximate, and all of the orbital elements associated with an orbiting body will vary with time. In this case, the state of the system defined at one particular epoch is referred to as a set of *osculating* orbital elements.

For simulations that are tied to actual observations of either solar system bodies or extrasolar planets, time is referenced to a continuous sequence of days called the Julian date (JD). A Julian day is defined as 86,400 seconds and the Julian date (or day number) is the number of such days that have elapsed since noon (universal time [UT]) on January 1, 4713 BC. Noon (UT) on New Year's Day 2000 corresponded to JD 2451545. The peculiar starting time of the Julian day sequence arises from the so-called Julian Cycle of 7980 years invented by Joseph Scaliger in 1583. The Julian Cycle and its starting date have no physical or astronomical significance, but the Julian date is, however, very widely used in astronomy. The conversion between JD and the calendar date is done using formulae that take leap years (and the 10 days omitted in October 1582 due to the conversion from the Julian to the Gregorian calendar) into account (see Montenbruck, 1989).*

The discovery of extrasolar planets (see, e.g., Marcy et al., 2000) is providing an exciting influx of opportunities for numerical few-body simulations that use the techniques described in this chapter. New planetary systems are currently being announced at a rate of several per month, and the number of known systems containing multiple planets is increasing quickly. With each discovery comes dynamical questions. Could terrestrial planets on habitable orbits exist within a particular system? Are there configurations that are consistent with the data and yet dynamically unstable? Are there alternate configurations of planets, which also provide good fits to the data?

A number of techniques have been used and proposed for detecting extrasolar planets — so far, the most successful have been the radial velocity and transit methods. Transits are detected by monitoring the stellar light output for the characteristic periodic dips in flux that occur when a planet passes in front of the star. This effect usually leads to a 1 to 2% decrease in brightness. The probability of observing a transit depends on the size of the star and the orientation and period of the orbit. For typical short-period "Hot Jupiter"-type planets with orbital periods of 3 to 5 days, transit probabilities are on the order of 10%. Transit measurements give the inclination i, which is the angle between the plane of the sky and the orbital plane, as well as P and the ratio of the planetary radius to the stellar radius.

* For phenomena, such as extrasolar planetary transits that require precise timing, the Julian date on Earth can be converted to the heliocentric Julian date at the solar system barycenter. For a given line of sight, this conversion accounts for the varying light travel time across the Earth's orbit during different parts of the year.

The stellar radial velocity half-amplitude, K, induced by one planet, m_2, orbiting the star m_1 is defined by

$$K = \left(\frac{2\pi G}{P}\right)^{1/3} \frac{m_2 \sin(i)}{(m_1 + m_2)^{2/3}} \frac{1}{\sqrt{1-e^2}} . \tag{3.39}$$

The inclination i is generally unknown unless the planet transits. Very accurate measurements of the Doppler shift of the absorption lines in the star's spectrum allow the velocity component of the star's motion along the line of sight to be determined at different times. Using the equations for Keplerian orbital motion, the star's line-of-sight velocity induced by the planet, as a function of time, is given by

$$v_r = K[\cos(f + \omega) + e \cos \omega], \tag{3.40}$$

where the *true anomaly*, f, is defined by

$$\tan(f/2) = \sqrt{\frac{1+e}{1-e}} \tan E/2 \tag{3.41}$$

and the *eccentric anomaly*, E, is computed by solving Kepler's equation at time t

$$\frac{2\pi}{P}(t - T_{peri}) = M = E - e \sin E . \tag{3.42}$$

Thus, from the observations of v_r vs. t, one obtains K, e, and P. The solution for E, along with the known e, allows one to determine the actual position of the planet in its orbit, relative to its periastron position, at a given time t. Equation (3.39) then gives the minimum mass of the planet, $m_2 \sin(i)$, as long as one has knowledge of the mass of the star and can assume that $m_2 << m_1$.

It is useful to remark on units. Imagine two 1-centimeter cubes of ice in space separated by 1 centimeter. A simple dimensional argument shows that if the cubes start from rest, they take approximately $\sqrt{\frac{1}{G}} \sim 1$ hour to come together. The cgs system of units (centimeters-grams-seconds) is not optimally tuned for a numerical solution of the problem, since 1 second is considerably smaller than the hour-long characteristic time. Indeed, for an N-body problem involving gram-sized masses separated by distances of a centimeter, an hour is a perfectly "natural" unit for measuring time, as it is the characteristic timescale over which significant evolution occurs.

In general, in an N-body problem, it is traditional to work in units where the gravitational constant

$$G = 1 \frac{(LU)^3}{(TU)^2 MU} \tag{3.43}$$

where LU is the unit in which length is measured (i.e., cm, pc, AU, etc.), TU is the unit of time, and MU is the unit of mass. In general, one is free to choose two of the three measurement units. The demand that $G = 1$, then automatically selects the third unit. For example, to express the integration of a planetary system, we might choose $MU = 1 M_\odot = 1.98911 \times 10^{33}$ gm and $TU = 1$ yr $= 31{,}557{,}600$ s. In this case, the unit of length is $LU = 5.0939 \times 10^{13}$cm.

3.4 THE FEW-BODY PROBLEM: BULIRSCH–STOER INTEGRATION

The need to integrate several bodies for hundreds to millions of orbits often arises in astronomy. One might wish to see, for example, whether a particular model planetary system is participating in a resonance. Or one might be interested in learning how the orbits of the planets would be affected if the solar system suffered an encounter with a passing star. Alternately, it might be of interest to study the dynamics of orbits within a specified galactic potential. For these types of problems, one requires numerical accuracy, flexibility, and speed. Bulirsch–Stoer integration (Stoer and Bulirsch, 1980, Section 7.2.14) performs very well on all three counts.

In the gravitational few-body problem, a single Bulirsch–Stoer integration step advances $6N$-coupled differential equations over a nonnegligible time interval while maintaining very high accuracy. In a simulation of a planetary system, a typical time step is often of the order of a tenth of an orbit of the shortest period planet in the system, and the fractional system energy accuracy across the time step is of order $\Delta E < 10^{-11}$.

The basic integration scheme for solving $d\mathbf{x}/dt = f(t, \mathbf{x})$ by Bulirsch–Stoer is the simple *modified midpoint* method for advancing a vector $\mathbf{x}(t)$ over a full time step $H = Nh$ by stringing together N equal substeps

$$\mathbf{x}_0 = \mathbf{x}(t) \tag{3.44}$$

$$\mathbf{x}_1 = \mathbf{x}_0 + hf(t, \mathbf{x}_0) \tag{3.45}$$

$$\mathbf{x}_n = \mathbf{x}_{n-2} + 2hf(t + [n-1]h, \mathbf{x}_{n-1}) \quad \text{for } n = 2, \ldots, N \tag{3.46}$$

$$\mathbf{x}(t + H) = \frac{1}{2}[\mathbf{x}_N + \mathbf{x}_{N-1} + hf(t + H, \mathbf{x}_N)]\,. \tag{3.47}$$

The subscripted $\mathbf{x}_n$s are approximate intermediate values at times $t + nh$. The error incurred by estimating $\mathbf{x}(t + H)$ with the above formulas can be expressed as a power series that contains terms to only even powers of h. This special property means that successively refined estimates of $\mathbf{x}(t + H)$ from Equation (3.47) (if they are done using values of N divisible by 2) can be combined in a weighted average to form a final estimate for $\mathbf{x}(t + H)$ that has a very high ratio of accuracy to computational effort.

The heart of the Bulirsch–Stoer method consists of generalizing this weighted average to identify it with a polynomial (a rational function could also be used) extrapolation from successive estimates of $\mathbf{x}(t+Nh)$, with the same value of H, but with different values of N. Thus, we evaluate $\mathbf{x}(t + H)k$ times, with $N = 2, 4, 6, 8, \ldots,$ fit the results to a polynomial of order $k - 1$, and then extrapolate to the value, $\mathbf{x}_{N\to\infty}$, that would occur if the step-size h were zero (Figure 3.4). This process is known as Richardson extrapolation.

The polynomials are given by

$$\mathbf{x}_{t+H}(h) = \mathbf{a}_0 + \mathbf{a}_1 h + \mathbf{a}_2 h^2 + \cdots + \mathbf{a}_k h^{k-1}. \tag{3.48}$$

The polynomial coefficients $\mathbf{a}_0 \ldots \mathbf{a}_{k-1}$ can be extracted from Lagrange's well-known formula that gives the unique polynomial of degree $k - 1$ passing through points

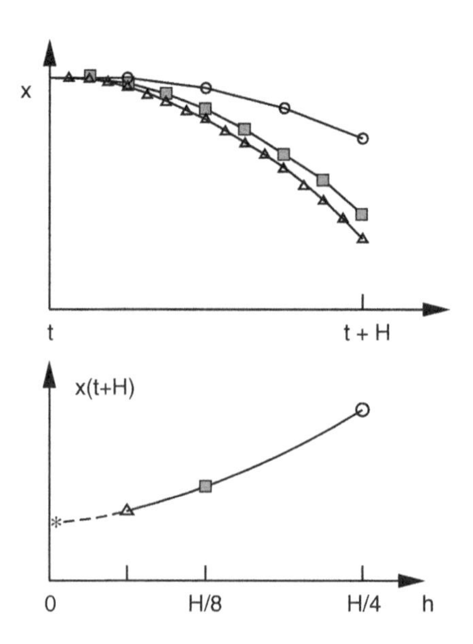

FIGURE 3.4 Schematic diagram for Richardson extrapolation. *Upper diagram*: the function x is integrated from time t to time $t + H$ using a subinterval $h = H/4$ (*open circles*), $h = H/8$ (*filled squares*), and $h = H/16$ (*open triangles*). *Lower diagram*: the values of $x(t + H)$ are plotted as a function of h (*solid line*) and then extrapolated to $h = 0$ (*dashed line*).

$h_1 \ldots h_k$ with values $\mathbf{x}_{t+H}(h_1), \mathbf{x}_{t+H}(h_2), \ldots, \mathbf{x}_{t+H}(h_k)$

$$\begin{aligned} \mathbf{x}_{t+H}(h) = {} & \frac{(h-h_2)(h-h_3)\ldots(h-h_k)}{(h_1-h_2)(h_1-h_3)\ldots(h_1-h_k)}\mathbf{x}_{t+H}(h_1) \\ & + \frac{(h-h_1)(h-h_3)\ldots(h-h_k)}{(h_2-h_1)(h_2-h_3)\ldots(h_2-h_k)}\mathbf{x}_{t+H}(h_2) \\ & + \cdots + \frac{(h-h_1)(h-h_2)\ldots(h-h_{k-1})}{(h_k-h_1)(h_k-h_2)\ldots(h_k-h_{k-1})}\mathbf{x}_{t+H}(h_k). \end{aligned} \tag{3.49}$$

Once the coefficients in Equation (3.48) have been determined, the expression is simply evaluated at $h = 0$ to obtain $\mathbf{x}_{t+H}(0)$.

The final step of the Bulirsch–Stoer process is the determination of numerical convergence. The whole procedure is redone with a different value of k, e.g., 10 evaluations of $\mathbf{x}(t + H)$ by the modified midpoint scheme instead of 8.[†] The two estimates of $\mathbf{x}_{N\to\infty}$ are compared. If the fractional difference is less than a given convergence criterion, for example 10^{-13}, then the solution for time step H is considered converged. If not, k is increased again until convergence is obtained. If the value of k becomes unacceptably large and convergence has still not been obtained, it may be necessary to reduce the basic time step H. Note that this convergence procedure is done at every time step H, so a single Bulirsch–Stoer time step requires a very large

[†] In the *Numerical Recipes in Fortran: The Art of Scientific Computing* routine *bsstep*, the sequence of k values is 2, 4, 6, ..., 16.

number of numerical operations. Why then use Bulirsch–Stoer? It turns out that for a given required accuracy, the Bulirsch–Stoer integration is several times faster than a fourth-order Runge–Kutta method.

For practical work, Equation (3.49) can be replaced by a more straightforwardly coded recursive formula, such as Neville's algorithm (Press et al., 1992, Sections 3.1, 16.4). Here we simply emphasize that once the polynomial has been constructed at $h_1 \ldots h_k$ using the known values $\mathbf{x}_{t+H}(h_1)$, $\mathbf{x}_{t+H}(h_2)$, . . ., $\mathbf{x}_{t+H}(h_k)$ returned from successive applications of the modified midpoint method, we can simply evaluate the value of the polynomial at $h = 0$, which provides our estimate $\mathbf{x}_{N\to\infty}$.

In summary, Bulirsch–Stoer is the preferred method if high accuracy in orbital integrations is required. It is able to handle situations where two orbiting planets make a close encounter. If computational speed is the prime requirement, then the symplectic technique (Section 3.6) can be used, but in general it cannot handle close encounters. In most situations, in fact, the Bulirsch–Stoer should be used for few-body integrations.

3.5 LYAPUNOV TIME ESTIMATION

A persistently remarkable phenomenon in the gravitational N-body problem is the presence of chaos. Chaotic behavior (for a detailed introduction, see Gleick, 1987) stems from the fact that the trajectories of bodies in the N-body problem, which are started very close to one another in phase space (i.e., with very similar position and velocity initial conditions), can, in many cases, diverge exponentially in terms of their phase space distance as time goes on. Eventually, this divergence causes the trajectories to develop complete independence. In the gravitational N-body problem, the timescale that is required for the trajectories to diverge depends sensitively on the details of all the initial conditions (particle masses, velocities, and positions.) Chaos can be observed in any system having more than two bodies, but it is important to remember that not every N-body system will exhibit measurable chaos. Furthermore, among systems that are chaotic, there is a wide range of severity in the chaotic behavior. In some cases, the trajectories, while divergent, will be confined to a narrowly defined region of phase space. Such motion is termed "weakly chaotic." In other cases, initially neighboring orbits are completely dispersed after a period of time. This represents the regime of strong chaos.

As an example, consider the following situation. We adopt our standard initial model containing the Sun and the eight planets in the present-day solar system and, on a somewhat mischievous whim, we insert a clone of the Earth into an orbit whose osculating orbital elements are identical to the Earth's apart from a single exception. We advance the mean anomaly, M, of the Earth's twin by π, which places it initially on the other side of the Sun from the true Earth. We then integrate this augmented version of the solar system forward in time for a million years using the Bulirsch–Stoer program *integrator.f*.

Contrary to what one might expect, the respective motions of the Earth and its twin are not immediately mutually unstable. Rather, the two planets participate in a variation of the horseshoe orbit, in which, during their periodic close encounters, they exchange energy and angular momentum. After any given time, the planet with the smaller osculating semimajor axis, a_1, will approach and will attempt to pass the planet with instantaneously larger semimajor axis, a_2. In the rotating frame, the inner planet is pulled forward, increases its energy and angular momentum, and, hence, increases its semimajor axis. The outer planet experiences the reverse behavior, winds up with a smaller semimajor axis, and the process repeats for the million-year duration of the simulation.

The chaotic trajectories arise largely from the hair-trigger sensitivity within the dynamics of the close encounter. Any slight difference in momentum prior to the encounter is amplified by the encounter itself. Successive differences between the two orbital paths are thereby magnified as the two planets pass energy and angular momentum back and forth, and the compounding of amplifications leads naturally to an exponential divergence in the two trajectories.

To quantify the orbital divergence, we can keep track of some measure of the phase space distance between two orbits. There are many potential definitions of the phase space distance d. The physical distance between the particles could be used, for example, or alternately, one could define $d = e_2 - e_1$. An exponential increase in phase space separation between the two trajectories can be written

$$d(t) = d(0)e^{\gamma(t-t_0)}, \tag{3.50}$$

where the constant γ is called the maximum Lyapunov characteristic exponent. (For a more detailed discussion of the characteristic exponents in gravitational few-body systems, see Lecar et al., 1992.) The exponent γ is inversely related to the characteristic (or e-folding) time for neighboring trajectories to drift apart, and if an orbit is chaotic, $\gamma > 0$. (A negative value for γ indicates converging trajectories.)

A practical scheme for measuring γ needs to maintain a local comparison of two trajectories. Once the phase space distance d is large, the accumulation of further phase space separation is not a measure of local conditions. A simple way to keep the estimation process local is to use the so-called shadowing method. A test particle is integrated from the initial starting condition that one wishes to test for chaos. A second shadow particle with slightly different initial conditions (in which the eccentricity, say, differs by one part in 10^6) is integrated alongside for a time Δt. At the close of the time interval, one estimates the exponent γ from

$$\gamma_{est}^1 = \frac{\ln[d(\Delta t)/d(0)]}{\Delta t}. \tag{3.51}$$

At the end of the time step, the shadow particle is returned to a phase space distance $d(0)$ away from the test particle, and the process is repeated. After n trials, the estimated value of the Lyapunov exponent is:

$$\gamma_{est}^n = \sum_{i=1}^{n} \frac{\ln[d(\Delta t)/d(0)]}{n\Delta t}. \tag{3.52}$$

As a specific example, Figure 3.5 and Figure 3.6 show the results of the integration of a chaotic system using a Bulirsch–Stoer procedure (*integrator.f*). Two planets are assumed to be in orbit around a star of 1 solar mass. Their initial orbital periods are

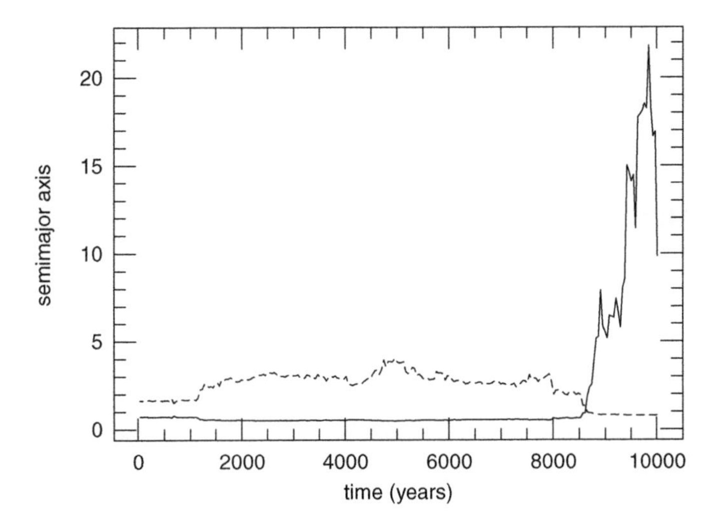

FIGURE 3.5 Semimajor axis (in astronomical units) as a function of time from an integration of the orbits of two planets around a star of 1 solar mass. *Solid curve*: Planet 1 whose initial orbital period was 250 days. *Dashed curve*: Planet 2 whose initial orbital period was 660 days. Note that the orbits cross at about 8600 years.

250 and 660 days, their orbital eccentricities are .28 and .27, their inclinations are zero (that is, they are in the same plane), and their masses are 1.89 and 3.75 Jupiter masses, for the inner and outer planet, respectively. Figure 3.5 shows the semimajor axes of the two planets as a function of time with the standard orbital parameters.

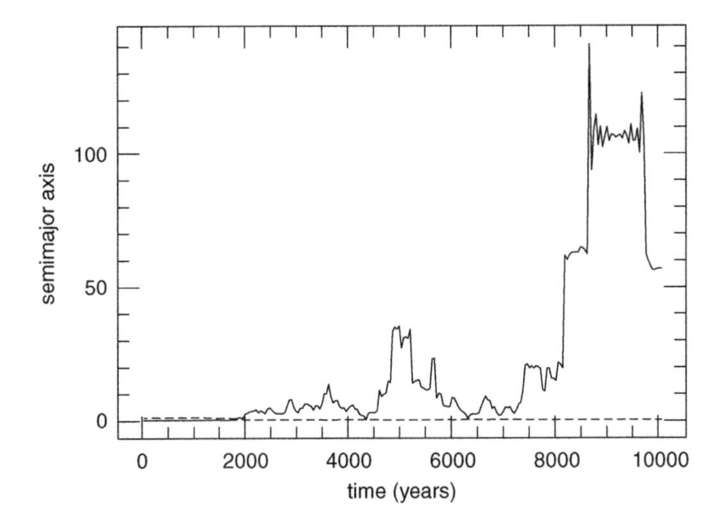

FIGURE 3.6 This figure is the same as Figure 3.5 except that the initial eccentricity of Planet 1 is changed from 0.28 to 0.28001. Note that the first orbit crossing occurs just before 2000 years.

In contrast, Figure 3.6 shows the results from a calculation in which all parameters are the same as the standard values, except that the eccentricity of the inner planet is increased from 0.28 to 0.28001. Clearly a change in the system that is smaller than the probable error of measurement of an eccentricity leads to a widely diverging result.

3.6 SYMPLECTIC INTEGRATION

The discussion in the previous sections has emphasized that many problems of astronomical interest involve the long-term behavior of small bodies orbiting a much larger parent body. The solar system provides a perfect example. While the orbit of a planet (the Earth, say) is well-described by a Keplerian ellipse over long timescales, the gravitational perturbations from other planets cannot be neglected. Indeed, the question of how the planets affect one another, and how the mutual interactions affect the long-term stability of the solar system has been a topic of intense interest from the time of Sir Isaac Newton onward.

The Newtonian gravitational interaction between two arbitrary point masses has an exact solution. This situation does not generally hold for systems containing three or more bodies. The underlying reason for the inherent difficulty in obtaining the motion for three gravitating objects was established by Jules Henri Poincaré in 1888 (see Murray and Dermott, 1999, Section 3.1 and 9.1), who showed that the three-body problem in nonintegrable. That is, it is formally impossible to write down a simple closed-form analytic solution for the future motion of a system of planets. Poincaré's work also emphasized the essential nonlinearity of the few-body problem in which the smallest change in initial conditions leads to completely different motions after long periods of time. Any solution for the motion of objects in the gravitational few-body problem, therefore, must be regarded as a sample from a statistical ensemble of equivalent solutions.

The advent of fast computers has revolutionized the problem of computing long-term planetary dynamics. At the time of this writing, a fast desktop computer, running for months on end, is capable of integrating the motion of the planets of the solar system for the entire lifetime of the Sun. Continued relentless increases in processor speed indicate that such calculations will eventually become interactive. The ability to do extremely long-term integrations has also benefited from so-called symplectic map algorithms that take specific advantage of the properties of near-Keplerian motion. The essence of the symplectic map is that the motion of a planet is divided into a Keplerian part and a non-Keplerian perturbative "kick" resulting from the interaction with bodies in the system other than the central body. Because the Keplerian part of the motion has an analytic solution, the computer can avoid wasting its time rediscovering the Keplerian ellipse on every orbit.

The symplectic algorithm relies on Hamilton's formulation of Newton's laws of dynamics (for a review, see Goldstein, 1980). The time development of the positions $\mathbf{q_i}$ and momenta $\mathbf{p_i}$ of the particles are written as a set of coupled first-order differential equations

$$\frac{dq_i}{dt} = \frac{\partial H}{\partial p_i} \tag{3.53}$$

$$\frac{-dp_i}{dt} = \frac{\partial H}{\partial q_i} \tag{3.54}$$

where H is the Hamiltonian for the N-body problem

$$H = \sum_{i=0}^{N-1} \frac{p_i^2}{2m_i} - \sum_{i<j} \frac{Gm_i m_j}{r_{ij}}. \tag{3.55}$$

Note that H is simply the total energy and, thus, is conserved. The idea behind the symplectic map is to separate the total Hamiltonian into a part corresponding to the Keplerian motion and a part that arises from the gravitational interactions with the other bodies

$$H = H_{Kepler} + H_{interaction}. \tag{3.56}$$

A Keplerian Hamiltonian is one that can be written in the form

$$H_{Kepler} = \frac{p^2}{2m} - \frac{GMm}{r}. \tag{3.57}$$

By contrast, in Equation (3.55), the definitions of p_i relative to a fixed origin, and the r_{ij}s in terms of particle–particle separations means that when there are more than two particles in the system, the terms in Equation (3.55) cannot immediately be read off in the form required to compose H_{Kepler}. That is, the momenta in the N-body Hamiltonian (Equation 3.55) refer to the center of mass, and the particles are not executing Keplerian motion about the center of mass.

Fortunately, however, a set of Keplerian Hamiltonians can be split off from H if the motion of the bodies is cast into Jacobi coordinates (see Wisdom and Holman, 1991). The system consists of a heavy mass m_0 and $N-1$ light masses m_i orbiting it. The center of mass of the entire system is taken to be the first Jacobi coordinate. The second Jacobi coordinate is the separation vector between the first planet and the central mass. Subsequent coordinates correspond to the separation vectors of the ith body relative to the center of mass of the $i-1$ preceding bodies. That is, if the first Jacobi coordinate (the COM) is labeled $\mathbf{x}'_0$, then the subsequent Jacobi coordinates are given by

$$\mathbf{x}'_i = \mathbf{x}_i - \frac{1}{\eta_{i-1}} \sum_{j=0}^{i-1} m_j \mathbf{x}_j, \tag{3.58}$$

where $\eta_i = \sum_{j=0}^{i} m_j$ is a running sum of the masses.

The Jacobi momenta are given by

$$\mathbf{p}'_i = \left(\frac{\eta_{i-1}}{\eta_i}\right)\mathbf{p}_i - \left(\frac{m_i}{\eta_i}\right)\sum_{j=0}^{i-1} \mathbf{p}_j \tag{3.59}$$

and

$$\mathbf{p}'_0 = \sum_{j=0}^{N-1} \mathbf{p}_j. \tag{3.60}$$

These are formed by multiplying the mass factors

$$m'_0 = M_{\text{tot}} = \eta_{N-1} \quad \text{for } i = 0 \tag{3.61}$$

and

$$m'_i = \eta_{i-1} m_i / \eta_i \quad \text{for } 0 < i < N \tag{3.62}$$

by $d\mathbf{x}'_i/dt$ (see Saha and Tremaine, 1994).

The essential utility of the Jacobi coordinates in describing a star surrounded by more than one planet dates back to Newton (1687), who was the first to consider how the gravitational forces between the planets perturb their Keplerian orbits around the Sun. In Book I, Section II, proposition 69, of the Principia, he wrote:

> *And hence, if several lesser bodies revolve about a greatest one, it can be found that the orbits described will approach closer to elliptical orbits, and the description of the areas will become more uniform* [...] *if the focus of each orbit is located in the common center of gravity of all the inner bodies.*

In modern terms, one says that to first order, orbits in a multiple-planet system are Keplerian when written in terms of Jacobi coordinates. Using the Jacobi coordinates, one finds that the full N-body Hamiltonian H can be written (see Murray and Dermott, 1999) in the desired form $H = H_{Kepler} + H_{interaction}$, where

$$H_{Kepler} = \sum_{i=1}^{N-1} \left(\frac{(p'_i)^2}{2m'_i} - \frac{G m_i m_0}{r'_i} \right) \tag{3.63}$$

and

$$H_{interaction} = \sum_{i=1}^{N-1} G m_i m_0 \left(\frac{1}{r'_i} - \frac{1}{r_{i0}} \right) - \sum_{0<i<j} \frac{G m_i m_j}{r_{ij}}, \tag{3.64}$$

where r_{i0} is, as usual, the distance between particle i and the massive object.

As outlined by Wisdom and Holman (1991), the basic idea behind the symplectic integrator is to imagine that the Hamiltonian Equation (3.56) represents a map that steps the system from an initial state at $t = t_0$ to an evolved state at a time $t = t_0 + \Delta t$. Operationally, the map can be carried out in two distinct steps. First, the Jacobi coordinates of the planets are advanced through the Keplerian portion of their trajectories. This can be done rapidly by taking advantage of the analytic nature of the solution. Numerically, this is accomplished by noting that the position and velocity vectors for a planet can be efficiently advanced through Keplerian motion using Gauss' f and g functions (see, e.g., Danby, 1988), discovered by Gauss in 1801 when working on the recovery of the asteroid Ceres. As we explain below, the computation of $\mathbf{r}(t)$, and $\mathbf{v}(t)$ can be done very quickly

$$\mathbf{r}(t) = f(t, t_0)\mathbf{r}(t_0) + g(t, t_0)\mathbf{v}(t_0), \tag{3.65}$$

$$\mathbf{v}(t) = \dot{f}(t, t_0)\mathbf{r}(t_0) + \dot{g}(t, t_0)\mathbf{v}(t_0), \tag{3.66}$$

$$f(t, t_0) = \frac{a}{r_0}[\cos(E - E_0) - 1] + 1, \tag{3.67}$$

$$g(t, t_0) = (t - t_0) + \frac{1}{n}[\sin(E - E_0) - (E - E_0)], \tag{3.68}$$

where r_0 is the magnitude of the position vector at time t_0 and E is the solution to Kepler's Equation (3.42) at time t.

The Keplerian steps for each particle are bracketed by first-order course corrections, which account for the planet–planet perturbations. Somewhat surprisingly, these course corrections, or "kicks," can be applied accurately in a simple impulsive fashion, in which the changes to the Jacobian vector velocities of the particles are given by

$$\Delta \mathbf{v}'_i = \Delta t \left(\frac{d\mathbf{v}'_i}{dt}\right)_{\text{interaction}}, \tag{3.69}$$

where

$$\left(\frac{d\mathbf{v}'_i}{dt}\right)_{\text{interaction}} = \frac{1}{m'_i}\left(-\frac{\partial H_{\text{interaction}}}{\partial \mathbf{r}'_i}\right). \tag{3.70}$$

$H_{\text{interaction}}$ is written most compactly in the form shown in Equation (3.64), which mixes ordinary interparticle distances, r_{ij} and Jacobi coordinates, r'_i.

The derivatives of the interaction Hamiltonian with respect to the Jacobi position coordinates are complicated, as they require the interaction Hamiltonian to first be written entirely in Jacobi coordinates. Murray and Dermott (1999) report the analytic formula

$$\left(\frac{d\mathbf{v}'_i}{dt}\right)_{\text{interaction}} = GM'_i\left[\frac{\mathbf{r_i}'}{r'^3_i} - \frac{\mathbf{r}_{0i}}{r^3_{0i}}\right] - \left(\frac{\eta_i}{\eta_{i-1}}\right)\sum_{j=1}^{i-1}\frac{Gm_j}{r^3_{ji}}\mathbf{r}_{ji}$$
$$+ \sum_{j=i+1}^{N-1}\frac{Gm_j}{r^3_{ij}}\mathbf{r}_{ij} - \frac{1}{\eta_{i-1}}\sum_{j=0}^{i-1}\sum_{k=i+1}^{N-1}\frac{Gm_jm_k}{r^3_{jk}}\mathbf{r}_{jk}, \tag{3.71}$$

where $M'_i = (\eta_i/\eta_{i-1})m_0$.

The code *symplectic.f* utilizes the foregoing ingredients to implement a simplified routine for integrating few-body problems. The code is best suited to situations in which one desires to understand the long-term effect of planet–planet perturbations in the absence of close encounters. An excellent example is the evolution of the orbital elements of the planets in the solar system over timescales of millions to billions of years. This version of the code breaks down when two masses come within 3 R_H of each other, where R_H is the "Hill sphere" radius of the larger object

$$R_H = a_i\left(\frac{m_i}{3m_0}\right)^{1/3}, \tag{3.72}$$

where a_i is the usual semimajor axis. Under such circumstances, the code must be augmented by a "regularization" technique (Section 3.8) or a switch to a Bulirsch–Stoer routine.

The input to the code is made in terms of the Keplerian orbital elements (P,M, i, e, ϖ, and Ω), along with the planetary and stellar masses. The code first transforms the orbital elements into astrocentric (star-centered) Cartesian coordinates (Murray and Dermott, 1999, Section 2.8). It then builds transformation matrices M and $\mathrm{M^t}$, which enable rapid conversion back and forth between Jacobi coordinates and ordinary coordinates

$$[\mathbf{r}'_\mathbf{j}] = [\mathrm{M}][\mathbf{r}_j], \tag{3.73}$$

with

$$[\mathrm{M}][\mathrm{M^t}] = 1. \tag{3.74}$$

Once the transformation matrices have been precomputed, the routine is ready to enter its main loop. Within this main loop, the Jacobian coordinates of the bodies are evaluated, and the bodies are advanced through the Keplerian portion of their orbits for time step Δt. The Keplerian advances are performed in Cartesian coordinates by making use of Gauss' f and g functions (Equation 3.67 and Equation 3.68). The Gauss f and g functions rely on knowing the eccentric anomaly $E(t)$. The eccentric anomaly is related to the mean anomaly, M, through Kepler's Equation (3.42), which can be solved using a Newton–Raphson root-finding scheme. Following the Keplerian step, the interaction Hamiltonians are used to further update the Jacobian velocities in accordance with Equation (3.69). This completes the time step. The method is readily converted to second-order by using an interval $\Delta t/2$ for the first and last time steps and Δt for all the others.

Again it should be emphasized that the symplectic technique works well for planetary orbits that are near Keplerian, i.e., there is little gravitational interaction between the planets themselves. The big advantage to the technique is that it can run up to 10 times faster than the Bulirsch–Stoer method for a given number of planetary orbits. The disadvantage is that once the orbits of two planets begin to approach each other, the method becomes inaccurate and breaks down. In this situation use of a Bulirsch–Stoer code is highly preferable.

3.7 *N*-BODY CODES FOR LARGE *N*

Our discussion so far has focused on the evolution of configurations such as planetary systems that have only a few bodies. When studying the evolution of few-body systems, one is often interested in the subtle interplay of perturbations that unfold over thousands or even millions of orbits. In such a case, accuracy is the paramount concern. The goal is to keep the conservation of energy and angular momentum in as strict order as possible. For larger systems, however, such as galaxies, one is faced with trying to understand the collective behavior of enormous numbers of particles. In the case of a galaxy, the number of orbits that one needs to follow is relatively small. Take the Sun as an example. The Sun's orbital period around the center of the galaxy is roughly 250 million years, meaning that the Sun has executed roughly 18 orbits since its formation. Clearly, the relevant problem for the collective dynamics of a galaxy concerns integrating an enormous number of particles, albeit for a relatively small number of orbits.

Two factors conspire to make the large N-body problem a challenge. The first concerns the number of interactions. The gravitational force on an individual body is the sum of the attractions of all the other bodies. Computing the force for all N bodies at a single time requires the evaluation of $\frac{1}{2}N(N-1)$ square roots. If forces are computed individually, this means that the computer time required to solve an N-body problem scales as N^2. No matter how fast the processor, N^2 algorithms are bound to become prohibitive for most astrophysical systems of interest.

A second problem is that of multiple time scales. In a self-gravitating system, such as a star cluster, stars that enter into tightly bound binary orbits or are in a region dense with other stars experience large gravitational forces. To accurately follow these stars, one is required to use a small time step. Such a small time step, however, when applied to the system as a whole, leads to disaster, as the vast majority of particles are updated far more often than necessary. Much of the art in N-body methods consists of implementing clever schemes to avoid these difficulties.

Perhaps the simplest strategy for integrating the motion of an assortment of particles with widely varying accelerations is to use the so-called "leapfrog" method in conjunction with time step doubling. The appropriate time step for each particle is first estimated using the criterion

$$\Delta t_i \simeq \eta \sqrt{\frac{1}{a_i}}, \tag{3.75}$$

where a_i is the magnitude of the acceleration of the ith particle and η is a small multiplicative factor. In order to give each particle an appropriate time step, the largest time step Δt_s, estimated from Equation (3.75), is subdivided into multiples of two

$$\Delta t_i = \frac{\Delta t_s}{2^{n_i}}, \tag{3.76}$$

with n_i chosen to be the smallest integer value for which $\Delta t_i < \eta\sqrt{1/a_i}$ is satisfied. For a particular particle, i, the leapfrog method consists of a simple second-order integration procedure to advance the positions and velocities

$$\mathbf{r}_i^{n+1/2} = \mathbf{r}_i^{n-1/2} + \Delta t_i \mathbf{v}_i^n + \mathcal{O}(\Delta t_i{}^3), \tag{3.77}$$

$$\mathbf{v}_i^{n+1} = \mathbf{v}_i^n + \Delta t_i \mathbf{a}_i^{n+1/2} + \mathcal{O}(\Delta t_i{}^3). \tag{3.78}$$

Note that in this pair of equations, the velocities and positions are defined at intervals separated by $\Delta t_i/2$ in order to maintain second-order accuracy, while maintaining the overall simplicity of a first-order algorithm, whereas, in general, a set of initial conditions will specify all of the velocities and positions at a single point in time. A series of integration steps can be initialized by

$$\mathbf{r}_i^{n+1/2} = \mathbf{r}_i^n + \frac{1}{2}\Delta t_i \mathbf{v}_i^n + \frac{1}{8}\Delta t_i{}^2 \mathbf{a}_i^n, \tag{3.79}$$

which preserves second-order accuracy.

To see how multiple time steps are handled, consider two particles, "A" and "B," within the overall aggregate of bodies, whose individual time steps have been found to be $\frac{\Delta t}{2}$ and Δt. At the beginning of the first time step ($t = 0$), we know both $\mathbf{r}$ and $\mathbf{v}$ for both particles. The startup formula (Equation 3.79) is used to advance the position vectors to $\frac{1}{4}\Delta t$ for "A," and $\frac{1}{2}\Delta t$ for "B." Once the particles are in place, the accelerations are computed based on the position at $\Delta t/4$ for "A" and $\Delta t/2$ for "B" (note the time asymmetry inherent in this force calculation). The acceleration felt by particle "A" is used to advance "A's" velocity to $t = \Delta t/2$. This velocity is in turn used to advance the position of "A" to $t = (3/4)\Delta t$. With the time asymmetry between "A" and "B" thus reversed, the velocity is advanced again to $t = \Delta t$ and, in doing so, second-order accuracy is recovered.

The original and updated positions for particle "A" are averaged together to obtain a position estimate for "A" centered at $t = \Delta t/2$. This position, as well as the original position for particle "B," can be used to compute an acceleration on "B," which is used to advance its velocity across the entire interval Δt, upon which, the velocities of both particles are set at $t = \Delta t$. This centered velocity can then be used to make a final advance of the position of particle "A" to $t = (\frac{5}{4})\Delta t$, at which point the cycle is ready to be run again. A simple generalization of this procedure up through the hierarchy of time step widths allows all of the particles to be updated in a synchronous fashion that maintains second-order accuracy of the method.

The second-order leapfrog method is often used for systems that have large numbers of particles and are subject to a large degree of internal dynamical dissipation. An excellent example would be a simulation of a collision between galaxies or even the evolution of a rich cluster of galaxies. When two galaxies approach each other, the large-scale kinetic energy associated with the motion of their individual centers of mass is transferred into small scale random motion of the individual particles within each galaxy. This process of draining energy from the largest scales into the smallest causes the orbits of binary galaxies to coalesce over periods of at most a few tens of orbits. To survey the overall structure of the final (generally ellipsoidal) remnant galaxy, one requires neither a fully exact rendering of the potential nor a high-order integration of the positions and velocities. On the other hand, for systems that encompass fewer objects, but many dynamical timescales, accuracy, and a lack of numerical dissipation are paramount.

If the number of particles that one wishes to follow is not too large, say $20 < N < 10^4$, then it is most advisable to adopt a technique that computes gravitational forces via the direct summation of the terms in Equation (3.1), but which integrates the particles themselves with less precision than either the symplectic integrator or the Bulirsch–Stoer scheme. Problems in this regime might include, for example, the dynamical evolution of clusters of stars or the tidal disruption of dwarf galaxies. For cases of this sort, we suggest adopting a straightforward scheme for direct integration that combines the method of Aarseth (1985) for advancing particles in their trajectories, and the algorithm of Ahmad and Cohen (1973) for controlling the frequency with which the full sum of $\frac{1}{2}N(N-1)$ interactions in Equation (3.1) is updated. An optimized version of this method is available as the NBODY2 code written by Sverre Aarseth.

Aarseth's method is a version of a predictor–corrector scheme, which in its simplest form can be described as follows: Suppose the acceleration on a particle $\mathbf{a}_i(t_0)$ has been calculated from the right-hand side of Equation (3.1). Then at an advanced time $t = t_0 + \Delta t$, the velocity and position can be obtained from

$$\mathbf{v}_i(t) = \mathbf{a}_i(t_0)\Delta t + \mathbf{v}_i(t_0)$$
$$\mathbf{r}_i(t) = \frac{1}{2}\mathbf{a}_i(t_0)(\Delta t)^2 + \mathbf{v}_i(t_0)\Delta t + \mathbf{r}_i(t_0). \tag{3.80}$$

One makes a provisional estimate of the radius at time t using the accelerations at t_0, recomputes the accelerations at t, and calculates an average acceleration $\bar{\mathbf{a}}_i = 0.5[\mathbf{a}_i(t_0) + \mathbf{a}_i(t)]$. This average is then used in Equations (3.80) to get the corrected positions and velocities at time t.

Aarseth's direct summation N-body algorithm hinges on representing the acceleration on a particular body of index i at time t in terms of a fourth-order polynomial based on knowledge of the acceleration at four previous times in the past, t_0, t_1, t_2, and t_3, with t_0 being the most recent

$$\mathbf{a}(t) = (((\mathbf{D}_4(t - t_3) + \mathbf{D}_3)(t - t_2) + \mathbf{D}_2)(t - t_1) + \mathbf{D}_1)(t - t_0) + \mathbf{a}_0, \tag{3.81}$$

where $\mathbf{a}_0$ is the acceleration at time t_0. This equation is a variant of the Lagrange interpolation formula. We identify $\mathbf{D}_0(t') = \mathbf{a}(t')$, where t' can be any past or future value of t. The "divided differences" $\mathbf{D}_1$, $\mathbf{D}_2$, $\mathbf{D}_3$, and $\mathbf{D}_4$ are defined as follows:

$$\mathbf{D}_1[t_j, t_k] = \frac{\mathbf{D}_0(t_j) - \mathbf{D}_0(t_k)}{t_j - t_k} \quad (j = k - 1)$$
$$\mathbf{D}_2[t_j, t_k] = \frac{\mathbf{D}_1[t_j, t_{k-1}] - \mathbf{D}_1[t_{k-1}, t_k]}{t_j - t_k} \quad (j = k - 2)$$
$$\mathbf{D}_3[t_j, t_k] = \frac{\mathbf{D}_2[t_j, t_{k-1}] - \mathbf{D}_2[t_{k-2}, t_k]}{t_j - t_k} \quad (j = k - 3)$$
$$\mathbf{D}_4 = \frac{\mathbf{D}_3[t, t_2] - \mathbf{D}_3[t_0, t_3]}{t - t_3}. \tag{3.82}$$

Note that $\mathbf{D}_4$ involves the acceleration at the *advanced* time t, where information is not initially available. This point is needed because the expansion (Equation 3.81) is fourth order and, thus, five points are needed to define it. The predictor–corrector method is used to obtain quantities at this point.

The fourth-order polynomial for the acceleration, **a**, can be converted into the first four terms of a Taylor series approximation of the acceleration in the neighborhood of t_0. When the acceleration of a particle is expressed as a Taylor series, one can immediately integrate the terms once with respect to time to obtain the particle velocity, and then again to obtain the position. The key idea underlying the method is that the instantaneous positions of particles at time t_i that are necessary to compute the acceleration of the particle i under consideration can be estimated at arbitrary t_i by integrating their acceleration polynomials. Each particle can then maintain a separate time step that is appropriate to its dynamical environment, allowing for a large increase in efficiency.

Equating the acceleration polynomial to a Taylor series, we find the following expressions for the successive derivatives of the acceleration at time $t = t_0$

$$\frac{d\mathbf{a}}{dt} = ((\mathbf{D}_4(t_0 - t_3) + \mathbf{D}_3)(t_0 - t_2) + \mathbf{D}_2)(t_0 - t_1) + \mathbf{D}_1 \tag{3.83}$$

$$\frac{d^2\mathbf{a}}{dt^2} = 2!(\mathbf{D}_4((t_0 - t_1)(t_0 - t_2) + (t_0 - t_2)(t_0 - t_3) + (t_0 - t_1)(t_0 - t_3)) + \mathbf{D}_3((t_0 - t_1) + (t_0 - t_2)) + \mathbf{D}_2) \tag{3.84}$$

$$\frac{d^3\mathbf{a}}{dt^3} = 3!(\mathbf{D}_4((t_0 - t_1) + (t_0 - t_2) + (t_0 - t_3)) + \mathbf{D}_3) \tag{3.85}$$

$$\frac{d^4\mathbf{a}}{dt^4} = 4!\mathbf{D}_4, \tag{3.86}$$

where the $\mathbf{D}_k$ are evaluated at $[t_0, t_k]$. The accelerations are now expressed in the form

$$\mathbf{a}(t) = a(\Delta t)^4 + b(\Delta t)^3 + c(\Delta t)^2 + d(\Delta t) + e. \tag{3.87}$$

The first time the acceleration is computed, the $\mathbf{D}_4$ terms are left out. Then, once a provisional acceleration at t has been obtained, it is used in a second pass to get a corrected acceleration. Once the positions and velocities have been updated by direct integration of Equation (3.87) for all particles, the actual acceleration on each particle is calculated by direct summation over all particles.

The time step required for any particular particle i depends on its acceleration. Particles accelerating rapidly require a small time step to adequately resolve the motion, whereas particles on nearly straight-line trajectories can be accurately integrated with large time steps. A simple method for estimating the time step for a particle involves the distance D_m and the relative velocity v_m with respect to its nearest neighbor. The travel time then is $\tau_1 \approx D_m/v_m$. The freefall time $\tau_2 \approx D_m^{3/2}$ because $\tau_{ff} \approx 1/\sqrt{G\rho}$ and G is usually set to 1. Thus, we can construct a timescale that is appropriate for each particle, that incorporates both τ_1 and τ_2

$$\Delta t_i = \frac{D_m^{3/2}}{\eta(1 + v_m D_m^{1/2})}, \tag{3.88}$$

where η is a dimensionless number less than 1. In practice, the time step for an individual particle is determined through the more sophisticated relation

$$\Delta t_i = \left[\frac{\eta(|\mathbf{a}||\mathbf{a}^{(2)}| + |\mathbf{a}^{(1)}|^2)}{(|\mathbf{a}^{(1)}||\mathbf{a}^{(3)}| + |\mathbf{a}^{(2)}|^2)}\right]^{1/2} \tag{3.89}$$

where the superscript in parentheses is the order of the derivative (i.e., $\mathbf{a}^{(1)} = d\mathbf{a}/dt$). The reader can verify that the right-hand side of Equation (3.89) indeed has the units of time. In this expression, the enthusiasm of the rate of change of the acceleration for decreasing the time step is judiciously tempered by contributions from lower-order derivatives of $\mathbf{a}$ appearing in the numerator. The time steps are adjusted during the calculation according to Scheme (3.76), so that only a discrete set of time steps is used.

3.8 CLOSE ENCOUNTERS AND REGULARIZATION

Close encounters pose a problem for N-body codes. During a close encounter between two point masses, the quantity $1/r_{ij}^2$ can become arbitrarily large, leading to arbitrarily large accelerations, even for particles with small masses. To follow large accelerations accurately, one requires small time steps, which slows down the algorithm and encourages the buildup of truncation error.

If a close encounter is a one-time event, as would be the case, for instance, with a moderately hyperbolic encounter between two stars, the overall performance of a code will not be severely compromised by an occasional downward spike in time step. Encounters become a continual problem, however, when a tight binary pair forms. Somewhat surprisingly, a chance encounter between three unbound stars can result in one member of the trio being ejected at high speed with the remaining pair stuck in a bound orbit. Once a binary forms, subsequent encounters with other stars in the cluster tend to sap energy from the binary, resulting in a decrease in the semimajor axis and an increase in orbital velocity. This process of "binary hardening" is essential to the overall dynamics of globular clusters, in which a small collection of tight binaries near the core of the cluster is able to release energy and support the core against catastrophic collapse, in much the same way that nuclear reactions between atoms provide the energy input (and attendant pressure) that supports a conventional star.

The simplest and least accurate, but still useful method for protecting an N-body algorithm against the catastrophic time sink provided by hard binaries is to include a softening term within the $\frac{1}{r^2}$ dependence of the gravitational force law. If we write, for example, the force exerted by particle j on particle i as

$$\mathbf{F}_{ij} = \frac{Gm_i m_j(\mathbf{r}_j - \mathbf{r}_i)}{\left(\epsilon^2 + |\mathbf{r}_i - \mathbf{r}_j|^2\right)^{3/2}}, \tag{3.90}$$

then the gravitational acceleration between the two particles saturates at a finite maximum magnitude of

$$|\mathbf{a}_j| = \frac{2Gm_i}{3^{3/2}\epsilon^2}, \tag{3.91}$$

which occurs when the particles reach a separation of $\frac{1}{\sqrt{2}}\epsilon$. By limiting the acceleration, the time step is not allowed to go to arbitrarily low values, but at the cost of not following the orbit of a tight binary star. Therefore, if softening is used, or more precisely, if the softening length ϵ is larger than the actual physical size of the particles that one is modeling, then tight binary orbits and detailed encounters cannot be followed.

When is it safe to use softening? It is safe whenever the dynamics of the system that one is trying to model are not driven by close encounters. That is, the use of softening is appropriate when a system is collisionless. A galaxy, for example, provides an excellent example of a collisionless system. The typical star has a radius 5×10^{10}cm, and the typical separation between stars is of order 5×10^{18}cm. The galaxy contains $\sim 10^{11}$ stars and is approximately 1.5×10^{23} cm in diameter. One can get a better intuitive sense for these numbers by imagining the construction of a scale model. Take a suitcase-sized box full of fine sand (a suitcase can easily hold 10^{11} sand grains) and spread the sand over a disk having a diameter roughly 100 times the width of North

America. At this level of rarefaction, the average separation between the sand grains is several miles. Because the average star in a galaxy must travel for vastly longer than the current age of the universe before running into another star, we say that the distribution is collisionless. Stars feel only the smooth global potential of the entire galaxy and remain unaffected by individual encounters. In the simulation of a galaxy, each particle represents not one star, but rather a bundle of stars on equivalent orbits. Stars on equivalent orbits can easily interpenetrate one another, hence, the use of a softened potential is well motivated.

However, in other problems, such as the evolution of a small cluster of stars, encounters and close binary formation cannot be neglected, and special treatment is needed. The idea underlying regularization is to introduce a coordinate transformation, which replaces the ordinary physical time with a regularized time. By effectively replacing the nonuniform oscillation of an unperturbed elliptical orbit by simple harmonic motion, the transformation allows the integration to be carried through the expensive moment of close approach between the two bodies with economy and accuracy. Regularization will even resolve the direct collision between two idealized point mass particles (i.e., two-body motion with zero angular momentum), although, in practice, the planets or stars in a real astrophysical simulation have finite radii, and the results of a true collision are very different from the mathematical trajectory of point masses.

The acceleration of the separation vector $\mathbf{R}$ for the components of a particular pair of particles within a larger N-body simulation is given by

$$\frac{d^2\mathbf{R}}{dt^2} = -G(m_1 + m_2)\frac{\mathbf{R}}{|\mathbf{R}|^3} + \mathbf{F}_{12} \tag{3.92}$$

where $\mathbf{F}_{12} = \mathbf{F}_1 - \mathbf{F}_2$ is the net external force per unit mass arising from the other bodies in the overall system. The regularization of Equation (3.92) involves transformation of both the time and the space coordinates. As the first step, we proceed by introducing a transformed, regularized time, τ, that is related to the ordinary time, t, by

$$dt = R^n d\tau \tag{3.93}$$

so that

$$\frac{d^2}{dt^2} = \frac{1}{R^{2n}}\frac{d^2}{d\tau^2} - \frac{n}{R^{2n+1}}\frac{dR}{d\tau}\frac{d}{d\tau}. \tag{3.94}$$

The equation of motion (Equation 3.92) becomes

$$\frac{d^2\mathbf{R}}{d\tau^2} = \frac{n}{R}\frac{dR}{d\tau}\frac{d\mathbf{R}}{d\tau} - G(m_1 + m_2)\frac{\mathbf{R}}{R^{3-2n}} + R^{2n}\mathbf{F}_{12} \tag{3.95}$$

For $n = 1$, which is the most practically useful choice, the separation between the particles is directly proportional to the rate at which real time passes in comparison to regularized time. In this case, by regularizing the time, we have removed the R^{-2} singularity in Equation (3.92) while introducing a term $\mathbf{R}/R$ that is indeterminate as the separation between the particles goes to zero. Thus, a second transformation is required to make a successful two-body regularization. This second ingredient involves transforming the space coordinates to get rid of the indeterminacy. This transformation gets successively tougher as one goes from one to two to three dimensions. Therefore, we look at each case in turn. (For details concerning the steps that we skip, see Aarseth (2003).)

The one-dimensional spatial transformation was first introduced by Euler. In one dimension, there is no distinguishing between **R** and R. In the absence of an external force, Equation (3.95) becomes

$$\frac{d^2R}{d\tau^2} = \frac{1}{R}\left(\frac{dR}{d\tau}\right)^2 - G(m_1 + m_2). \tag{3.96}$$

This equation still runs into difficulties in numerical integrations. However, we can use energy conservation to improve things. The binding energy per reduced mass $\mu = m_1 m_2/(m_1 + m_2)$, is:

$$h = \frac{1}{2}\left(\frac{dR}{dt}\right)^2 - \frac{G}{R}(m_1 + m_2) \tag{3.97}$$

and is fixed for any given unperturbed binary orbit. Using the fact that

$$\frac{dR}{dt} = \frac{1}{R}\frac{dR}{d\tau}, \tag{3.98}$$

we have

$$\frac{d^2R}{d\tau^2} = 2hR + G(m_1 + m_2), \tag{3.99}$$

which is free of any problems as R goes through zero. If we further write $u^2 = R$, then the equation of motion reduces to a simple harmonic oscillator

$$\frac{d^2u}{d\tau^2} = \frac{1}{2}hu. \tag{3.100}$$

Equation (3.100) is readily integrated using the techniques discussed earlier in this chapter.

In numerical practice, then, one could handle the development of an effectively straight-line collision between two particles in a simulation by switching from (x, t) to (u, τ) when the bodies reach a prespecified minimum distance, using Equation (3.100) to resolve the collision and then switching back after the bodies are safely separated from one another.

A workable coordinate transformation for handling close binary encounters in two dimensions was originally described by Levi-Civita (1904). As in one dimension, the problem consists of eliminating the indeterminate term $\mathbf{R}/R$ from Equation (3.95). The regularization proceeds by first rewriting the components of $\mathbf{R} = (x, y)$ in terms of new variables u_1 and u_2

$$x = u_1^2 - u_2^2 \tag{3.101}$$

$$y = 2u_1u_2, \tag{3.102}$$

so that in matrix notation

$$\mathbf{R} = \mathcal{L}\mathbf{u}, \tag{3.103}$$

where

$$\mathcal{L}(\mathbf{u}) = \begin{bmatrix} u_1 & -u_2 \\ u_2 & u_1 \end{bmatrix}. \tag{3.104}$$

The transformation has the following mathematical properties (for arbitrary $\mathbf{u},\ \mathbf{v}$):

1. $\mathcal{L}^T(\mathbf{u})\mathcal{L}(\mathbf{u}) = RI$, where I is the unit matrix

$$I = \begin{bmatrix} 1 & 0 \\ 0 & 1 \end{bmatrix}. \tag{3.105}$$

2. $\frac{d}{d\tau}\mathcal{L}(\mathbf{u}) = \mathcal{L}\left(\frac{d\mathbf{u}}{d\tau}\right)$
3. $\mathcal{L}(\mathbf{u})\mathbf{v} = \mathcal{L}(\mathbf{v})\mathbf{u}$
4. $\mathbf{u}\cdot\mathbf{u}\mathcal{L}(\mathbf{v})\mathbf{v} - 2\mathbf{u}\cdot\mathbf{v}\mathcal{L}(\mathbf{u})\mathbf{v} + \mathbf{v}\cdot\mathbf{v}\mathcal{L}(\mathbf{u})\mathbf{u} = 0$

With these transformations, the equation of motion (Equation 3.95) can be rewritten in terms of the new variable $\mathbf{u}$, using properties two and three of Equation (3.105)

$$\begin{aligned} \frac{d\mathbf{R}}{d\tau} &= 2\mathcal{L}(\mathbf{u})\frac{d\mathbf{u}}{d\tau} \\ \frac{d^2\mathbf{R}}{d\tau^2} &= 2\mathcal{L}(\mathbf{u})\frac{d^2\mathbf{u}}{d\tau^2} + 2\mathcal{L}\left(\frac{d\mathbf{u}}{d\tau}\right)\frac{d\mathbf{u}}{d\tau}. \end{aligned} \tag{3.106}$$

When these expressions are substituted into the equation of motion, using property four and $n = 1$, plus some algebra, then

$$2\mathbf{u}\cdot\mathbf{u}\mathcal{L}(\mathbf{u})\frac{d^2\mathbf{u}}{d\tau^2} - 2\frac{d\mathbf{u}}{d\tau}\cdot\frac{d\mathbf{u}}{d\tau}\mathcal{L}(\mathbf{u})\mathbf{u} + G(m_1+m_2)\mathcal{L}(\mathbf{u})\mathbf{u} = (\mathbf{u}\cdot\mathbf{u})^3\mathbf{F}_{12}. \tag{3.107}$$

After still more manipulation, one obtains the equation in a form that contains neither singularities nor indeterminacies as R goes to 0

$$\frac{d^2\mathbf{u}}{d\tau^2} = \frac{1}{2}h\mathbf{u} + \frac{1}{2}R\mathcal{L}^{\mathrm{T}}(\mathbf{u})\mathbf{F}_{12}. \tag{3.108}$$

As was the case in 1-D, if one has only an isolated binary with no external perturbation, then h is exactly conserved, and the equation of motion reduces once again to the readily solvable simple harmonic oscillator. For the general case, however, the perturbation $\mathbf{F}_{12}$ leads to a nonconservation of binding energy and h must be expressed as

$$h = \left[2\frac{d\mathbf{u}}{d\tau}\cdot\frac{d\mathbf{u}}{d\tau} - G(m_1+m_2)\right]\Big/R. \tag{3.109}$$

Note that in ordinary coordinates

$$\frac{d}{dt}\left[\frac{1}{2}\left(\frac{dR}{dt}\right)^2 - \frac{G}{R}(m_1+m_2)\right] = \frac{d\mathbf{R}}{dt}\cdot\mathbf{F}_{12}, \tag{3.110}$$

which can be used, along with the coordinate transformations, to write

$$\frac{dh}{d\tau} = 2\frac{d\mathbf{u}}{d\tau} \cdot \mathcal{L}^T(\mathbf{u})\mathbf{F}_{12}. \tag{3.111}$$

Equation (3.108) and Equation (3.111) can be smoothly integrated as ordinary differential equations through $R = 0$.

An interesting application of the Levi-Civita transformation in a numerical N-body simulation was given by Szebehely and Peters (1967) in their solution of the long-standing Pythagorean three-body problem. In this problem, which idealizes the situation that can occur during an encounter between a binary star and a single star in an evolving star cluster, three point masses of masses $m_1 = 3$, $m_2 = 4$, and $m_3 = 5$ are placed at the vertices opposite the respective sides of a 3-4-5 right triangle. The three particles are initially at rest and the problem is to completely describe their subsequent motion.

As is the case for many bound encounters between a single star and a binary, the motion is extremely complex. The zero-velocity initial condition ensures that the total angular momentum $\mathbf{L} = 0$. This means that in theory, a triple collision is possible, in which case a solution for all time would not exist. Numerical integration using regularization (Figure 3.7), however, shows that no triple collision takes place. During the

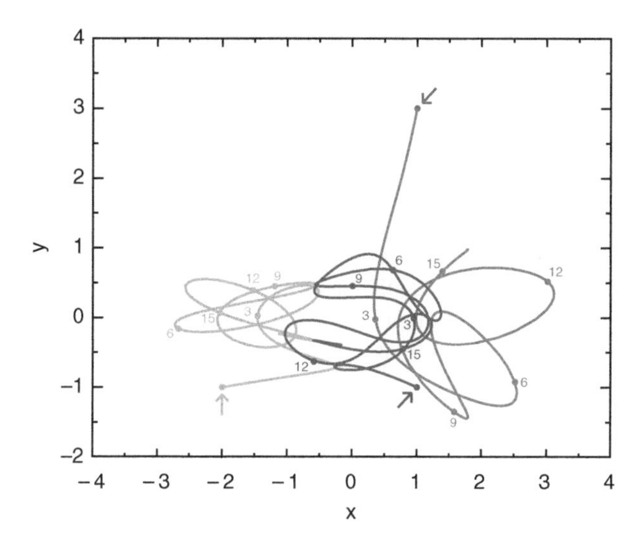

FIGURE 3.7 ***A color version of this figure follows page 212*** Integration of the three-body Pythagorean problem. The initial positions of the particles (indicated by arrows) are at the vertices of a right triangle. The particles have masses of 3, 4, and 5 grams. Positions in the (x, y) plane are expressed in cm. In these units, the unit of time is 3872 s. The equations are integrated for 16 time units, and the positions at selected times are marked on the curves. An initial close encounter between mass 4 and mass 5 occurs at about $x = -0.25$, $y = -0.75$. The end result of the simulation (not shown), which is determined at about $t = 60$, is the formation of a binary by masses 4 and 5 and the ejection of mass 3 from the system. (After Szebehely and Peters (1967). Figure courtesy of Evan Kirby.)

course of the calculation, the distances r_{ij} between the particles are monitored. When two bodies come closer than a threshold value r_{min}, the Levi-Civita transformation is applied to regularize the motion, with the third body playing the part of the perturber, which drives the harmonic oscillator equation. The two bodies are integrated through the close encounter using the regularized equation and are swapped back into Cartesian coordinates once $r_{ij} > r_{min}$. Despite the overhead associated with switching back and forth, the calculation is completed much more quickly and with higher accuracy than if regularization is not used.

The Levi-Civita procedure for regularizing two-dimensional motion transfers in a complicated way to three dimensions, the details of which are not treated here. The equation of motion (Equation 3.92) for a binary pair in three dimensions must first be transformed to a *four-dimensional* coordinate system before it can be regularized. In the context of the numerical N-body problem, this was first done by Kustaanheimo and Stiefel (1965), who gave the transformation

$$\begin{aligned} R_1 &= {u_1}^2 - {u_2}^2 - {u_3}^2 + {u_4}^2 \\ R_2 &= 2(u_1 u_2 - u_3 u_4) \\ R_3 &= 2(u_1 u_3 + u_2 u_4) \\ R_4 &= 0, \end{aligned} \tag{3.112}$$

so that if

$$\mathbf{R} = \mathcal{L}(\mathbf{u})\mathbf{u}, \tag{3.113}$$

one has

$$\mathcal{L}(\mathbf{u}) = \begin{bmatrix} u_1 & -u_2 & -u_3 & u_4 \\ u_2 & u_1 & -u_4 & -u_3 \\ u_3 & u_4 & u_1 & u_2 \\ u_4 & -u_3 & u_2 & -u_1 \end{bmatrix}. \tag{3.114}$$

The analysis outlined by Aarseth (2003; Section 4.4) shows, however, that the equations of motion reduce to exactly the same form as for the two-dimensional case, namely Equation (3.108) and Equation (3.111). Thus, the solution reduces to one that resembles a forced harmonic oscillator, and there is no problem as $R \to 0$.

3.9 FORCE CALCULATION: THE TREE METHOD

We first consider strategies for reducing the $\sim \mathcal{O}(N^2)$ dependence of computational time on the number of particles, which would result if the right-hand side of Equation (3.1) were simply summed over all particles. In any aggregate of gravitating bodies, the particles that are closest to the body being considered will exert the largest contributions to the instantaneous gravitational acceleration, while distant particles will have relatively little individual effect. Therefore, one can drastically speed up an N-body code by retaining careful consideration of nearby neighbors, while treating the more numerous distant bodies as simpler aggregates. This is the basic idea behind the hierarchical tree method.

The tree algorithm (as implemented, for example, by Barnes and Hut 1986, 1989) works by tessellating the volume of space that contains particles into a hierarchy of nested cubic cells, or nodes, all of which contain at least one particle. The aggregate of cells is called the tree, and the finest subdivisions, those containing only a single particle, are the leaves.

Construction of the tree proceeds recursively from the top down. A large cube is first drawn that encompasses all of the particles in the simulation. This lowest-order node is then split into eight equal subvolumes. Those containing no particles are discarded, while those containing multiple particles are subject to further subdivision. As the tree is constructed, the total mass, the center-of-mass location, and the quadrupole moment of each node is computed and stored.

The quadrupole moment is derived from the $n = 2$ term in the multipole expansion of the gravitational potential arising from a distribution of mass

$$\Phi(\mathbf{R}) = -G \sum_{n=0}^{\infty} \frac{1}{|\mathbf{R}|^{n+1}} \int |\mathbf{r}|^n P_n(\cos\theta)\rho(\mathbf{r})d^3\mathbf{r}, \tag{3.115}$$

where the geometry used in the expansion is shown in Figure 3.8. In Equation (3.115), $P_n(x)$ is the nth Legendre Polynomial, and θ is the angle between the radius vectors of a mass element located at position $\mathbf{r} = (x_1, x_2, x_3)$ (inside the node) and the location $\mathbf{R} = (X_1, X_2, X_3)$ at which the potential is to be evaluated (outside the node). The $n = 0$ (the monopole) term corresponds to the potential $(-GM/R)$ that would occur

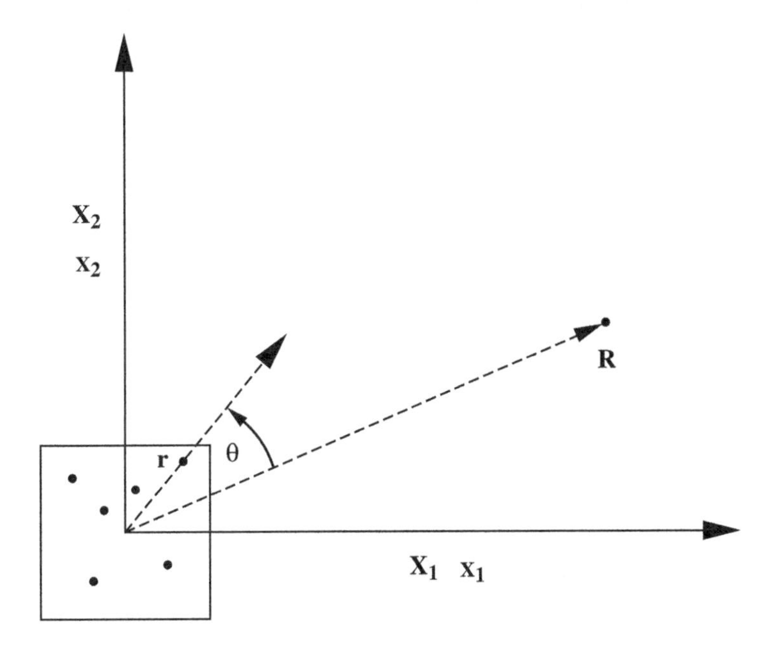

FIGURE 3.8 Geometry for the Legendre expansion in Equation (3.115), simplified to two dimensions. A node of the tree is indicated by the box. A particle i in the box has a radius vector $\mathbf{r}_i$. As long as the box is sufficiently far away from an external particle at $\mathbf{R}$, the contribution of the node to the potential at $\mathbf{R}$ is obtained from the sum of the monopole and quadrupole terms of the Legendre expansion of the potential from the particles in the box. The origin of the coordinate system is the center of mass of the box.

if all of the mass (M) in the node under consideration were concentrated at the center of mass. Mass cannot be negative, hence, the $n = 1$ (the dipole term) vanishes by symmetry. The $n = 2$ term gives the quadrupole contribution to the potential. It can be rewritten in the form

$$\Phi(\mathbf{R})_2 = -\frac{G}{2|\mathbf{R}|^3} \sum_{i,j=1}^{3} \frac{X_i X_j}{R^2} Q_{ij}\,, \tag{3.116}$$

where

$$Q_{ij} = \int \rho(\mathbf{r})(3x_i x_j - r^2 \delta_{ij}) d^3\mathbf{r} \tag{3.117}$$

is the quadrupole moment tensor. For a collection of point particles within a node of the tree, the integral in Equation (3.117) reduces to a sum.

After the tree has been constructed, which is redone for every time step, the gravitational force on each particle is evaluated. These forces are computed by working systematically down through the tree. At each level, the width, w, of a node is compared to the distance, R, between the node's center of mass and the location of the current particle. If $w/R < \delta$, where δ is a specified fractional distance, then the attraction between all the particles in the node and the current particle is evaluated using the monopole and quadrupole contributions that have been stored for the node, and no further descent into the node's subdivisions is necessary. Alternately, if $w/R > \delta$, the node must be further subdivided. As this procedure is repeatedly applied, all of the nodes in the tree are either added to a monopole–quadrupole approximate contribution or are found to contain a single particle. For nodes containing single particles, the contribution to the gravitational force is obtained through direct summation.

With the switching criterion between direct summation and multipole expansion set to the form $w/R < \delta$, the sizes of cells contributing approximate monopole–quadrupole-based forces increases in direct proportion to the distance from the current particle, and the sum over N particles is replaced by a sum containing only $\sim \log N$ terms. For simulations in which δ is a small fraction of the total physical domain, the computational time required to obtain all the accelerations scales as $\mathcal{O}(N \log N)$. For large N, the difference between N^2 and $N \log N$ is enormous.

For a large aggregate of particles, essentially nothing is lost by approximating the direct interactions with the first terms of the gravitational multipole expansion. Any benefit from finely resolving distant aggregates of mass is often outweighed by the attendant accumulation of truncation errors and the errors arising from the use of finitely resolved time steps. Indeed, in the few-body techniques considered in previous sections, the planets, which are physically extended objects, were reduced to point masses, which amounts to truncating their multipole expansions at a single term. For a refined investigation of the overall dynamics of the solar system, one might, for example, approximate the Earth–Moon binary by both the total point mass contribution as well as the quadrupole contribution.

For one million particles, the factor of increase in speed provided by shifting from direct summation to the tree method can approach 70,000. One might ask, is there a way to further speed up the computation of the gravitational forces so that finding the accelerations becomes an $\mathcal{O} \sim (N)$ process?

Remarkably, an order N method was employed by Erik Holmberg of the Lund Observatory in Sweden in 1941. Instead of integrating the equations of motion with a computer, Holmberg modeled a two-dimensional system of gravitating particles as an actual physical distribution of movable light bulbs laid out on a gridded sheet of dark paper. Because the intensity of light from a point source diminishes as $1/r^2$, one can directly relate the intensity of the light at a particular spot to the gravitational acceleration. The N^2 process of computing the gravitational force on a given particle from all of the other particles reduces to a measurement of the total intensity of light in two perpendicular directions using a photocell and a galvanometer. Since one set of measurements is required for the location of each light bulb, the method scales as $\mathcal{O}(N)$.

With his intensity measurement analog method for computing the net gravitational acceleration on each of his light bulb "point masses," Holmberg could compute the change in trajectories that would occur over a time interval using a simple integration scheme such as Euler's method. A time step would then be completed by moving all of the light bulbs to their updated positions, at which point a new estimate of the gravitational acceleration could be made. Holmberg's scheme allowed him to gain a better understanding of important aspects of the dynamics of close encounters between disk galaxies, including the phenomena of orbital decay and the formation of tidal tails. Viewed from today, experiments such as this one seem quaint, but there is an important lesson to be drawn: Use of an analog method reduces an $\mathcal{O}(N^2)$ computation to $\mathcal{O}(N)$, foreshadowing a time when quantum computation will similarly reduce the computational time required from $\mathcal{O}(N^2)$ to $\mathcal{O}(N)$.

3.10 FORCE CALCULATION: FAST FOURIER TRANSFORMS

In its most familiar guise, the technique of Fourier analysis allows one to express a continuous time signal in terms of an infinite set of frequency components. That is, the distribution of frequencies, $H(\omega)$, corresponding to a time series $h(t)$ is given by

$$H(\omega) = \int_{-\infty}^{\infty} h(t) \exp(i\omega t) dt. \tag{3.118}$$

Even after learning to manipulate complex variables, some readers feel uncomfortable with the fact that the time series $h(t)$ is real, whereas the Fourier transform, $H(\omega)$, is a complex number: "Isn't a frequency a number of Hertz — a real quantity?"

$H(\omega)$ is complex because it has to express two pieces of information for every frequency; at each ω, we can write $H(\omega) = H_0(\omega) \exp[i\theta_0(\omega)]$. $H_0(\omega)$, which is purely real, gives the amplitude of the sinusoidal contribution for ω, whereas $\theta_0(\omega)$ gives its phase. The phase information allows us to correctly position the sinusoid relative to all the others so that when we make an inverse Fourier transform, the original (purely real) time-dependent signal

$$h(t) = \frac{1}{2\pi} \int_{-\infty}^{\infty} H(\omega) \exp(-i\omega t) dt \tag{3.119}$$

is recovered.

The utility of Fourier transforms in the context of the N-body problem stems from a remarkable result known as the convolution theorem. If $g(t)$ and $h(t)$ are functions, then their convolution is:

$$g \star h = \int_{-\infty}^{\infty} g(\tau)h(t-\tau)d\tau, \tag{3.120}$$

and the Fourier transform of the convolution is $G(\omega)H(\omega)$, where G and H are the Fourier transforms of g and h, respectively.

The transform of the convolution of two functions, thus, is equal to the product of their transforms, and if these transforms are known, we can avoid explicitly computing the integral. To motivate this, note that the gravitational potential of a continuous field of sources

$$\Phi(\mathbf{x}) = -G\int \frac{\rho(\mathbf{x}')}{|\mathbf{x}-\mathbf{x}'|}d^3\mathbf{x}' \tag{3.121}$$

has the basic structure of the Expression (3.120), with $\mathbf{x}'$ playing the role of τ. This connection allows us to use Fourier transforms to do potential calculations.

The first step in making a simple Fourier transform potential solver is to define a three-dimensional Cartesian grid that encompasses all the particles. One then loops through the N particles and computes an index

$$ix = \mathrm{A_{int}}\left[\frac{(x_i - x_{min})}{(x_{max} - x_{min})}Nc_x\right] \tag{3.122}$$

for each particle, along with analogous indices iy and iz. In the above formula, x_{max} and x_{min} are the predefined x-direction limits of the grid, and Nc_x is the number of cells assigned to the x-direction. The $\mathrm{A_{int}}$ function indicates truncation to an integer value (e.g., $\mathrm{A_{int}}[2.647] = 2$). Within the loop, as the cell indices of each particle are computed, the mass of each cell, $M_{ix,iy,iz}$, can be tallied, along with the three coordinate arrays, x_{ix}, y_{iy}, z_{iz}, for the cell centers. For example,

$$x_{ix} = x_{min} + \frac{(ix - 0.5)}{Nc_x}(x_{max} - x_{min}). \tag{3.123}$$

The softened gravitational potentials at the cell centers are given by

$$\Phi_{jx,jy,jz} = -\sum_{ix=0}^{Nc_x-1}\sum_{iy=0}^{Nc_y-1}\sum_{iz=0}^{Nc_z-1} M_{ix,iy,iz}\frac{G}{D},$$

where

$$D = \sqrt{\epsilon^2 + |x_{jx} - x_{ix}|^2 + |x_{jy} - x_{iy}|^2 + |x_{jz} - x_{iz}|^2}. \tag{3.124}$$

With this discretization, we have replaced the integral in Equation (3.121) with a sum over ix, iy, and iz. The Fourier transform and inverse transforms (Equation 3.118) and (Equation 3.119) can be similarly discretized through the use of multidimensional Fourier transforms, which map one set of $N = N_1N_2N_3$ numbers arranged in a three-dimensional grid $h(k_1, k_2, k_3)$ to a second set of N numbers in an identically arrayed grid $H(n_1, n_2, n_3)$

$$H(n_1, n_2, n_3) = \sum_{k_3=0}^{N_3-1} \sum_{k_2=0}^{N_2-1} \sum_{k_1=0}^{N_1-1} h(k_1, k_2, k_3) \exp(E_1) \exp(E_2) \exp(E_3) \quad (3.125)$$

with $E_1 = 2\pi i k_1 n_1/N_1$, $E_2 = 2\pi i k_2 n_2/N_2$, and $E_3 = 2\pi i k_3 n_3/N_3$. Likewise, the discrete three-dimensional inverse Fourier transform is given by

$$h(k_1, k_2, k_3) = \frac{1}{N} \sum_{n_3=0}^{N_3-1} \sum_{n_2=0}^{N_2-1} \sum_{n_1=0}^{N_1-1} H(n_1, n_2, n_3) \exp(-E_1) \exp(-E_2) \exp(-E_3). \quad (3.126)$$

Note that multidimensional Fourier transforms are computed by taking successive one-dimensional transforms, so that the overall effort involves a comparable number of operations to taking a one-dimensional transform of an equivalently long ($N_{\text{tot}} = N_1 N_2 N_3$) sequence of numbers. That is, we can make a minor alteration in the way that Equation (3.125) is expressed

$$H(n_1, n_2, n_3) = \sum_{k_1=0}^{N_1-1} \left[\sum_{k_2=0}^{N_2-1} \left[\sum_{k_3=0}^{N_3-1} h(k_1, k_2, k_3) \exp(E_3) \right] \exp(E_2) \right] \exp(E_1), \quad (3.127)$$

which makes it clear that we can compute the one-dimensional transforms from the inner nesting to the outer nesting, thus giving us the capability of computing the Fourier transform of a set of numbers defined on a grid.

In order to actually find the potentials on the grid that has been superimposed on the mass distribution, we leverage the fact that the Fourier convolution theorem has a discrete analogy. That is, if

$$Z_k = \frac{1}{\sqrt{2K}} \sum_{k'=-K}^{K-1} Y_{k-k'} X_{k'} \quad (3.128)$$

defines a convolution, then the Fourier transform $\hat{Z}_p$ is obtained from

$$\hat{Z}_p = \hat{X}_p \hat{Y}_p. \quad (3.129)$$

(For proof see Binney and Tremaine [1987, Section 2.8].) From the form of Equation (3.118), it is clear that a Fourier integral transform applies over an unbounded domain of time (or of space, when we identify t with a position coordinate). A localized function that goes to zero beyond a certain boundary can still be represented by periodic functions because an infinite number of frequency components are being called upon to provide perfectly destructive interference within the entire infinite region beyond the realm where $h(t)$ is nonzero. With a finite and discretized distribution, however, we have no such luxury. The "frequency components" $\hat{X}_p$ constitute a finite set and lead to a representation X_k from the inverse transformation that must be understood to be periodic.

The equation for the potential (Equation 3.124) has essentially the form required to implement the discrete Fourier transform. If we scale the size of the cells so that each cell is cubic with sides of unit length, then we have

$$\Phi_{ijk} = -\sum_{i'=0}^{K-1}\sum_{j'=0}^{K-1}\sum_{k'=0}^{K-1} X(i-i', j-j', k-k')M_{i',j',k'}, \tag{3.130}$$

where $X = G/D$. The range of the indices that are being summed over in Equation (3.130) is half of the range of the indices in the sums in the expression for the discrete Fourier convolution. In order to apply the Fourier convolution theorem to compute the potential, we need to augment the sums in Equation (3.130) so that they run over the full range of the indices. This is done by defining the mass cube $M_{i',j',k'}$ to be twice as large in every direction, but with all of the new cells occupied with zero particles. The geometric array $X(i-i', j-j', k-k')$ is also expanded so that the indices run from $-K$ to $K-1$. (All of these geometric terms can be precomputed and then reused every time the mass distribution changes.)

Given this simple modification, the Fourier convolution theorem can be applied to yield an estimate of the potential at each point in the grid. One obtains the three-dimensional Fourier transforms, $\hat{X}$, and $\hat{M}$ of the matrices X and M, and multiplies each individual term, $\hat{X}_{ijk}$ with its counterpart $\hat{M}_{ijk}$. The resulting matrix is the Fourier transform, $\hat{\Phi}$, of the potential. The potentials Φ themselves are then obtained by computing the discrete inverse Fourier transform.

With what has been discussed so far, the only advantage in computational speed from the Fourier method arises from the fact that X, with its profusion of square roots, can be precomputed. The Fourier transforms, if done in the naïve way by working term-by-term through the triple sums, are an order N^2 operation and, as such, add no benefit beyond a simple, direct summation over all grid cells to obtain the potential. The advantage of the Fourier transform comes from the fact that the discrete Fourier transform can be computed as an $N \log N$ operation, which, for large N, offers a gargantuan improvement in speed.

The order $N \log N$ *fast Fourier transform* (FFT) exploits the fact that the Fourier transform of N discrete points, whose indices run from 0 to $N-1$ can be rewritten as a sum of two transforms of half the length, whose indices run from 0 to $N/2-1$. That is, assume that the transform length N is an integer power of two, $N = 2^m$. The kth component of the (1-D) transform

$$H_k = \sum_{j=0}^{N-1} h_j \exp(2\pi ijk/N) \tag{3.131}$$

can be expressed as the addition of a series of even terms with length $N/2$ to a series of odd terms of length $N/2$

$$H_k = \sum_{j=0}^{N/2-1} h_{2j} \exp[2\pi ik(2j)/N] + \sum_{j=0}^{N/2-1} h_{2j+1} \exp[2\pi ik(2j+1)/N], \tag{3.132}$$

which, with a slight rearrangement, are seen themselves to be two length $N/2$ individual Fourier series, with the second multiplied by the complex factor $\exp(2\pi ik/N)$

$$H_k = H_{ek} + \exp(2\pi ik/N)H_{ok} = \sum_{j=0}^{N/2-1} h_{2j}\exp[2\pi ikj/(N/2)]$$
$$+ \exp(2\pi ik/N)\sum_{j=0}^{N/2-1} h_{2j+1}\exp[2\pi ikj/(N/2)]. \quad (3.133)$$

In order to evaluate the components of H_k that have $k > N/2$, we operationally require the sums in Equation (3.133) to run from 0 to $N-1$. This demand is met by recognizing that the sums are periodic. Thus, the full 0 to $N-1$ range of k can be produced by augmenting each series on the right-hand side of Equation (3.132) by a repeated copy of itself for $k > N/2$.

The FFT draws its speed from the recursive applications of Equation (3.133). Using Equation (3.132), the even series, H_{ek} can be further subdivided to form

$$H_{ek} = H_{eek} + \exp[2\pi ik/(N/2)]H_{eok} \quad (3.134)$$

and the odd series becomes

$$\exp(2\pi ik/N)H_{ok} = \exp(2\pi ik/N)[H_{oek} + \exp[2\pi ik/(N/2)]H_{ook}]\ . \quad (3.135)$$

One proceeds through m subdivisions until one is left with N *one-point* Fourier transforms, each multiplied by the individual complex factor, $\exp(i\phi_k)$, built up through the m subdivisions.

Remarkably, the N one-point Fourier transforms (call them H_{1k}) that remain after m recursive applications of Equation (3.134) and Equation (3.135) have an exact (yet scrambled) one-to-one correspondence with the N terms of the original input function h. The mapping $k \to l$, which equates each term h_k with a one-point transform H_{1l} proceeds by bit-reversing the indices k (see, Press et al, 1992). That is, one first writes the integer k as a binary number with m digits, and then determines the integer l by reading off the binary digits of k from right to left, as opposed to the usual left to right. For example, for $N = 32$, $k = 19$ has binary representation 10011, indicating that the binary representation for l is 11001, i.e., $l = 25$.

To compute the transform, one first loops through the indices k to get the bit reversed values l and the corresponding array of one-point transforms $H_{1k} = h_l$. One then uses Equation (3.134) and Equation (3.135) to combine each pair of one-point transforms to form two-point transforms. Both terms in each transform are computed. The sets of two-point transforms are then combined to form four-point transforms, and the process continues a total of $m = \log_2 N$ times to form the final FFT. At each level, forming the combinations requires order N operations, making the entire procedure order $N \log N$.

As a concrete example, consider a four-point FFT of a discrete signal that has been sampled four times

$$h = [h_0, h_1, h_2, h_3]. \quad (3.136)$$

The bit reversal step yields 00 → 00, 01 → 10, 10 → 01, and 11 → 11, so that the one-point transforms are given by, $H_{10} = h_0$, $H_{11} = h_2$, $H_{12} = h_1$, and $H_{13} = h_3$. The rule (Equation 3.134) is used to combine H_{10} and H_{11} into a two-point transform

$$[H_{20} = h_0 + h_2, H_{21} = h_0 + h_2 \exp(i\pi)], \tag{3.137}$$

and to combine H_{12} and H_{13} into a second two-point transform with terms $[h_1 + h_3, h_1 + h_3 \exp(i\pi)]$.

Finally, with $N = 4$, the components of the four-point FFT can be assembled from the two-point FFT

$$\begin{aligned} H_{40} &= h_0 + h_2 + h_1 + h_3 \\ H_{41} &= h_0 + h_2 \exp(i\pi) + \exp(i\pi/2)[h_1 + h_3 \exp(i\pi)] \\ H_{42} &= h_0 + h_2 + \exp(i\pi)[h_1 + h_3] \\ H_{43} &= h_0 + h_2 \exp(i\pi) + \exp(3\pi i/2)[h_1 + h_3 \exp(i\pi)]. \end{aligned} \tag{3.138}$$

Notice that each transformed component, H_{4k}, is itself formed from the sum of four terms, suggesting at first glance that order $N^2 = 16$ operations are required to compute the transform. Upon inspection, however, one sees that the terms are not fully independent, having been assembled from precomputed sums of length $N/2$. For a four-point FFT, the overhead involved in computing the terms of form $\exp(i\theta)$ effectively cancels the gain in efficiency from the FFT. As N increases, however, the $N \log N$ scaling of the FFT algorithm imparts an increasingly massive advantage.

Exercise

Verify Equation (3.138) by calculating its inverse discrete Fourier transform.

Once the potential has been computed by employing the FFT and the discrete convolution theorem, the individual potentials can be differenced across the grid cells to obtain the acceleration components a_x, a_y, and a_z.

We may summarize the situation for a large N system, where it is not practical to calculate the gravitational forces by direct summation over all particles. The Fourier method, also known as the *particle-mesh* (PM) method, may be better in a situation where the mass is distributed, more or less, uniformly. A grid is overlaid on the system of particles and a number of grid-related quantities can be precomputed once and then used over and over again. An example would be the case of very large-scale cosmological simulations. However, if the mass distribution becomes clumpy, with a lot of mass concentrated into a few grid cells with large almost empty spaces in between, then the Fourier method will not give an accurate potential in the dense regions unless the grid is adaptively refined. In this case, the tree method would be better, since it automatically puts the leaves (smallest subdivisions) where the mass is and, thus, is adaptive in a fully Lagrangian sense. An example of such a problem would be the encounter and collision between two galaxies. Another alternative is the so-called "particle-particle-particle-mesh" (P^3M) scheme (Hockney and Eastwood, 1981, Efstathiou and Eastwood, 1981) in which the force acting on a given particle is split into two parts, one involving long-range smoothly varying forces, which are calculated with the Fourier method on a grid, and the other involving short-range forces arising only from a small set of particles within a specified distance from the given particle. The short-range forces are calculated by direct summation. This type of code is widely used in cosmological simulations. Nevertheless a (gridless) tree code may still have an advantage, since any grid imposes a particular geometry on the problem and can introduce errors as a result of force interpolation.

REFERENCES

Aarseth, S. (1985) *Dynamics of Star Clusters* (Dordrecht: Reidel), p. 251.

Aarseth, S. (2003) *Gravitational N-Body Simulations* (Cambridge: Cambridge University Press).

Ahmad, A. and Cohen, L. (1973) *Astrophys. J.* **179**: 885.

Barnes, J. and Hut, P. (1986) *Nature* **324**: 446.

Barnes, J. and Hut, P. (1989) *Astrophys. J. Suppl.* **70**: 389.

Binney, J. and Tremaine, S. (1987) *Galactic Dynamics* (Princeton: Princeton University Press).

Danby, J. (1988) *Fundamentals of Celestial Mechanics,* 2nd ed. (Richmond, VA: Willmann-Bell).

Efstathiou, G. and Eastwood, J. W. (1981) *Monthly Notices of the Royal Astronomical Society* **194**: 503.

Gleick, J. (1987) *Chaos: Making a New Science* (New York: Viking Penguin).

Goldstein, H. (1980) *Classical Mechanics* (Reading, MA: Addison-Wesley).

Hockney, R. W. and Eastwood, J. W. (1981) *Computer Simulation Using Particles* (New York: McGraw-Hill).

Holmberg, E. (1941) *Astrophys. J.* **94**: 385.

Kustaanheimo, P. and Stiefel, E. (1965) *J. Reine Angew. Math.* **218**: 204.

Lecar, M., Franklin, F., and Murison, M. (1992) *Astron. J.* **104**: 1230.

Levi-Civita, T. (1904) *Ann. Mat. Ser.* 3, **9**: 1.

Marcy, G. W., Cochran, W. D. and Mayor, M. (2000) *Protostars and Planets IV,* V. Mannings, A. P. Boss and S. S. Russell, Eds. (Tucson: University of Arizona Press), p. 1285.

Montenbruck, O. (1989) *Practical Ephemeris Calculations* (Heidelberg: Springer-Verlag).

Murray, C. and Dermott, S. (1999) *Solar System Dynamics* (Cambridge: Cambridge University Press).

Newton, I. (1687) *Philosophiae Naturalis Principia Mathematica* (London: Royal Society).

Press, W. H., Teukolsky, S. A., Vetterling, W. T. and Flannery, B. P. (1992) *Numerical Recipes in Fortran: The Art of Scientific Computing,* 2nd ed. (Cambridge: Cambridge University Press).

Saha, P. and Tremaine, S. (1994) *Astron. J.* **108**: 1962.

Stoer, J. and Bulirsch, R. (1980) *Introduction to Numerical Analysis* (New York: Springer–Verlag).

Szebehely, V. and Peters, C. F. (1967) *Astron. J.* **72**: 876.

Wisdom, J. and Holman, M. (1991) *Astron. J.* **102**: 1528.

4 Smoothed Particle Hydrodynamics

The major hydrodynamical problems in astrophysics, for example, galaxy formation and evolution, the collapse of interstellar clouds in star formation, and the evolution of disks, have been attacked numerically through two major approaches. The first is the grid-based method, which was introduced in Chapter 2 and will be elaborated upon in Chapter 6. In one space dimension, grids can be either Lagrangian or Eulerian in nature, but in two or three space dimensions, a fixed Eulerian grid is generally used, with the option of subdividing the grid as the calculation proceeds. However, particularly in two or three space dimensions, it has also been found productive to use a Lagrangian method that is more closely related to N-body methods, known as smoothed particle hydrodynamics (SPH). This chapter provides an introduction to that method. Section 4.1 considers the very basic equations, in their simplest form, that must be put together to perform hydrodynamical calculations in three space dimensions using the SPH technique. Section 4.2 shows the results of a simple test calculation. Section 4.3 then discusses improvements needed in various aspects of an SPH code to bring it up to research grade. The techniques have been expertly reviewed by Benz (1990), Monaghan (1992), and Price (2004). At the end of the chapter appears a summary of the advantages and disadvantages of SPH as compared with grid-based codes.

4.1 RUDIMENTARY SPH

A gas dynamical system differs fundamentally from a gravitational N-body system because gas experiences pressure. At the microscopic level, pressure is the cumulative effect of collisions of individual atoms or molecules that continuously frustrate the ballistic trajectories of the particles within the time- and space-dependent gravitational field. At the macroscopic level, the sum of all the tumult amounts to a force on each fluid element that is proportional to the local pressure gradient. The equations describing the motion of an individual packet of fluid are extended to read

$$\frac{d\mathbf{r}_i}{dt} = \mathbf{v}_i, \tag{4.1}$$

$$\frac{d\mathbf{v}_i}{dt} = -\frac{1}{\rho_i}\nabla P_i - \nabla\Phi_i, \tag{4.2}$$

where P_i, ρ_i, and Φ_i are the pressure, density, and gravitational potential, respectively. In Chapter 6, we will describe in considerable detail how these equations of motion for the fluid elements can be cast onto a fixed grid. With this method, the so-called "Eulerian" approach, one tracks the properties of the fluid at the locations of the grid cells. The individual parcels of fluid are not followed. This approach differs fundamentally from the N-body codes, where each particle represents a particular discrete bundle of mass. We say that an N-body code functions in a "Lagrangian" framework. In simplest terms, the SPH method is an extension of an N-body particle method (with large N!) to include the properties of gases.

Eulerian methods tend to perform better in situations that have a basic underlying symmetry, or for problems in which one needs to simulate only part of a larger system. For example, if one wishes to study the fate of small disturbances in an equilibrium disk, the use of a cylindrical grid allows one to exactly capture the underlying equilibrium that constitutes the bulk of the motion. Likewise, a grid would be the natural choice for simulating a relatively small volume of a global atmosphere. One cares about the properties of the gas localized at a particular spot. There is no concern for the detailed fate of the temporary confederation of molecules that make up a weather front.

The Lagrangian view is most useful for situations that are highly dynamic, and which do not conform well to an underlying symmetry. Arbitrary geometries are allowed, and there is no need to define a spatial "edge" to the problem, as is required for a fixed grid. Collisions between stars or planets provide good examples of intrinsically Lagrangian problems. Also, the collapse of a mass of gas under gravity from arbitrary initial conditions is a problem well suited to a Lagrangian approach. The major applications of SPH in astrophysics involve problems in three space dimensions.

Much of the machinery of the previous chapter seems poised to tackle Equation (4.1) and Equation (4.2), save for the $\nabla P_i/\rho_i$ term. The essential motivation reduces to: How does one accurately represent the gradient of the pressure at a single point? The idea behind smoothed particle hydrodynamics is to imagine that each particle represents not a point mass, but rather, a smeared-out distribution of density

$$\rho_j(\mathbf{r}) = m_j W(|\mathbf{r} - \mathbf{r}_j|; h), \tag{4.3}$$

where $\rho_j(\mathbf{r})$ is the contribution to the density at point $\mathbf{r}$ arising from a particle j at $\mathbf{r}_j$ and W is the "smoothing kernel" that describes the form of the mass distribution associated with a particular particle of mass m_j. When thinking about smoothed particle hydrodynamics, it helps to use the example of a Gaussian (spherically symmetric) kernel

$$W(|\mathbf{r} - \mathbf{r}_j|; h) = \frac{1}{h^3 \pi^{3/2}} \exp[-(|\mathbf{r} - \mathbf{r}_j|/h)^2]. \tag{4.4}$$

The half-width h of the density distribution associated with the particle is called the smoothing length; the kernel, therefore, has dimensions of $(\text{volume})^{-1}$. The physical density at a given point, or at the location of a particle i, is obtained by summing the overlapping contribution from all of the particles (j)

$$\rho(\mathbf{r}) = \sum_{j=1}^{N} \rho_j(\mathbf{r}) \tag{4.5}$$

where particle i itself is included in the summation. With the Gaussian kernel, along with other kernels that are typically used, the dominant contribution to $\rho(\mathbf{r})$ is from the nearby particles. The kernel must be defined so that

$$\int_0^\infty W(|\mathbf{r}-\mathbf{r}_j|;h)d\mathbf{r} = 1 \tag{4.6}$$

and

$$\lim(h \to 0)W(|\mathbf{r}-\mathbf{r}_j|;h) = \delta_D(\mathbf{r}-\mathbf{r}_j) \tag{4.7}$$

where δ_D is the Dirac delta function, which goes to infinity at $\mathbf{r} = \mathbf{r}_j$ in such a way as to maintain the normalization of Equation (4.6) and is otherwise zero.

Other physical quantities associated with the fluid (for instance, the pressure or the temperature) are estimated using density-weighted interpolations between the known particle values

$$A(\mathbf{r}) = \sum_{j=1}^{N} m_j \frac{A_j}{\rho_j} W(|\mathbf{r}-\mathbf{r}_j|, h) \tag{4.8}$$

where $\mathbf{r}$ refers, for example, to the position of a particular particle i and $\rho_i/m_i = n_i$, the number of particles per unit volume. The Lagrangian nature of SPH is cemented by the fact that the derivatives of physical quantities can be obtained directly from derivatives of the kernel itself (Hernquist and Katz, 1989). For example, the pressure gradient $\nabla P(\mathbf{r})$ which is needed in Equation (4.2) is obtained most simply as

$$\nabla P(\mathbf{r}) = \sum_{j=1}^{N} m_j \frac{P_j}{\rho_j} \nabla W(|\mathbf{r}-\mathbf{r}_j|, h) \tag{4.9}$$

where the derivative is calculated at $\mathbf{r}$.

One can employ these ideas to quickly write a basic SPH code. The elements of this code must include:

- An initial condition that provides positions, velocities, and masses for all particles. Then the density ρ_i assigned to each particle is obtained from Equation (4.5).
- A method for calculating the gravitational forces $\nabla\Phi_i$.
- A physical law relating P_i and ρ_i.
- A time integration scheme, such as those described in the previous chapter, for advancing the positions and velocities of all particles according to Equation (4.1) and Equation (4.2).
- A scheme for determining $\nabla P_i/\rho_i$, for example, the combination of Equation (4.3), Equation (4.4), Equation (4.5), and Equation (4.9).
- A method for determining the smoothing length h. It scales most appropriately as

  $$h = \frac{h_0}{<\rho>^{1/3}}, \tag{4.10}$$

  where h_0 is an adjustable constant and

  $$<\rho> = \frac{1}{N}\sum_{i=1}^{N} \rho_i \tag{4.11}$$

  is the average density for the entire aggregate of N particles.

In the absolutely most rudimentary instance, the positions and velocities can be advanced using Euler's method for time stepping, the gravitational forces are obtained by direct summation, and h is the same for all particles, with h_0 chosen so that a reasonable number of particles (20 to 50) are contained within a smoothing length. In an N-body code, taking a large time step leads merely to a decrease in accuracy. In SPH, the presence of a sound speed $c_s^2 = \frac{\partial P}{\partial \rho}$, through which the effects of pressure are transmitted, requires that each particle must obey the CFL condition

$$\Delta t_i < \frac{h_i}{c_s}. \tag{4.12}$$

A time step limit based on fluid velocity is not needed in this type of Lagrangian scheme. The calculation then proceeds in two separate steps. First, the density is updated for all particles from Equation (4.5), then other particle properties are calculated, including the pressure gradient, the gravitational force, and the new position and velocity.

4.2 COLLIDING PLANETS: AN SPH TEST PROBLEM

A code based on these concepts is extremely simple and, indeed, the first implementations of SPH, by Lucy (1977) and by Gingold and Monaghan (1977), were not much more complicated than the version we discuss here. In order to get a feeling for how the code works (and in order to get a feeling for its shortcomings), we look at how the rudimentary code handles an astrophysical test problem of the type most generally suitable to the SPH method — a collision of two planets.

Collisions between planets sound rather melodramatic, but in fact, they are likely a fairly important occurrence in the evolution of planetary systems. The Earth–Moon system is thought to have arisen from the collision of a Mars-size terrestrial planet with the proto-Earth during the first 100 million years following the formation of the solar system. The large range in eccentricities among the observed extrasolar planets hints at histories of orbital instability and collisions.

In order to run the test problem, we need a reasonable model for the equilibrium structure of a giant planet — we need to know where to position the particles initially. The relationship between temperature, density, and pressure within an astrophysical object is called the equation of state. The ideal gas law, $P = nk_BT$, where n is the number of free particles per unit volume or, more usefully,

$$P = \frac{R_g}{\mu}\rho T \tag{4.13}$$

where $R_g = k_B/m_u$ is the gas constant, m_u is the atomic mass unit, and μ is the mean atomic weight per free particle, is a familiar example. Given any two quantities from the P, ρ, T trio, one immediately knows the third. In trying to use the ideal gas law with the SPH equations that we have formulated so far, we run into the difficulty of determining the temperature. We, therefore, need to introduce the complication of an energy equation, but before doing so, we note that there are many situations in astronomy where the pressure depends only on the density (known as a *barotropic* equation of state). The most important example is the isothermal gas. If a gas can cool off very efficiently when it is compressed, then the temperature will stay constant, and the ideal gas law reduces to $P \propto \rho$. The dynamics of molecular clouds, for instance, often obey the isothermal approximation. Other examples of barotropic equations of state include those for white dwarfs (Chapter 5). When a white dwarf is of low mass, the electrons are moving at nonrelativistic speeds, and the equation of state is $P \propto \rho^{5/3}$. As white dwarfs approach the Chandrasekhar mass, the electrons obtain velocities that approach the speed of light. When this happens, the equation of state approaches the limiting form $P \propto \rho^{4/3}$.

The equation of state for a giant planet is approximated by an ideal gas only in the outermost atmospheric layers. Deeper down, the material is either in the form of a dense molecular fluid or liquid metallic hydrogen, whose physics is very complicated and beyond the scope of this book (see de Pater and Lissauer, 2001, Chap. 6). However, detailed models of a planet like Jupiter have shown that the structure can be approximated by a polytrope $P = K\rho^{\gamma}$ with $\gamma = 2$. Such models for Jupiter (e.g., de Pater and Lissauer, 2001) give density $\rho_c = 5$ g cm^{-3}, pressure $P_c = 6.5 \times 10^{13}$ dyne cm^{-2}, and temperature $T_c \approx 22{,}000$ K near the center of the planet, just outside a small core composed of heavy elements. These data allow us to fix a reasonable value for $K = 2.6 \times 10^{12}$ dyne g^{-2} cm^4.

One consequence of this relation is that the radius of the planet is very insensitive to the mass. A planet with 10 times the mass of Jupiter is only marginally larger than Jupiter in physical size. With $\gamma = 2$, one can build an equilibrium model in spherical symmetry, which does a reasonable job of approximating Jupiter's interior structure by balancing the pressure produced by this equation of state against the force of gravity at every radial location within the star. This equilibrium is defined by the following equations, which are discussed in more detail in the following chapter

$$\frac{dP}{dr} = -\frac{GM_r\rho}{r^2} \tag{4.14}$$

and

$$\frac{dM_r}{dr} = 4\pi r^2 \rho, \tag{4.15}$$

where M_r is the mass enclosed within radius r.

Detailed studies of this system of equations, in connection with the polytropic equation of state (Chandrasekhar, 1939), show that for $\gamma = 2$ there is an analytic solution

$$\frac{\rho}{\rho_c} = \frac{\sin(\xi)}{\xi}, \tag{4.16}$$

where ρ_c is the central density and

$$\xi = \frac{r}{\alpha} \quad \text{where} \quad \alpha = \left(\frac{K}{2\pi G}\right)^{1/2}. \tag{4.17}$$

A model for the planet is specified if one provides the two parameters ρ_c and K.

Exercise

Calculate the outer radius of Jupiter (where the density $\rho \to 0$) according to the polytropic model and compare with its actual radius.

An SPH realization of a $\gamma = 2$ polytrope is produced by distributing the particles so that their density profile recovers Equation (4.16) above. In normalized coordinates $\alpha = 1$ and $\rho_c = 1$, the total mass $M = 4\pi^2$. The $N_{Total} = M/m_i$ equal mass particles assigned to constitute the planet (in three space dimensions) must be allocated within radial cells of width $\delta\xi = \xi_2 - \xi_1$ according to

$$N(\delta\xi) = \frac{(\sin\xi_2 - \xi_2\cos\xi_2 - \sin\xi_1 + \xi_1\cos\xi_1)}{\pi} N_{Total}. \tag{4.18}$$

The distribution over the longitudinal and latitudinal angles ϕ and θ can be made by placing the particles randomly within $0 < \phi < 2\pi$ and $-1 < \cos(\theta) < 1$. (Note that one distributes particles through $\cos(\theta)$ rather than θ; there is much more surface area near the equator than near the poles.)

Exercise

Derive Equation (4.18).

Collisions between planets generally occur when the planets are moving roughly at their mutual escape velocity. For an initial condition, we construct two 2500-particle planets and place them in the x-y plane, initially separated by three planetary radii (R_p) in the x-direction, and 1 R_p in the y-direction. The planets are given velocities in the x-direction of $\mathbf{v} = \pm 1/2\sqrt{2GM/3R_p}$, so that they encounter each other in an off-center collision at the origin.

Even with only 5000 particles, the basic sequence of events in the collision is clear. The angular momentum of the two-planet system relative to the center of mass is transported into a pair of spiral waves, which serve to transport angular momentum away from the center of the newly merged bodies. The bulk of the mass of the two planets settles into a distended and rapidly rotating central body.

4.3 NECESSARY IMPROVEMENTS TO RUDIMENTARY SPH

Despite being both crude and inefficient, the rudimentary SPH implementation manages to capture the essential dynamical features of the collision between Jovian planets (Figure 4.1). This is no small accomplishment. It is remarkable that such a small programming effort can produce results that are qualitatively reasonable. Often, when you are starting out in research, your advisor will give you a copy of a state-of-the-art numerical code, and your task will be to use (and usually modify) this code so that it can be used to simulate a cutting-edge problem. Such an assignment can seem daunting to say the least. Highly competitive codes often contain a lot of physics and they are always finely tuned to maximize speed and minimize memory usage. Before you use the high-octane code, try spending a few days writing and

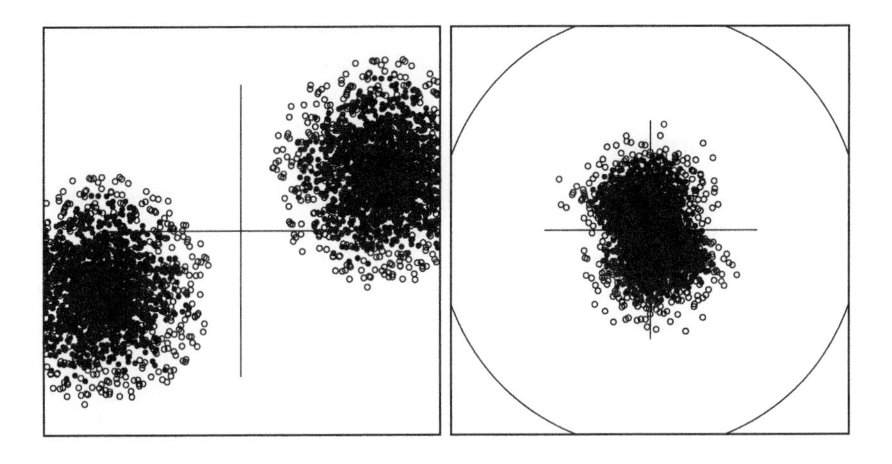

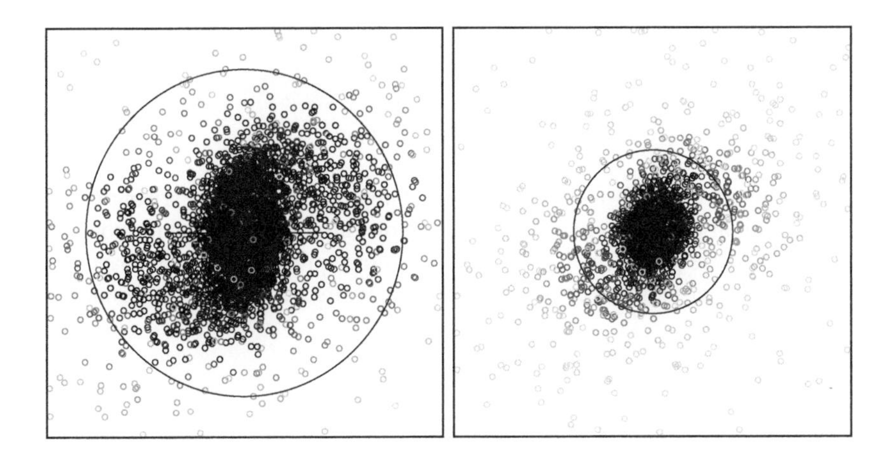

FIGURE 4.1 Rudimentary SPH model of the collision between two polytropes. Particle positions, projected onto the (x, y) plane, are shown at four different times. The final configuration is rotationally flattened and has a disk around it. The circle indicates the scale; it has the same physical radius in all frames.

running the simplest possible program that could be used on the problem. In doing this, the overall picture of what you are supposed to accomplish will invariably make more sense. You will gain a feel for why sophisticated techniques are needed to get a research-grade result and, most importantly, you will gain confidence that you can correctly modify the state-of-the-art code.

The performance of the rudimentary SPH code on the planet collision problem immediately suggests many directions in which improvements can be made. It is useful to look at these successively. Our discussion is designed so that each improvement can be added in order (or turned on or off) to make the code increasingly sophisticated.

4.3.1 Initial Conditions

The random assignment of the particles in the spherical shells, as described for the test problem, leads to large local fluctuations in the density from point to point. These large density fluctuations in turn imply large pressure gradients that will cause the supposed equilibrium of particles in the initial model to experience rather violent agitation. This situation can be partially alleviated by placing the particles equidistant from each other (note, however, that the problem of covering a sphere with equidistant particles is a hard one) or, alternatively, one can introduce an initial frictional damping to bring the model into a stable equilibrium. The friction can be implemented by evolving the particles that have been laid down to form the planet with the following equations of motion describing a simple leapfrog-type scheme. This method for quiet starts was suggested by Lucy (1977) in his original paper on SPH

$$\mathbf{v}_i^{n+1} = \left[\mathbf{v}_i^{n-1} + \Delta t \left(-\frac{1}{\rho_i}\nabla P_i - \nabla\Phi_i\right)^n\right] \Big/ (1+\delta), \tag{4.19}$$

and

$$\mathbf{r}_i^{n+2} = \mathbf{r}_i^n + (1+\delta)\Delta t \mathbf{v}_i^{n+1}. \tag{4.20}$$

If $\delta > 0$, the effect of the $1+\delta$ term acts to damp out the motions arising from the stochastic initial pressure gradients. For example, after about five pulsations of an $N = 2500$ particle planet (with $\delta = 0.3$), the average pressure gradients have nearly vanished, and a far more acceptable equilibrium configuration has been produced.

4.3.2 Kernels with Compact Support

The next point concerns the assumed structure of the kernel. With a Gaussian kernel, each particle normally contributes a finite density to every point in space. The computation of interpolated quantities, thus, requires summing over all particles, most of which have absolutely negligible contributions. This situation is resolved by introducing a smoothing kernel covering finite volume (that is, one with "compact support"). A given particle then interacts only with a more limited subset of nearest neighbors, which allows for an evaluation of the pressure gradient that involves only nonnegligible contributions. The spline kernel of Monaghan and Lattanzio (1985) is a popular choice; in three space dimensions, it reads

$$\begin{aligned} W(r,h) &= \frac{1}{\pi h^3}\left[1 - \frac{3}{2}(r/h)^2 + \frac{3}{4}(r/h)^3\right], \quad 0 \le r/h \le 1 \\ W(r,h) &= \frac{1}{4\pi h^3}[2-(r/h)]^3 \quad 1 \le r/h \le 2 \\ W(r,h) &= 0 \quad r/h \ge 2 \end{aligned} \tag{4.21}$$

where $r = |\mathbf{r} - \mathbf{r}_j|$. This kernel goes to zero at a distance of $2h$ from a given particle j, and is readily differentiated in the x-, y-, and z-directions to build ∇W. From a Taylor expansion, it can be shown that the dominant error term is of $O(h^2)$.

Having a kernel with compact support allows for a speedup of the operations which require summing over particles, since only nearby neighbors contribute. In practice, this requires knowing the continuously evolving "interaction list" for each particle.

Brute force computation of the interaction lists scales as N^2, which certainly defeats the purpose of the finite kernel. One way to speed up the search for neighbors is to use grids with linked lists. One first determines the three-dimensional volume and, hence, the three intervals $\Delta x = x_{max} - x_{min}$, $\Delta y = y_{max} - y_{min}$, and $\Delta z = z_{max} - z_{min}$, which span the current SPH particle distribution. One can then construct three one-dimensional grids for the x-, y-, and z-directions, respectively. The x grid, for example, is assigned a total width Δx, and is subdivided into $\Delta x/(2h)$ cells. With the grids in place, one loops through the particles, computing their cell locations in the x, y, and z grids. The loop over the particles allows each populated grid cell to be labeled with a list of contained particles. That is, each cell has an associated array of "pointers" to the particles that it contains. The neighbor list for any particle can then be assembled from those pointers that separately match its x, y, and z positions.

The use of a grid for particle neighbor searches is very efficient (see Hockney and Eastwood, 1981), especially when the particle smoothing lengths are constant, but a grid compromises the fully Lagrangian nature of SPH. If the particle distribution is highly clustered, then the grid will have to contain a large number of cells, which may strain the memory allocation. Furthermore, the use of a variable smoothing length (see below) reduces the effectiveness of a grid search with linked lists. An alternative neighbor-finding scheme is to use a tree.

4.3.3 Combining SPH with a Tree Code

Both SPH and the tree method are fully Lagrangian and, when intermeshed, they can work together extremely effectively (Hernquist and Katz, 1989). The use of trees to compute gravitational forces was discussed in the previous chapter. The root cell of the tree is the cubic volume which contains all of the particles in the simulation. At each time step, the tree is built (in three dimensions) by recursively subdividing the root cell into octants. If an octant contains no particles, it is abandoned. The subdivision is continued until each subcell contains only one particle. These fundamental subcells are the leaves of the tree.

The gravitational potential is generally softened by introduction of a softening length ϵ, which is not necessarily the same as the smoothing length h (Equation 3.90). For SPH, the softening length can be combined with the spline kernel (Equation 4.21) as described by Hernquist and Katz (1989). In this case, it is best to use $\epsilon = h$, which implies that no softening occurs outside $2h$, and that the numerical resolution for the calculation of gravitational forces and of pressure forces is the same.

Once constructed, a tree representation of the particle distribution performs the double duty of providing both the gravitational potential and the near-neighbor list. The neighbor search consists of finding all the particles whose kernels impinge on the sphere of radius $2h_i$ enveloping the target particle i. This sphere is circumscribed by a cube with side length $4h_i$, and the three Cartesian ranges of the edges of this box are noted. The neighbors are logged by probing recursively through the cells of the tree. When the volume of a cell under consideration fails to overlap the cube circumscribing particle i, no further descent into the cell is made. When a descent into a cell reaches a cell containing one particle j, final admission of j to i's neighbor list is made if j is within $2h_i$ of i. In this manner, large regions of the search space can be discarded with relatively little effort. The tree method is particularly appropriate if the SPH uses variable smoothing lengths, a standard feature in most modern codes.

4.3.4 Variable Smoothing Lengths

Many problems in astrophysics span a wide range of densities, and SPH will not perform well with a uniform smoothing length if the density has a considerable dynamic range. The solution is to give each particle its own smoothing length, h_i, which is determined at each time step by demanding that each particle interact with approximately the same number of neighbors (N_{target}, typically about 50). It is important to avoid time-consuming efforts to adjust the smoothing length to capture the target quantity of neighbors. The standard method is simply to link changes in h for a given particle to changes in density

$$\frac{dh_i}{dt} = -\frac{h_i}{3\rho_i}\frac{d\rho_i}{dt}. \tag{4.22}$$

Another viable method is to extrapolate smoothing lengths from time step $n-1$ (when particle i had N_i^{n-1} neighbors) to time step n, based on the deviation of N_i^{n-1} from the target number and then, for stability, to average the old and new smoothing lengths

$$h_i^n = h_i^{n-1}\frac{1}{2}\left[1+\left(\frac{N_{target}}{N_i^{n-1}}\right)^{1/3}\right]. \tag{4.23}$$

As a second step, the number of neighbors N_i^n at the advanced time can be compared against the number N_{target} desired. If the agreement is worse than a specified tolerance, then h_i is increased or decreased until the agreement is sufficient. This iterative procedure is not overly time-consuming and it guarantees that all points in the fluid are treated with comparable sampling.

Introduction of the variable smoothing length presents an ambiguity in the interpretation of the fundamental SPH interpolation equation. On the one hand, one could consider a given particle i as a passive marker, whose physical quantities are determined by those nearby particles whose smoothing lengths (h_j) overlap the position of the particle

$$A_i = \sum_j m_j \frac{A_j}{\rho_j} W(r_{ij}, h_j) \tag{4.24}$$

where $r_{ij} = |\mathbf{r}_i - \mathbf{r}_j|$.

On the other hand, one could consider the particle i as playing the active role of collecting the contributions of other particles only out to *its own* smoothing length h_i

$$A_i = \sum_j m_j \frac{A_j}{\rho_j} W(r_{ij}, h_i). \tag{4.25}$$

If either one or the other interpretation is used, then the angular and linear momentum of a system of particles will not be conserved, leading to spurious generation of a system velocity and rotation for an aggregate of particles that are initially at rest. In other words, to conserve momentum, the force of particle j on particle i must be the same as that of i on j. The solution is to symmetrize the smoothing interaction between particles by employing the smoothing lengths of both the particle itself and its neighbors in a linear combination

$$A_i = \frac{1}{2} \sum_j m_j \frac{A_j}{\rho_j} [W(r_{ij}, h_j) + W(r_{ij}, h_i)] \tag{4.26}$$

and

$$\rho_i = \frac{1}{2} \sum_j m_j [W(r_{ij}, h_j) + W(r_{ij}, h_i)]. \tag{4.27}$$

The expression for the pressure gradient must also be modified into a symmetrized form to conserve linear and angular momentum. Various expressions are possible, one of which uses the relation

$$\nabla P = 2\sqrt{P}\nabla\sqrt{P} \tag{4.28}$$

in conjunction with Equation (4.26) to construct a symmetrized expression for the acceleration resulting from the pressure gradient

$$\frac{\nabla P_i}{\rho_i} = \sum_{j=1}^{N} m_j \frac{\sqrt{P_i P_j}}{\rho_i \rho_j} [\nabla_i W(r_{ij}, h_i) + \nabla_i W(r_{ij}, h_j)], \tag{4.29}$$

where ∇_i refers to the derivative with respect to the position of particle i. Note that a geometric mean for ρ^2 and P is used, for reasons of numerical stability. However, in some applications, the arithmetic mean of P/ρ^2 is used. In this case one uses

$$\frac{\nabla P}{\rho} = \nabla\left(\frac{P}{\rho}\right) + \frac{P}{\rho^2}\nabla\rho \tag{4.30}$$

and the pressure gradient term becomes

$$\frac{\nabla P_i}{\rho_i} = \sum_{j=1}^{N} \frac{m_j}{2}\left(\frac{P_i}{\rho_i^2} + \frac{P_j}{\rho_j^2}\right) [\nabla_i W(r_{ij}, h_i) + \nabla_i W(r_{ij}, h_j)]. \tag{4.31}$$

Exercise

Derive Equation (4.31). What approximations, if any, are used?

The error introduced into SPH by including the variable smoothing length is of order h^2, the same order as in SPH in general. Such errors show up as nonconservation in, for example, entropy or energy in a flow (Hernquist, 1993). With a sufficient number of particles, the errors are relatively small, and almost all codes in use have the feature of variable smoothing length, as its advantages far outweigh its disadvantages. The errors can be corrected by introducing additional terms involving derivatives of h (Springel and Hernquist, 2002; Nelson and Papaloizou, 1994). In certain test problems, such as the collision of two polytropes, inclusion of these terms improves energy conservation significantly; nevertheless most codes do not include them.

4.3.5 A Resolution Requirement

In problems involving self-gravity, a sufficient number of particles must be used so that the Jeans mass is resolved. This requirement is necessary so that (1) elements of material that should collapse under their own gravity actually do collapse, and (2) elements of material that do not have sufficient self-gravity to collapse actually do not collapse. For a uniform-density gas of a given temperature T and a given density ρ, the Jeans mass

$$M_J = \left(\frac{5R_gT}{2G\mu}\right)^{3/2}\left(\frac{4\pi\rho}{3}\right)^{-1/2} \tag{4.32}$$

is the minimum mass needed for the parcel of gas to collapse under its own gravity. In SPH, the minimum resolved mass $M_{\rm min}$ must be less than M_J (Bate and Burkert, 1997) at all times and locations. An alternate statement of the resolution requirement is that the Jeans *length*

$$R_J \approx \left(\frac{3M_J}{4\pi\rho}\right)^{1/3} = \left(\frac{5R_gT}{2\mu}\right)^{1/2}\left(\frac{3}{4\pi G\rho}\right)^{1/2} \tag{4.33}$$

must contain several smoothing lengths.

The minimum resolved mass in SPH is:

$$M_{\rm min} \approx (1.5 - 2)mN_{\rm target} \approx (75 - 100)m, \tag{4.34}$$

where m is the mass of an individual particle. For example, suppose a cloud of 200 $M_\odot$ collapses isothermally at 100 K ($\mu = 2$) from a density of 10^{-20} g cm^{-3} to a density of 10^{-14} g cm^{-3}. As the density increases, M_J decreases to a minimum value of 3×10^{32} g. To resolve this mass, one needs 100 particles, so the mass of each must be $m = 3 \times 10^{30}$ g. Assuming the cloud of 4×10^{35} g is composed of equal-mass particles, the total number of particles needed to represent the collapse is at least 133,333. The specification of this minimum mass implicitly assumes that the calculation is carried out with a variable smoothing length h, adjusted for each particle so that $N_{\rm target} \approx 50$. It is also assumes that the gravitational softening length ϵ varies in time for each particle, such that $\epsilon \approx h$.

Exercise

A cloud of 1 solar mass, temperature 12 K, and mean molecular weight 2 collapses from an initial density of 10^{-18} g cm^{-3}. The collapse is isothermal up to a density of 10^{-13} g cm^{-3}, then it is adiabatic with $P \propto \rho^{5/3}$ up to a final density of 10^{-10} g cm^{-3}. Assuming that each particle has 50 nearest neighbors, how many particles are required to resolve the Jeans mass?

If this requirement is not met, one possible consequence is inhibition of fragmentation, i.e., a region that is gravitationally unstable ($M > M_J$) does not actually collapse. This unphysical result could occur if the unresolved region has $\epsilon > h$, so that pressure effects, which tend to make the region expand, become more important than gravitational forces. On the other hand, if $\epsilon < h$, a possible consequence is artificial fragmentation, that is, collapse of a region that is not gravitationally unstable ($M < M_J$). In this case, gravitational forces are artificially enhanced relative to pressure forces. Note, however, that a region that is marginally Jeans unstable ($M \approx M_J$), is marginally resolved, and has $\epsilon = h$ will collapse, but the collapse will be slower than it should be because both gravitational forces and pressure forces are weakened on the smallest scales. It should be emphasized that all of these problems are minimized if the resolution requirement is met. It must be adhered to in all problems of galaxy formation, star formation, or planet formation that involve gravitational collapse.

4.3.6 Introducing an Energy Equation into SPH

So far, our formulation has relied on barotropic equations of state, with $P = P(\rho)$, and for some astrophysical applications, especially cases of exploratory or "back-of-the-envelope" computing, this is a reasonable approximation. In general, however, the pressure depends on two thermodynamic state variables. The ideal gas law, $P = (\gamma - 1)\rho\epsilon$, where ϵ is the internal energy per unit mass of the gas, is the most important example. The dependence of the pressure on two variables complicates the simplest SPH formulation by requiring that a third equation be advanced in time along with the particles' positions (Equation 4.1), and velocities (Equation 4.2).

The first law of thermodynamics couples the specific internal energy of a fluid to the density and pressure. For situations in which the gas is adiabatic (where the timescale of interest is short enough so that there is no net inflow or outflow of heat from the SPH particles), the first law is the Lagrangian Equation (1.58)

$$\frac{d\epsilon}{dt} = -\left(\frac{P}{\rho}\right)\nabla \cdot \mathbf{v}. \tag{4.35}$$

The thermal energy, therefore, can be advanced along with the positions and velocities of the particles, thereby allowing the pressure to be computed at every time step. This allows the simulation to consistently account for the transfer of bulk kinetic energy into internal energy

$$\frac{d\epsilon_i}{dt} = \sum_{j=1}^{N} m_j \left(\frac{\sqrt{P_i P_j}}{\rho_i \rho_j} \right) \mathbf{v}_{ij} \cdot \frac{1}{2}[\nabla_\mathbf{i} W(r_{ij}, h_i) + \nabla_\mathbf{i} W(r_{ij}, h_j)] \tag{4.36}$$

where $\mathbf{v}_{ij} = \mathbf{v}_i - \mathbf{v}_j$. A potential difficulty with this difference scheme is that the time derivative of ϵ could be so large that ϵ_i could become negative (Hernquist and Katz, 1989). This situation can occur, for example, if a single particle moves rapidly away from the rest and becomes relatively isolated. It helps in this case to substitute

$$\frac{1}{2}\left(\frac{P_i}{\rho_i^2} + \frac{P_j}{\rho_j^2} \right) \quad \text{or} \quad \left(\frac{P_i}{\rho_i^2} \right) \quad \text{for} \quad \left(\frac{\sqrt{P_i P_j}}{\rho_i \rho_j} \right). \tag{4.37}$$

However, it is clear that widely separated particles cannot represent fluid flow correctly.

4.3.7 Heat Transfer in SPH

Equation (4.36) cannot account for heat that leaks into or out of the simulation volume or is transferred between particles. One way for this to happen is if the fluid can efficiently conduct heat from one spot to another. This process of conduction is described by a diffusion equation

$$\frac{d\epsilon}{dt} = \frac{1}{\rho} \boldsymbol{\nabla} \cdot (k \boldsymbol{\nabla} T) \tag{4.38}$$

where k is the heat conduction coefficient and for an ideal gas $\epsilon = c_V T$. An example for the case of radiation is given by Equation (1.127).

Diffusion of heat depends on second spatial derivatives of the internal energy. Because the information regarding the variation of the fluid internal energy is contained only at a discrete set of points (the particles), it is tricky to get an accurate value of the rate of change of the rate of change. A way around this is to imagine that heat transfer occurs macroscopically between pairs of SPH particles and to construct an integral approximation to Equation (4.38), as derived by Jubelgas et al. (2004). They suggest modeling diffusion's contribution with

$$\frac{d\epsilon_i}{dt} = -\sum_{j=1}^{N} m_j \frac{(k_j + k_i)(T_i - T_j)(\mathbf{r}_{ij} \cdot \boldsymbol{\nabla}_i W_{ij})}{(\rho_i \rho_j)|\mathbf{r}_{ij}^2|}, \tag{4.39}$$

although other forms are possible (Monaghan, 1992). Equation (4.39) conserves thermal energy. Here the notation $\boldsymbol{\nabla}_i W_{ij}$ can be interpreted to represent the gradient of the kernel $W(r_{ij}, h_{ij})$, where $h_{ij} = 0.5(h_i + h_j)$, an alternative method for symmetrizing the kernel to that given in Equation (4.26).

In the case of radiation diffusion, the change in internal energy caused by transport, in the one-temperature approximation where the radiation temperature $T_{\rm rad}$ and the gas temperature $T_{\rm m}$ are the same, is given by

$$\frac{d\epsilon}{dt} = \frac{c}{\rho} \boldsymbol{\nabla} \cdot \left(\frac{\lambda}{\kappa \rho} \boldsymbol{\nabla} u \right), \tag{4.40}$$

where $u = aT^4$ is the radiation energy density, κ is the mean opacity, and λ is the flux limiter (Equation 1.120). In optically thick regions $\lambda = 1/3$ and the quantity $c/(3\kappa\rho)$ is known as the radiative diffusivity. An appropriate SPH formulation of this equation is:

$$\frac{d\epsilon_i}{dt} \approx \sum_{j=1}^{N} \frac{4cm_j}{\rho_i \rho_j} \left(\frac{q_i q_j}{q_i + q_j} \right) (u_i - u_j) \left(\frac{\mathbf{r}_{ij} \cdot \nabla_i W_{ij}}{|\mathbf{r}_{ij}|^2} \right) \tag{4.41}$$

where

$$q_i = \frac{\lambda_i}{\kappa_i \rho_i}. \tag{4.42}$$

Again, energy exchange takes place between pairs of particles and energy is conserved. An implicit method for solution of this equation, in the more general case where $T_{\rm m} \neq T_{\rm rad}$, is given by Whitehouse et al. (2005). An example of the solution of the equations of radiation hydrodynamics (Chapter 9, Section 9.4) with the SPH technique is shown in Figure 4.2. The starting point of this calculation is a rotating cold cloud (10 K) with size 3.2×10^{16} cm. It collapses to form a binary protostar, with the components separated by about 10^{16} cm. The result in the figure is shown at a time of 1.3 initial free-fall times, where the maximum density has increased to 10^{-1} g cm^{-3} from an initial value of 1.44×10^{-17} g cm^{-3}, an extremely demanding calculation for any hydrodynamics code.

4.3.8 Shocks in SPH

When parcels of gas collide with relative velocity larger than the sound speed, sound waves are unable to carry all of the collision energy away, and the entropy of the gas increases. An accurate SPH formulation, therefore, requires a method to deal with the occurrence of shocks. Physically, a shock can be thought of as a compression of a gas that exceeds an "elastic limit." A shocked gas cannot return the work that was done to compress it; shocks provide a viscous drag on the gas, which turns kinetic energy irreversibly into heat. For the treatment of shocks, SPH generally uses an artificial viscosity that has a similar effect to that used in grid-based codes (Chapter 6, Section 6.1.4). The momentum equation is then written

$$\frac{d\mathbf{v}_i}{dt} = -\frac{1}{\rho_i} \nabla P_i - \mathbf{a}_i^{\rm visc} - \nabla \Phi_i, \tag{4.43}$$

where

$$\mathbf{a}_i^{\rm visc} = \frac{1}{2} \sum_{j=1}^{N} m_j \prod_{ij} [\nabla_i W(r_{ij}, h_i) + \nabla_i W(r_{ij}, h_j)]. \tag{4.44}$$

The artificial viscosity tensor Π_{ij} needs to translate the locally sampled density and velocity of the particles into a deceleration in a way that correctly captures the net macroscopic effect of shocks. Π_{ij} must be constructed so that the total linear and angular momenta of the particles are conserved. It should, for example, vanish if the particles are participating in rigid body rotation. A number of different prescriptions

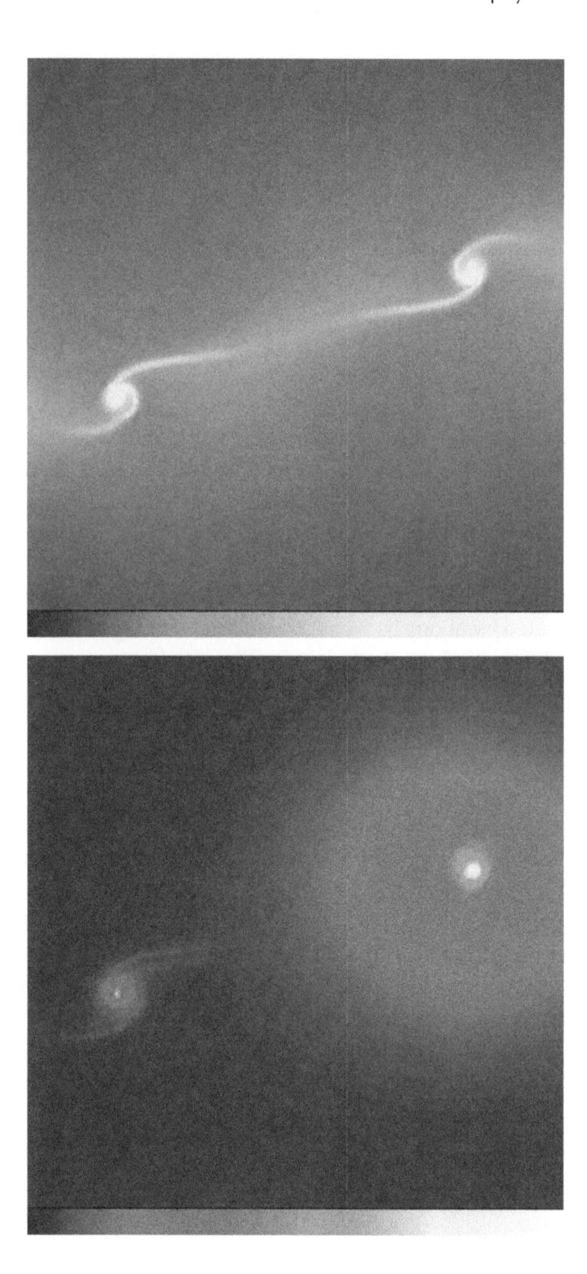

FIGURE 4.2 ***A color version of this figure follows page 212*** An SPH model, with 150,000 particles, including radiation transport, of the collapse of a protostellar cloud. *Upper panel:* column density integrated parallel to the rotation axis, on a logarithmic grey scale with log N ranging from -1 to $+3.5$, where N is given in g cm^{-2}. *Lower panel:* gas temperature, weighted along each column according to the local mass, on a logarithmic grey scale ranging from log $T = 1$ (black) to log $T = 3$ (white). Because of slight asymmetries in the SPH initial conditions, the protostar on the right has evolved slightly farther and has heated the surrounding gas more than the protostar on the left. [From Whitehouse, S. C. and Bate, M. R. (2006). *Monthly Notices, Royal Astronomical Society* **367**:32. With permission.)

are possible. Hernquist and Katz (1989) suggest a symmetrized version of Π_{ij} that is based on an earlier prescription by Monaghan (1988)

$$\prod_{ij} = \frac{-0.5\alpha\mu_{ij}(c_i + c_j) + \beta\mu_{ij}^2}{0.5(\rho_i + \rho_j)}, \tag{4.45}$$

with

$$\mu_{ij} = \frac{\mathbf{v}_{ij} \cdot \mathbf{r}_{ij}}{h_{ij}\left(r_{ij}^2/h_{ij}^2 + \eta^2\right)} \quad \text{for} \quad \mathbf{v}_{ij} \cdot \mathbf{r}_{ij} < 0, \tag{4.46}$$

and $\mu_{ij} = 0$ for $\mathbf{v}_{ij} \cdot \mathbf{r}_{ij} \geq 0$. Here, c_i is the sound speed of particle i, h_{ij} is $0.5(h_i + h_j)$, and $\mathbf{v}_{ij}$ is the velocity difference between a pair ij. The quantity $\eta^2 \approx 0.01$ is a softening parameter to prevent problems if r_{ij} happens to be very small. The α term in Equation (4.45) dissipates kinetic energy when particles are converging on one another and, hence, plays the role of a bulk viscosity (think of squeezing a sponge cake); it also has a component of shear viscosity. Its purpose is mainly to suppress post-shock oscillations. The β term in Equation (4.45) contributes strongly when particles are crowding at high velocity and, thus, allows the SPH method to handle development of shocks, in a manner analogous to the von Neumann–Richtmyer artificial viscosity that is ubiquitous in grid-based hydrodynamics schemes. Typical values for the viscosity coefficients are $\alpha \sim 1$ and $\beta \sim 2$ (Monaghan, 1992). Note that the quantity $\mathbf{v}_{ij} \cdot \mathbf{r}_{ij}$ is used instead of the more usual $\nabla \cdot \mathbf{v}$ to determine whether the flow is compressive and, therefore, an artificial viscosity should be used.

In general, the energy lost in the viscous deceleration of a particle must reappear as a viscous heating contribution in the internal energy equation, which is added to the term involving $\nabla \cdot \mathbf{v}$

$$\frac{d\epsilon_i}{dt} = \sum_{j=1}^{N} \frac{m_j}{2} \prod_{ij} \mathbf{v}_{ij} \cdot \frac{1}{2}[\nabla_i W(r_{ij}, h_i) + \nabla_i W(r_{ij}, h_j)]. \tag{4.47}$$

An analogous contribution to the internal energy equation appears in grid-based formulations of the equations of hydrodynamics, as described in Chapter 6, Section 6.1.4.

An alternate form is suggested by Hernquist and Katz (1989), which does in fact use $\nabla \cdot \mathbf{v}$. The form given by Equation (4.45) has been found to generate considerable viscosity in shear flows, such as differentially rotating disks, even if no shocks are present (the artificial viscosity vanishes for rigid body rotation, but not for differential rotation). The alternate form has less shear viscosity, but is less accurate in the treatment of the flow near the shock.

$$\prod_{ij} = \frac{q_i}{\rho_i^2} + \frac{q_j}{\rho_j^2} \tag{4.48}$$

$$q_i = \alpha h_i \rho_i c_i |\nabla \cdot \mathbf{v}|_i + \beta h_i^2 \rho_i |\nabla \cdot \mathbf{v}|_i^2 \tag{4.49}$$

for $\nabla \cdot \mathbf{v} < 0$, and $q_i = 0$ in the opposite case. The velocity divergence is given by an expression closely related to Equation (4.36)

$$\nabla \cdot \mathbf{v} = -\frac{1}{\rho_i}\sum_{j=1}^{N} m_j \mathbf{v}_{ij} \cdot \frac{1}{2}[\nabla_{\mathbf{i}} W(r_{ij}, h_i) + \nabla_{\mathbf{i}} W(r_{ij}, h_j)]. \tag{4.50}$$

Other devices have been employed in order to preserve the more accurate shock treatment of Equation (4.45), but to reduce the shear viscosity effect caused by the α term. Morris and Monaghan (1997) give each particle its own parameter α_i. Each parameter is evolved in time according to

$$\frac{d\alpha_i}{dt} = -\frac{\alpha_i - \alpha_{\min}}{\tau_i} + S_i, \tag{4.51}$$

where the first term causes α_i to decay to $\alpha_{\min}$ on a timescale τ_i, while the second term is a source term, which causes α_i to increase near shocks. The time scale is generally taken to be $\tau_i = h_i/(S_s c_i)$, where c_i is the sound speed. If the arbitrary constant S_s is taken to be 0.2 to 0.1, the viscosity is reduced over 2 to 5 smoothing lengths to $\alpha_{\min}$, which typically is taken to be 0.1, rather than the standard value of $\alpha = 1$. The source term must give rapid increase in dissipation in regions of strong compression; the following expression has been shown to give good results

$$S_i = \max(-\nabla \cdot \mathbf{v}, 0)(2.0 - \alpha_i). \tag{4.52}$$

Another approach for reducing the effects of artificial viscosity away from shocks is the so-called Balsara switch (Balsara, 1995). The artificial viscosity tensor Π_{ij} has the same form as in Equation (4.45), but the quantity μ_{ij} has an additional factor

$$\mu_{ij} = \frac{\mathbf{v}_{ij} \cdot \mathbf{r}_{ij}}{h_{ij}\left(r_{ij}^2/h_{ij}^2 + \eta^2\right)} \frac{f_i + f_j}{2} \quad \text{for} \quad \mathbf{v}_{ij} \cdot \mathbf{r}_{ij} < 0, \tag{4.53}$$

and $\mu_{ij} = 0$ for $\mathbf{v}_{ij} \cdot \mathbf{r}_{ij} \geq 0$. This form is similar to Equation (4.46) except for the f factor. This factor is designed to be close to unity in regions of strong compression, but zero in regions of strong vorticity ($\nabla \times \mathbf{v}$)

$$f_i = \frac{|\nabla \cdot \mathbf{v}|_i}{|\nabla \cdot \mathbf{v}|_i + |\nabla \times \mathbf{v}|_i + 0.0001 c_i/h}. \tag{4.54}$$

While an artificial viscosity prescription, such as Equation (4.44), allows SPH to give fairly reasonable and stable results in the presence of strong shocks, SPH nevertheless is not the best technique to use for problems where strong shocks are a decisive factor in the dynamics. For example, it is often remarked that particles tend to penetrate artificially through shocks at high Mach number. If one is faced with a problem that is highly Lagrangian and where shocks are important, the specialized technique of Godunov-SPH (also known as GPH) is showing great potential. Godunov methods, whether grid-based or SPH-based, take advantage of the fact that the one-dimensional shock tube problem, in which two initially separated parcels of gas at different pressures are brought into contact, has an analytic solution (for further explanation, see Chapter 6, Section 6.1.3). For each pair of interacting particles, at each time step, Godunov-SPH treats the particle–particle interaction as a shock-tube problem along the direction of the separation vector $\mathbf{r}_{ij}$. Thus, the treatment is one-dimensional even in an overall three-dimensional problem. No artificial viscosity is needed. The method essentially eliminates the problem of particle penetration and significantly reduces the problem of dissipation in shear flows (Inutsuka, 2002; Cha and Whitworth, 2003). Another distinct advantage is that the time step limitations introduced by the artificial viscosity method — the α and β terms in Equation (4.59) and Equation (4.60) below — are removed. Even though the computation time per time step may be a factor of two longer for the Godunov method compared with the artificial viscosity method, the computer time required for a given total problem time may be significantly reduced. Nevertheless, the artificial viscosity method is convenient and is still widely used.

4.3.9 Time Integration

For the advancement in time of the radii and velocities of the particles according to Equation (4.1) and Equation (4.2), a Runge–Kutta integration or a predictor–corrector scheme can be used, but it has also been found adequate to use a leapfrog scheme, which is second-order accurate in Δt

$$\mathbf{r}_i^{n+1/2} = \mathbf{r}_i^{n-1/2} + \Delta t \mathbf{v}_i^n \tag{4.55}$$

$$\mathbf{v}_i^{n+1} = \mathbf{v}_i^n + \Delta t \mathbf{a}_i^{n+1/2}. \tag{4.56}$$

This scheme runs into difficulty only if artificial viscosity is used, in which case the acceleration $\mathbf{a}_i$ depends on the velocity. However, one can use an estimate for the velocity at $n + 1/2$

$$\hat{\mathbf{v}}_i^{n+1/2} = \mathbf{v}_i^n + 0.5\Delta t \mathbf{a}_i^{n-1/2} \tag{4.57}$$

to use in the calculation of the acceleration $\mathbf{a}_i^{n+1/2}$ needed in Equation (4.56), which is then used to update $\mathbf{v}_i^{n+1}$.

If artificial viscosity is included, the time step limit must be modified from the simple criterion given in Equation (4.12), as discussed in Chapter 2, Section 2.4. There are a number of ways to limit the time step, with the overall goal of conservation of linear and angular momentum and total energy. Two alternate methods are outlined here.

The first method (Monaghan, 1992) is based on an assumed uniform time step for all particles and combines two time steps Δt_{f} and Δt_{cv} as follows

$$\Delta t_{\mathrm{f}} = \min \left(\frac{h_i}{|\mathbf{a}_i|} \right)^{1/2} \tag{4.58}$$

$$\Delta t_{\mathrm{cv}} = \min \left[\frac{h_i}{c_i + 0.6(\alpha c_i + \beta \max_j \mu_{ij})} \right]. \tag{4.59}$$

Then, the final $\Delta t = 0.25 \min(\Delta t_{\mathrm{f}}, \Delta t_{\mathrm{cv}})$. The maxima and minima are taken over all particles, and the numerical coefficients are adjusted to optimum values by experiment. Thus, the first criterion reduces the time step on the basis of strong forces, while the second reduces it on the basis of large sound speed or large artificial viscosity.

The second method (Hernquist and Katz, 1989) allows each particle to have its individual time step Δt_i, determined as follows for the case where the artificial viscosity is given by Equation (4.45)

$$\Delta t_i = \frac{C_0 h_i}{h_i |\nabla \cdot \mathbf{v}_i| + c_i + 1.2(\alpha c_i + \beta \max_j |\mu_{ij}|)} \tag{4.60}$$

where C_0 is the Courant number, which is taken to be, for example, 0.3. Note that there is an additional term $h_i|\nabla \cdot \mathbf{v}_i|$ not required by the CFL condition, which is included to improve the accuracy of the calculation. It is included in all cases, regardless of the sign of $\nabla \cdot \mathbf{v}_i$. Note also that $\mu_{ij} = 0$ for $\mathbf{v}_{ij} \cdot \mathbf{r}_{ij} > 0$. In practice, the time steps are divided into bins, with $\Delta t_n = \Delta t_{max}/2^n$, where Δt_{max} is the largest allowed time step. All particles within a given bin ($\Delta t_n - \Delta t_{n+1}$) are evolved with the same time step.

4.4 SUMMARY

A number of significant advantages and disadvantages of SPH methods are made evident as a result of numerical tests of such methods as compared with fixed-grid techniques.

- The code is self-adaptive and does not require extensive grid-refinement techniques to resolve regions of high density.
- If there are extensive regions that are almost empty, such as in collision problems, SPH is more efficient than grid codes.
- SPH is closely related to N-body methods for which there exists a vast literature as well as specialized hardware.
- SPH may be superior to grid codes in handling problems with free boundaries and geometries without significant symmetries.

However,

- SPH has more difficulties with the accurate treatment of shocks than do grid-based codes.
- SPH is less developed in treating the physical effects of radiation transfer and magnetohydrodynamics than are many grid-based codes.
- The calculation of numerical second spatial derivatives can be a problem in SPH, since the disorder of the particles can lead to errors in these derivatives. However, an approximation can be used to reduce terms that involve second derivatives to ones involving only first derivatives (Price, 2004).
- SPH is less suited to the treatment of problems with fixed boundaries than are grid-based codes.
- If variable smoothing lengths are used, then the gravitational softening length must also be adjusted in space and time so that numerical resolution is the same for gravitational and pressure forces. However, a variable

softening length can introduce energy nonconservation, if not properly corrected for.
- The controlled setup of initial conditions is more complicated in SPH than in grid-based codes, since particles can be set down in an infinite number of ways. It takes some care in setting up the initial conditions to make sure that these are not influencing the outcome of the simulation.

Finally it should be pointed out that the SPH method and the finite difference method on a grid both solve exactly the same hydrodynamic equations, only with quite different numerical difference schemes. Thus, on the same hydrodynamical problem, and with similar spatial resolution, the two types of codes should give qualitatively the same results, although details, especially on small scales, could be different. A few such tests have been carried out and confirm that similar results are obtained. Note in particular the pioneering study by Durisen et al. (1986). In SPH, to obtain reliable numerical results, one should compute the same problem with several different total numbers of particles to make sure that the result is numerically converged. In addition, it is valuable to compare the results of grid-based calculations and SPH calculations on the same problem to make sure that neither method is generating incorrect results.

As a final example, we show the utility of SPH in the investigation of a fundamental problem of solar system astrophysics: the origin of the Moon (Figure 4.3 to Figure 4.5). The impactor arrives from the right on a trajectory with constant $Y = 6000$ km, hitting the proto-Earth at escape velocity at an angle of 47° between the trajectory and the outward normal on the Earth. The end result of such a calculation is a system with total mass and angular momentum close to that of the Earth–Moon system, and

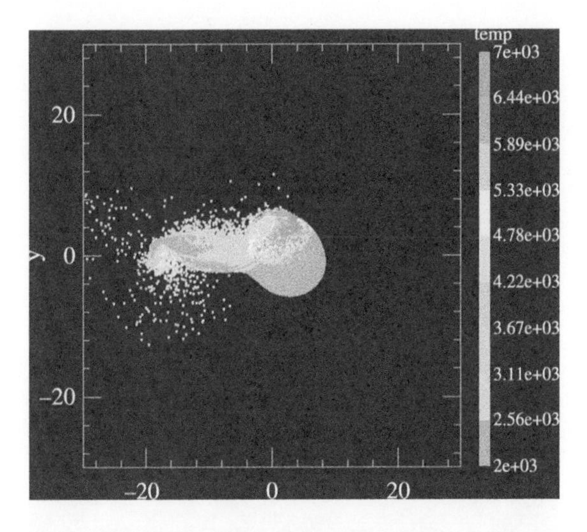

FIGURE 4.3 ***A color version of this figure follows page 212*** An SPH model of the collision between a Mars-sized object (0.14 $M_\oplus$) with the proto-Earth (0.81 $M_\oplus$) at a time of 0.86 hours. The grey scale gives the temperature from 7000 K to 2000 K. The particle positions are projected onto the (x, y) plane. The linear scale is given in units of 1000 km. The total number of particles is 120,000. (From Canup, R. (2004) *Icarus* **168**:433. With permission.)

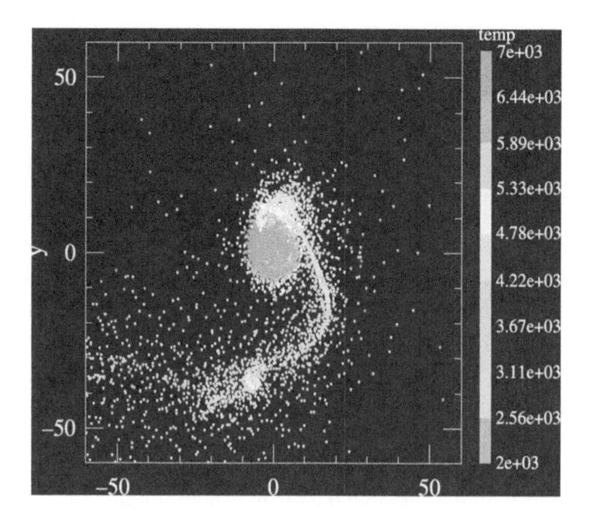

FIGURE 4.4 ***A color version of this figure follows page 212*** As in Figure 4.3, after a time of 5.89 hours. (From Canup, R. (2004) *Icarus* **168**:433. With permission.)

with an orbiting disk of approximately lunar mass. The iron core of the impactor joins with the initial iron core of the proto-Earth, and the disk is composed mainly of the silicate mantle material initially in the impactor. Thus, the calculation can explain the overall composition difference between Moon and Earth.

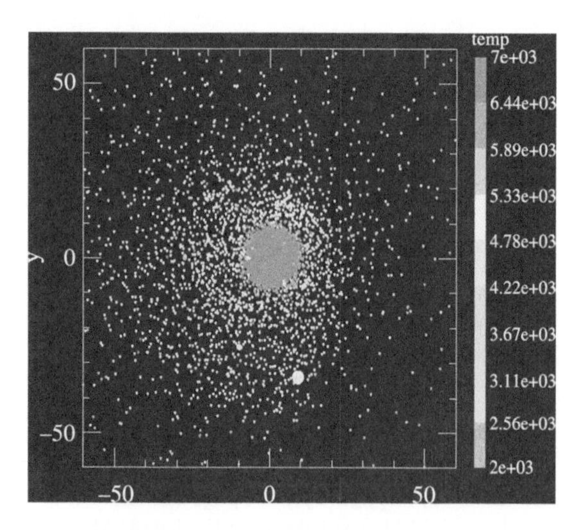

FIGURE 4.5 ***A color version of this figure follows page 212*** As in Figure 4.3 after a time of 21.9 hours. Retained material orbiting the Earth forms a disk with a clump with about 60% of a lunar mass. (From Canup, R. (2004) *Icarus* **168**:433. With permission.)

REFERENCES

Balsara, D. (1995) *J. Comp. Phys.* **121**: 357.

Bate, M. R. and Burkert, A. (1997) *Monthly Notices of the Royal Astronomical Society* **288**: 1060.

Benz, W. (1990) *The Numerical Modelling of Nonlinear Stellar Pulsations,* J. R. Buchler, Ed. (Dordrecht: Kluwer) p. 269.

Canup, R. (2004) *Icarus* **168**: 433.

Cha, S.-H. and Whitworth, A. P. (2003) *Monthly Notices of the Royal Astronomical Society* **340**: 73.

Chandrasekhar, S. (1939) *An Introduction to the Theory of Stellar Structure* (Chicago: University of Chicago Press).

Durisen, R. H., Gingold, R. A., Tohline, J. E., and Boss, A. P. (1986) *Astrophys. J.* **305**: 281.

de Pater, I. and Lissauer, J. J. (2001) *Planetary Sciences* (Cambridge: Cambridge University Press).

Gingold, R. A. and Monaghan, J. J. (1977) *Monthly Notices of the Royal Astronomical Society* **181**: 375.

Hernquist, L. (1993) *Astrophys. J.* **404**: 717.

Hernquist, L. and Katz, N. (1989) *Astrophys. J. Suppl.* **70**: 419.

Hockney, R. W. and Eastwood, J. W. (1981) *Computer Simulation Using Particles* (New York: McGraw-Hill).

Inutsuka, S. (2002) *J. Comp. Phys.* **179**: 238.

Jubelgas, M., Springel, V. and Dolag, K. (2004) *Monthly Notices of the Royal Astronomical Society* **351**: 423.

Lucy, L. (1977) *Astron. J.* **82**: 1013.

Monaghan, J. J. (1992) *Annu. Rev. Astron. Astrophys.* **30**: 543.

Monaghan, J. J. and Lattanzio, J. C. (1985) *Astron. Astrophys.* **149**: 135.

Morris, J. P. and Monaghan, J. J. (1997) *J. Comp. Phys.* **136**: 41.

Nelson, R. P. and Papaloizou, J. C. B. (1994) *Monthly Notices of the Royal Astronomical Society* **270**: 1.

Price, D. (2004) Ph.D. thesis, Cambridge: Cambridge University, Chap. 3: http://arXiv.org/abs/astro-ph/0507472.

Springel, V. and Hernquist, L. (2002) *Monthly Notices of the Royal Astronomical Society* **333**: 649.

Whitehouse, S. C. and Bate, M. R. (2006) *Monthly Notices of the Royal Astronomical Society* **367**: 32.

Whitehouse, S. C., Bate, M. R., and Monaghan, J. J. (2005) *Monthly Notices of the Royal Astronomical Society* **364**: 1367.

5 Stellar Evolution

The structural properties of a star, such as the variation of its temperature, density, and rate of energy production from center to surface, can be determined through a combination of theoretical calculations, based on known physical laws and observational data. Stellar evolution refers to the change in these physical properties with time, again as determined from both theory and observation. The first main phase of stellar evolution is the pre-main-sequence phase, during which the star is not hot enough to ignite nuclear reactions in the interior, so gravitational contraction provides most of the star's energy. Next comes the main-sequence phase, which takes up most of the lifetime of the star, in which nuclear fusion of hydrogen to helium in the central region provides the energy. Once hydrogen has been exhausted at the center, the post-main-sequence phase starts, in which hydrogen burning away from the center, as well as the burning of helium, carbon, or heavier elements, may provide the energy. The evolutionary properties are strongly dependent on the initial mass of the star and to some extent on its initial chemical composition. While the final phase of evolution of most stars is a white dwarf, the stars in the higher mass range (above about nine times the mass of our Sun) evolve to a supernova explosion and can end up as a neutron star or a black hole. The general definition of a star is a gaseous object that, at some point in its evolution, obtains 100% of its energy from the fusion of protons (^{1}H) to helium nuclei. The boundary in mass between stars and substellar objects (also known as brown dwarfs) falls at about 0.075 solar masses ($M_\odot$), below which the interior temperatures are not high enough to generate adequate energy by nuclear fusion to supply the radiation released at the surface.

The structure of a star is determined by the interaction of a number of basic physical processes, including nuclear fusion; energy transport by radiation, convection, and conduction; atomic physics involving, especially, the interaction of radiation with matter; and the equation of state and thermodynamics of a gas. These principles, combined with basic equilibrium relations and assumed mass and chemical composition, allow the construction of mathematical models of stars. During the pre-main-sequence phase, the structure evolves as a result of gravitational contraction and the resulting heating of the interior. On the main sequence, evolution is driven by the change in chemical composition, specifically the conversion of hydrogen to helium as a result of nuclear reactions. During the post-main-sequence phase, the evolution is driven by a combination of nuclear transformations as well as contraction and expansion of various layers of the star. To follow this evolution mathematically requires the solution of a complicated set of equations. Even though a star can, to a high degree of approximation, be regarded as a sphere, so that the equations can be set up in one space dimension, the time-dependent solution is complicated enough so that numerical methods are required. The goal of the calculations is to obtain a complete evolutionary history of a star, as a function of its initial mass and chemical composition, from its birth in an interstellar cloud to its final state as a compact remnant, and to predict the observable properties of the object as a function of time. This chapter discusses the equations that are to be solved, the numerical method most commonly used to solve them, and the physics that must be included in the calculation.

5.1 EQUATIONS FOR EQUILIBRIUM OF A STAR

In this chapter we describe the main phases of evolution, during which the star is in hydrostatic equilibrium. Thus, we do not consider some of the exotic phases, such as nova or supernova explosions, pulsars, or interaction of close binary stars. We assume that rotation and magnetic fields are unimportant, and that the force of gravity on a mass element is exactly balanced by the difference in pressure on its upper and lower surface. In spherical symmetry, an excellent approximation for most stars, the one-dimensional Lagrangian equation of motion (Equation 1.55) can be written

$$0 = -\frac{1}{\rho}\frac{dP}{dr} - \frac{GM_r}{r^2}. \tag{5.1}$$

where M_r is the mass within radius r, ρ the density, and P the pressure. Using the equation for the mass of a spherical shell of radial thickness dr

$$dM_r = 4\pi r^2 \rho dr, \tag{5.2}$$

we obtain the form of the hydrostatic equation where the Lagrangian coordinate is taken to be M_r

$$\frac{dP}{dM_r} = -\frac{GM_r}{4\pi r^4}. \tag{5.3}$$

A second equilibrium condition, mass conservation in a spherical shell, is obtained by rewriting Equation (5.2)

$$\frac{dr}{dM_r} = \frac{1}{4\pi r^2 \rho}, \tag{5.4}$$

where the dependent variables ρ and r are, in general, functions of M_r and time t, but the ordinary derivative is used to emphasize the fact that the system is Lagrangian.

The third condition is that of conservation of energy. The simplest situation is referred to as *thermal equilibrium*, which means that the star is producing enough energy by nuclear processes to exactly balance the loss of energy by radiation at the surface. Then, if we define L_r as the net amount of energy per unit time that crosses a sphere at radius r in the outward direction, this condition is written

$$\frac{dL_r}{dM_r} = E_{\rm nuc}, \tag{5.5}$$

where $E_{\rm nuc}$ is defined as the nuclear energy generation rate per unit mass (see Section 5.6.3). Integrating over the mass, we obtain

$$L = \int_0^M E_{\rm nuc}\, dM_r, \tag{5.6}$$

where M is the total mass and L is the total output of energy from the star per second. The condition of thermal equilibrium holds on the main sequence, but in many stages of stellar evolution, it does not because the star may be expanding, contracting, heating, or cooling. The more general equation of conservation of energy, effectively the first law of thermodynamics, then must be used. Combining Equation (1.124) and Equation (1.126), we obtain

$$\frac{dL_r}{dM_r} = E_{\rm nuc} - \frac{d\epsilon}{dt} - P\frac{dV}{dt}, \tag{5.7}$$

where $V = 1/\rho$, ϵ is the internal energy per unit mass, and $L_r = 4\pi r^2 F_r$, where the net outward energy flux F_r is the amount of energy per unit time per unit area crossing a spherical surface at radius r. If $E_{\rm nuc} = 0$, the star contracts and obtains its energy from the third term on the right-hand side. Using thermodynamic relations (Kippenhahn and Weigert, 1990, Chapter 4, Sections 4.1, 4.2) one can rewrite this equation in terms of time derivatives of P and the temperature T

$$\frac{dL_r}{dM_r} = E_{\rm nuc} - c_P\frac{dT}{dt} + \frac{\delta}{\rho}\frac{dP}{dt}, \tag{5.8}$$

where c_P, the specific heat at constant pressure, is given by

$$c_P = \left(\frac{\partial \epsilon}{\partial T}\right)_P + P\left(\frac{\partial V}{\partial T}\right)_P \tag{5.9}$$

and

$$\delta = -\left(\frac{\partial \ln \rho}{\partial \ln T}\right)_P. \tag{5.10}$$

5.2 RADIATIVE, CONDUCTIVE, AND CONVECTIVE ENERGY TRANSFER

To obtain a detailed solution for the structure and evolution of a star, one must solve Equation (5.3), Equation (5.4), and Equation (5.8) along with a fourth differential equation, which describes the energy transport.

The transport of energy outward from the interior of a star to its surface depends in general on the existence of a temperature gradient. Heat will be carried by various processes from hotter regions to cooler regions; the processes that need to be considered include (1) radiation transport, (2) convective transport, and (3) conductive transport. Neutrino transport must be considered only in exceptional circumstances when the density is above 10^{10} g cm^{-3}; at lower densities neutrinos produced in stars simply escape directly without interacting with matter. In each case, a relation must be found between the energy flux F_r and the temperature gradient dT/dr. The diffusion approximation for radiation transport (Equation 1.117) can be rewritten, with M_r as the independent variable, in a form that can be used for all three types of transport

$$\frac{dT}{dM_r} = -\frac{GM_rT}{4\pi r^4 P}\nabla \tag{5.11}$$

where, if the energy transport is by radiation,

$$\nabla = \nabla_{\rm rad} = \frac{3}{16\pi Gac}\frac{\kappa_R L_r P}{M_r T^4} = \left(\frac{\partial \ln T}{\partial \ln P}\right)_{\rm rad} \tag{5.12}$$

where the derivative refers to the actual temperature–pressure variation in the structure of the star, and κ_R is the Rosseland mean opacity.

Exercise

Derive the expression for $\nabla_{\rm rad}$ by combining Equation (1.117) and Equation (5.1).

The criterion for the onset of convection is:

$$\nabla_{\rm rad} > \nabla_{\rm ad} = (\partial \ln T/\partial \ln P)_{\rm ad} \approx \frac{\gamma - 1}{\gamma}, \tag{5.13}$$

where γ is the ratio of specific heats c_P/c_V and $\nabla_{\rm ad}$ is the so-called *adiabatic gradient*. Each point in the star must be tested to see if this condition is satisfied and, if so, ∇ in Equation (5.11) is replaced by the "convective gradient" ∇_{conv}. In most of the interior of a star, it can be shown that convection is very efficient and that only a very small excess of the actual temperature gradient over the adiabatic gradient is required for the transport of the entire flux by convection. In this case, $\nabla_{conv} = \nabla_{\rm ad}$ to a high degree of accuracy. However, if convection occurs in the surface layers of a star, for example, just below the photosphere as in the case of the Sun, then ∇_{conv} must be calculated in more detail. A simple one-dimensional hydrodynamic formulation known as the "mixing-length theory" (Böhm-Vitense, 1958; Kippenhahn and Weigert, 1990) is generally used, but it contains an arbitrary parameter, the ratio of the mixing length, which is effectively the mean free path of a convective element, to the local pressure scale height $H = P/|dP/dr| = P/(g\rho)$, where g is the acceleration of gravity. This ratio can be determined empirically by fitting a stellar model to the known properties of the Sun, or numerically through two-dimensional or three-dimensional simulations of the turbulent, convective motions in the outer layers of a star (Stein and Nordlund, 1998; Ludwig et al., 1999).

If transport is by conduction, an equivalent "conductive opacity" κ_{cond} can be obtained from the conductivity K, defined by

$$F_{r,cond} = -K\boldsymbol{\nabla}T, \tag{5.14}$$

where $F_{r,cond}$ is the net conductive flux crossing radius r. Conduction, which involves transfer of heat by material particles, can occur, for example, in the interior of a white dwarf where electrons move at very high speeds owing to the phenomenon known as "electron degeneracy" (Section 5.6.1). The radiative flux from Equation (1.117) can be written as

$$F_{rad} = -\frac{4acT^3}{3\kappa_R\rho}\boldsymbol{\nabla}T. \tag{5.15}$$

The conduction flux can be written in parallel form

$$F_{cond} = -\frac{4acT^3}{3\kappa_{cond}\rho}\boldsymbol{\nabla}T, \tag{5.16}$$

if κ_{cond} is defined to be

$$\kappa_{cond} = \frac{4acT^3}{3\rho K}. \quad (5.17)$$

Since the total flux $F_{tot} = F_{rad} + F_{cond}$, it can also be written in the same form

$$F_{tot} = -\frac{4acT^3}{3\kappa_{tot}\rho}\nabla T \quad (5.18)$$

if

$$\frac{1}{\kappa_{tot}} = \frac{1}{\kappa_R} + \frac{1}{\kappa_{cond}}. \quad (5.19)$$

Thus, radiative and conductive effects can be combined in Equation (5.11) if κ_{tot} is used in the expression for $\nabla_{\rm rad}$.

5.3 CHANGE IN CHEMICAL COMPOSITION

The mass fraction X_i of a nucleus with mass number i is related to N_i, the number of nuclei of i per unit volume, by the expression

$$X_i = \frac{m_i N_i}{\rho}, \quad (5.20)$$

where m_i is the mass of nucleus i and ρ is the total mass density of all species. For the evolution of moderate-mass stars, the main species that need to be considered are X_1, X_4, X_{12}, X_{14}, and X_{16}, referring to ^{1}H, ^{4}He, ^{12}C, ^{14}N, and ^{16}O, respectively. Until helium burning sets in, ^{1}H plus ^{4}He compose more than 98% of the mass in the case of solar composition. The change in composition from nuclear reactions for each species can be written, again as a Lagrangian derivative

$$\left(\frac{dX_i}{dt}\right)_{nuc} = \frac{m_i}{\rho}\left[\sum_j r_{ji} - \sum_k r_{ik}\right], \quad (5.21)$$

where r_{ji} is the reaction rate, that is, the number of reactions per unit volume per second, of a reaction that produces i, and where r_{ik} is the reaction rate of a reaction that destroys i. This expression must be evaluated for each species i at all layers of the star. Equation (5.21) can be reexpressed in terms of the energy release $E_{\rm nuc}$ per unit mass per unit time. If a particular reaction produces energy at a rate E_{mn} per unit mass, and the energy released per reaction is Q_{mn} then

$$\rho E_{mn} = r_{mn} Q_{mn}. \quad (5.22)$$

If we further define the energy per unit mass of nucleus m resulting from its transformation to nucleus n as $q_{mn} = Q_{mn}/m_m$ where m_m is the mass of nucleus m, then Equation (5.21) can be written

$$\left(\frac{dX_i}{dt}\right)_{nuc} = \sum_j \frac{E_{ji}}{q_{ji}} - \sum_k \frac{E_{ik}}{q_{ik}}. \quad (5.23)$$

A layer dM_r at mass fraction M_r changes its composition at a rate given by Equation (5.23) if the layer is radiative. However, if the layer is convective, the motions will mix the layers within the convection zone on timescales much shorter than the timescales of nuclear transformation (unless the evolutionary timescale for the star is very short, which may occur for a very massive star at some stages). Thus, in general, a convection zone can be assumed to be chemically homogeneous. If the convection zone extends from M_{r1} to M_{r2}, then the rate of change in composition in the convection zone for species i is:

$$\frac{dX_{i,conv}}{dt} = \frac{1}{M_{r2} - M_{r1}} \int_{M_{r1}}^{M_{r2}} \left(\frac{dX_i(M_r)}{dt} \right)_{nuc} dM_r. \tag{5.24}$$

This expression does not include corrections for motions of the convective zone boundaries (see Clayton, 1968, Section 6.3).

5.4 BOUNDARY CONDITIONS

At the center of the star $M_r = 0$, the radius r and luminosity L_r also must vanish. Because of the singularities in the equations at the center, one usually expands the equations in a Taylor series to first order in M_r and applies the boundary condition at a small value of M_r close to the center, where

$$r = \left(\frac{3}{4\pi \rho_c} \right) M_r^{1/3} \tag{5.25}$$

$$P = P_c - \frac{1}{2} \left(\frac{4\pi}{3} \right)^{1/3} G \rho_c^{4/3} M_r^{2/3} \tag{5.26}$$

$$L_r = \left(E_{\rm nuc} - \frac{d\epsilon}{dt} - P \frac{dV}{dt} \right)_c M_r \tag{5.27}$$

$$T = T_c - \frac{1}{2} \left(\frac{4\pi}{3} \right)^{1/3} G \left(\frac{\rho_c^{4/3}}{P_c} \right) \cdot \nabla_c \cdot T_c \cdot M_r^{2/3}, \tag{5.28}$$

where the subscript c indicates the central value. The gradient ∇_c is set to $\nabla_{rad,c}$ if the layer is radiative or conductive, or to $\nabla_{ad,c}$ if the layer is convective. At the center, $\nabla_{rad,c}$, which is a function of L_r/M_r, goes to a constant value, but it must be evaluated slightly off center. Note that ρ_c, T_c, and P_c are not boundary conditions; they are to be determined.

In the case of the surface layer, the boundary conditions are more complicated. A simple way to apply them is to take first the definition of the effective (or photospheric) temperature

$$L = 4\pi R^2 \sigma_B T_{\mathit{eff}}^4, \tag{5.29}$$

where L is the surface luminosity, σ_B is the Stefan–Boltzmann constant, and R is the surface radius. Second, one can obtain the equation of hydrostatic equilibrium for an atmosphere with a fixed value of the gravitational acceleration g by combining Equation (5.3) and Equation (5.4)

$$\frac{dP}{dr} = -\frac{GM_r \rho}{r^2} = -g\rho, \tag{5.30}$$

if $M_r = M$ in the thin atmospheric layer. Then from the definition of optical depth $d\tau = -\kappa_R \rho dr$, where κ_R is the radiative opacity, we obtain

$$\frac{dP}{d\tau} = \frac{g}{\kappa_R}. \tag{5.31}$$

Integrating this expression approximately from a small value of τ inwards to $\tau = 2/3$, which defines the photosphere

$$\kappa_{ph} P_{ph} = \frac{2}{3} g, \tag{5.32}$$

where the subscript ph refers to the photosphere. An adequate but still approximate representation of the two surface boundary conditions is given by Equation (5.29) and Equation (5.32).

In order to make detailed comparisons of models with observations, it is desirable to have a more accurate atmospheric calculation. In some cases, frequency-dependent radiation transport has been employed in the atmosphere so that colors and fluxes could be calculated to compare with observed objects (Baraffe et al., 1998). Another reason to make a detailed atmosphere model is that the dominant elements H and He tend to be only partially ionized in the outer layers, while in most of the deep interior they are fully ionized. Furthermore, the convective temperature gradient in the outer convection zone of cool stars is usually superadiabatic rather than adiabatic in a fairly thin surface layer. It is convenient to confine the complicating effects of partial ionization and nonadiabatic convection to the thin surface layer where they are important and thereby to allow the simplification of the physics in the deep interior. The strategy, therefore, is to apply atmospheric physics in the outermost mass zone, between the surface $M_r = M$ and a deeper layer $M_r = M_{\mathrm{atm}}$ and to apply the outer boundary condition for the interior at M_{atm}.

A simple integration for a frequency-independent (grey) radiative atmosphere would proceed as follows (see Chapter 9, Section 9.3 for a discussion of frequency-dependent atmospheres) :

1. Assume that $M_r = M$ and $L_r = L$: The energy generation in the atmosphere is expected to be negligible.
2. Provide input values for L and R as well as the chemical composition X_i; then T_{eff} and g are determined.
3. Start at a small value of τ, say 0.01. Integrate Equation (5.31) inwards in discrete steps in τ, supplementing it with the $T - \tau$ relation

$$T^4 = \frac{3}{4} T_{\mathit{eff}}^4 \left(\tau + \frac{2}{3} \right) \tag{5.33}$$

and the ideal-gas equation of state

$$P = \frac{R_g}{\mu} \rho T. \tag{5.34}$$

Equation (5.33) is derived from considerations of radiation transfer by Mihalas (1978; Chapter 3) and is equivalent to the radiation diffusion

Equation (5.15). In regions of partial ionization, the quantity μ, the mean atomic weight per free particle, must be obtained with the aid of the Saha ionization equation

$$\log\left(\frac{N_{r+1}}{N_r}P_e\right) = -\theta\chi_r + 2.5\log T + \log\frac{2u_{r+1}}{u_r} - 0.48, \quad (5.35)$$

where the electron pressure P_e is given in dyne cm^{-2}; N_{r+1} and N_r are, respectively, the numbers of ions of a given element in the $(r+1)$st and rth stages of ionization; χ_r is the ionization energy from r to $r+1$ in electron volts, $\theta = 5040/T$; and the us are the partition functions for stages $(r+1)$ and r

$$u_r = g_{r,0} + g_{r,1}\exp(-\psi_{r,1}/kT) + g_{r,2}\exp(-\psi_{r,2}/kT) + \cdots \quad (5.36)$$

where $g_{r,i}$ and $\psi_{r,i}$ are, respectively, the statistical weight and excitation potential of the ith excited state of an atom in ionization stage r. At each level τ, one normally provides a first guess for the free electron pressure P_e, calculates $T(\tau)$, determines the degree of ionization, and iterates on P_e until the total pressure from the equation of state agrees with that derived from the condition of hydrostatic equilibrium (Equation 5.31).

4. Integrate downward until the mass level M_{atm} is reached. Then provide P and T from the atmospheric integration for the surface boundary condition for the interior.

This procedure has to be modified if the layer turns out to be convective. Then the radiative equation (5.33) is replaced by

$$\frac{d\log T}{d\log P} = \nabla_{conv} \quad (5.37)$$

where the convective gradient is obtained through a theory or a simulation of convection. The procedure for obtaining ∇_{conv} from the mixing-length theory is described by Kippenhahn and Weigert (1990; Sections 7.1, 7.2). Thus, to summarize, Equation (5.31) is used to get the pressure as a function of depth, Equation (5.33) or Equation (5.37) is used to get the temperature, and Equation (5.34) is used to obtain the density. A calculation of a gray atmosphere is included in the program STELLAR (Chapter 10, Section 10.2).

The lower boundary of the atmosphere M_{atm} is determined when two conditions are satisfied: (1) if the layers are convective near the lower boundary, the temperature and density have to be high enough so that $\nabla_{conv} = \nabla_{ad}$ to, say, 0.1%, and (2) the temperature must be high enough so that hydrogen and helium are practically fully ionized. If the layer near the lower boundary of the atmosphere is radiative, only the second condition applies. It generally turns out that a temperature of about 10^5 K is required to satisfy both conditions. As a star evolves, the mass layer corresponding to this temperature changes. For example, in a large red giant with a low-density envelope, M_{atm} may include 10% of the total mass of the star, while for the main-sequence Sun that layer includes only 0.01% of the mass. Numerically, a procedure must be developed to allow the value of M_{atm} to change with time.

5.5 AN IMPLICIT LAGRANGIAN TECHNIQUE: HENYEY METHOD

The overall problem is to solve for the complete structure of the star, i.e., for the basic dependent variables P, r, L_r, and T as a function of M_r and time. Equation (5.3), Equation (5.4), Equation (5.8), and Equation (5.11) are solved from a specified initial condition with boundary conditions given in the previous section. Additional relations (see Section 5.6) are necessary to obtain $\rho(P, T, X_i)$, $E_{\text{nuc}}(P, T, X_i)$, $\epsilon(P, T, X_i)$, $\nabla_{\text{ad}}(P, T, X_i)$, and $\kappa_{tot}(P, T, X_i)$. The four differential equations are solved implicitly and simultaneously for a time $t^{n+1} = t^n + \Delta t$, assuming that a complete model at time t^n already exists. Usually, once the complete model at t^{n+1} is obtained, the composition variables X_i are then updated explicitly for the next time step according to Equation (5.23), using values for E_{nuc} obtained at t^{n+1}

$$X_i(t^{n+2}, M_r) = X_i(t^{n+1}, M_r) + \Delta t \left(\frac{dX_i}{dt}\right)_{nuc}^{n+1}, \tag{5.38}$$

in a radiative zone, while the analogous expression derived from Equation (5.24) is used in convection zones. Note the Lagrangian form of this equation, which is appropriate for composition changes. In situations for advanced stages of stellar evolution where complicated nuclear reaction networks are solved, the equation must be modified to include implicit solutions. The original Henyey method is described by Henyey, Forbes, and Gould (1964). A different version, using the same general principle, is described by Kippenhahn and Weigert (1990, Section 11.2.) Here we describe the procedure coded in the program STELLAR, which accompanies this volume.

To solve the structure equations, one sets up a Lagrangian grid of J mesh points, with $j = 1$ representing the center $M_r = 0$ and $j = J$ representing the surface $M_r = M$, where M is the total mass. Although there are phases of stellar evolution when stars lose mass, we assume here that $M(t) = \text{constant}$. The coordinate M_r, which will be referred to as M_j, is very nearly equal to the total mass over a large range of radii near the surface. Thus, when one is setting up the mass grid it is necessary to store the mass increments $dM_j = M_j - M_{j-1}$ in addition to the masses themselves, in order to avoid subtractions of very nearly equal quantities. This procedure is used in the code STELLAR. Another method is to use logarithmic variables (Schwarzschild, 1958; Kippenhahn et al., 1967), solving for $\log T$, $\log P$, etc. as a function of $\log(M_r/M)$. In any case, whatever form for M_r is used, the distribution of mesh points must be set up so that no calculated variable varies by more than a few percent between M_j and M_{j+1}. As the structure of the star changes with evolution, mesh points are added or deleted in order to take into account regions where physical quantities change rapidly in space. It is also possible to set up a grid, all of whose points are continuously redistributed during the evolution, concentrating the points where they are needed most (Eggleton, 1971). To simplify the description of the method, here we assume that we have a set of J fixed Lagrangian zones.

Even for the case of the Sun, which is not a particularly extreme case of stellar structure, the pressure varies by 12 orders of magnitude from center to surface. The atmosphere calculation covers several orders of magnitude in pressure even though its mass range is very small; thus the interior can be well represented by 70 to 100 zones.

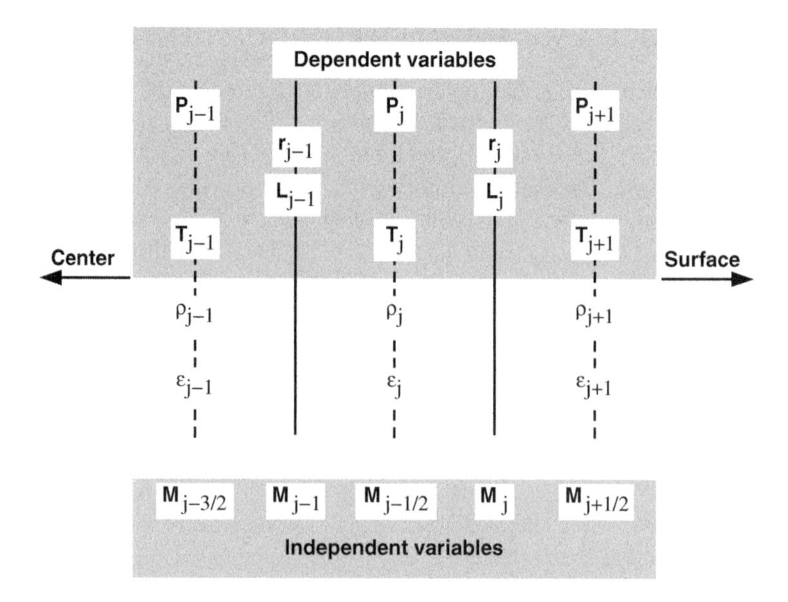

FIGURE 5.1 The setup of the grid for a stellar structure calculation, showing where the various variables are defined. The basic independent variable is the Lagrangian coordinate M_r. The four basic dependent variables are shown in the shaded region at the top. The functions ϵ and ρ, as well as $E_{\rm nuc}$ and κ_R, are derived as functions of P, T, and composition.

As the star evolves and the contrast between central and surface pressure becomes even greater, more zones are needed.

In the code STELLAR, a staggered grid is used to provide spatial centering of the equations (Figure 5.1). The four basic dependent variables that are solved for are $P,\ r,\ L_r,\ T$. The radius r_j and luminosity L_j are defined on the grid points M_j. The pressure P_j, temperature T_j, and density ρ_j are defined halfway between grid points, so that P_j lies between M_j and M_{j-1}; the corresponding mass coordinate is known as $M_{j-1/2}$. We reexpress Equation (5.3), Equation (5.4), Equation (5.8), and Equation (5.11) as follows:

$$P_{j+1} - P_j + \frac{(M_{j+1/2} - M_{j-1/2})GM_j}{4\pi r_j^4} = 0 \tag{5.39}$$

$$r_j^3 - r_{j-1}^3 - (M_j - M_{j-1}) \cdot \frac{3}{4\pi\rho_j} = 0 \tag{5.40}$$

$$L_j - L_{j-1} - (M_j - M_{j-1})\left[E_{\mathrm{nuc},j} - c_{P,j}(T_j - T_j^n)/\Delta t + \frac{\delta}{\rho_j}(P_j - P_j^n)/\Delta t\right] = 0 \tag{5.41}$$

$$T_{j+1} - T_j + \frac{(M_{j+1/2} - M_{j-1/2})GM_j}{4\pi r_j^4} 0.5(B_{j+1} + B_j) = 0. \tag{5.42}$$

Here, $B_j = T_j \nabla_j / P_j$. Equation (5.39) and Equation (5.42) are centered at M_j, while Equation (5.40) and Equation (5.41) are centered at $M_{j-1/2}$. A superscript n indicates that the quantity is evaluated at the previous time step, while a superscript $n+1$ is implied for all other quantities. Note that the time derivatives are evaluated by backward differences that greatly enhance the stability of the system. It is also important to note that for a given value of j, the above set of difference equations involves three grid points: $j-1, j, j+1$. Other Henyey schemes described in the literature are generally simpler in the sense that they define all variables on the same grid points and couple only the two neighboring points $j, j+1$ in the set of difference equations.

Because of the staggered grid, the inner boundary condition requires only slight modification. The first grid point M_1 does not lie at $M_r = 0$, but at a short distance out from the center. The central pressure P_1 and central temperature T_1 are actually defined at $M_{1/2}$. The first and fourth equations above, Equation (5.39) and Equation (5.42), are written exactly as above, with $j = 1$. In the second and third equations, j is set to 1, and r_{j-1}, L_{j-1}, and M_{j-1} are set to zero. Thus, these two equations reduce to boundary conditions (5.25) and (5.27), respectively.

At the surface, two boundary conditions are specified, which involve the input of the current values of the total radius and the total luminosity $R = r_J$ and $L = L_J$, and the output of the quantities P_{atm}, T_{atm}, and Δr_{atm}, where Δr_{atm} is the vertical thickness of the atmosphere. If the boundary conditions are evaluated at the photosphere (M_J), then Equation (5.29) and Equation (5.32) are used, and $\Delta r_{\text{atm}} = 0$. If a model atmosphere is used, then P_{atm} and T_{atm} refer to the lower boundary of the atmosphere, at $j = J$ and $M_{\text{atm}} = M_{J-1/2}$. Assuming that an atmosphere is used, then there will be three functions λ, which cannot be expressed analytically, that connect the input values and the output values

$$\lambda_1 = P_J - \phi(r_J, L_J) = 0 \tag{5.43}$$

$$\lambda_2 = T_J - \psi(r_J, L_J) = 0 \tag{5.44}$$

$$\lambda_3 = \Delta r_{\text{atm}} - \zeta(r_J, L_J) = 0. \tag{5.45}$$

(5.46)

The functions ϕ, ψ, and ζ are obtained numerically.

The boundary conditions at J then are:

$$P_J - P_{\text{atm}} = 0 \tag{5.47}$$

$$r_J - r_{J-1} - \Delta r_{\text{atm}} = 0 \tag{5.48}$$

$$L_J - L_{J-1} = 0 \tag{5.49}$$

$$T_J - T_{\text{atm}} = 0 \tag{5.50}$$

Equation (5.49) expresses the fact that no energy is generated or absorbed in the atmosphere. Note that P_J and T_J are evaluated at $M_{J-1/2}$, while Δr_{atm} extends farther inward to M_{J-1}.

The Henyey method involves the generalization of the Newton–Raphson method for finding the root of a function to the case of a large number of variables. Let us first illustrate the procedure for a function of one variable. Suppose, on an interval $a < x < b$, a function $y = f(x)$, and its derivative $f'(x)$ are continuous, and $f''(x)$ is finite. We search for zero points for $f(x)$ by an iterative procedure. Suppose the zero point is at x_0 so that $f(x_0) = 0$. We use the Taylor expansion

$$f(x_0) = f(x) + (x_0 - x) \cdot f'(x) + O[(x_0 - x)^2]. \tag{5.51}$$

We take a trial value of $x = x^{(1)}$ and rewrite the above equation

$$0 = f(x^{(1)}) + (x^{(2)} - x^{(1)}) \cdot f'(x^{(1)}). \tag{5.52}$$

That is, we solve for the value $x^{(2)}$, which would be the zero point if the function were linear. Thus, if $f'(x^{(1)}) \neq 0$, the first correction to x is:

$$\delta x^{(1)} = (x^{(2)} - x^{(1)}) = -\frac{f(x^{(1)})}{f'(x^{(1)})}. \tag{5.53}$$

In general, $f(x)$ is not linear, so $x^{(2)}$ is just an approximation to the root. Iterations proceed according to

$$\delta x^{(n)} = (x^{(n+1)} - x^{(n)}) = -\frac{f(x^{(n)})}{f'(x^{(n)})}. \tag{5.54}$$

Convergence occurs if $\delta x^{(n)} \to 0$. The initial guess $x = x^{(1)}$ must be close enough to the actual solution so that it is inside the region of convergence.

As an example, we wish to find the maximum of the function $f(x) = x^2/(e^x - 1)$. Thus, we need x such that

$$0 = f'(x) = [e^x(2 - x) - 2]\frac{x}{(e^x - 1)^2} \tag{5.55}$$

or, equivalently, we wish to find the zero point of

$$g(x) = e^x(2 - x) - 2. \tag{5.56}$$

Since $g'(x) = e^x(1 - x)$, the prescription for the corrections δx is:

$$\delta x^{(n)} = (x^{(n+1)} - x^{(n)}) = \frac{x^{(n)} - 2 + 2\exp(-x^{(n)})}{1 - x^{(n)}}. \tag{5.57}$$

Starting from an initial guess $x^{(1)} = 1.1$, the sequence of x values that is generated is 3.443, 2.826, 2.309, 1.921, 1.689, 1.604, 1.59377, 1.59362. Convergence to 1 part in 10^4 in x is achieved in 8 iterations. It could happen that the iterations oscillate about the true solution or do not converge at all, in which case it is often helpful to take only a fraction of the indicated correction $\delta x^{(n)}$.

For the system of stellar structure equations, we have a total of J mesh points, two independent variables M_r and t, and four differential equations, which we must solve for four unknown functions. The dependent variables are P, r, L, and T, and ρ can be obtained as a function of (P, T) through the equation of state. Other quantities, such as κ_R and $E_{\rm nuc}$, can be expressed as functions of (ρ, T). Thus, there are $4J$ unknowns to be solved for simultaneously: P_j, r_j, L_j, and T_j, for all j. We have available $4(J-1)$ difference equations, given by Equation (5.39) to Equation (5.42), evaluated from $j = 1$ to $J - 1$ and with $r_0 = L_0 = 0$ serving as the central boundary conditions. The four missing relations are provided by the surface boundary conditions, Equation (5.47) to Equation (5.50). We express the ith difference equation at the jth mesh point as

$$G_i^j(P_n, r_n, L_n, T_n,\ n = j-1, j, j+1) = 0, \tag{5.58}$$

where i runs from 1 to 4, and j from 1 to J. Equation (5.39) to Equation (5.42), as well as Equation (5.47) to Equation (5.50), are expressed in this form; the coefficients of some of the variables are clearly equal to zero. Thus, for example,

$$G_1^J = P_J - P_{\rm atm} \tag{5.59}$$

and

$$G_2^6 = r_6^3 - r_5^3 - (M_6 - M_5)\cdot\frac{3}{4\pi\rho_6}. \tag{5.60}$$

We then expand each function G in a Taylor series, in analogy to Equation (5.52). For simplicity, let $(P_j,\ r_j,\ L_j,\ T_j) = (x_1^j,\ x_2^j,\ x_3^j,\ x_4^j)$, so x_i^j is the ith dependent variable at the jth grid point. At a given value of j, each G_i^j is in general a function of up to eight variables. The expansion becomes

$$\begin{aligned} G_i^j &+ \frac{\partial G_i^j}{\partial x_1^{j-1}}\cdot\delta x_1^{j-1} + \frac{\partial G_i^j}{\partial x_2^{j-1}}\cdot\delta x_2^{j-1} + \frac{\partial G_i^j}{\partial x_3^{j-1}}\cdot\delta x_3^{j-1} + \frac{\partial G_i^j}{\partial x_4^{j-1}}\cdot\delta x_4^{j-1} \\ &+ \frac{\partial G_i^j}{\partial x_1^{j}}\cdot\delta x_1^{j} + \frac{\partial G_i^j}{\partial x_2^{j}}\cdot\delta x_2^{j} + \frac{\partial G_i^j}{\partial x_3^{j}}\cdot\delta x_3^{j} + \frac{\partial G_i^j}{\partial x_4^{j}}\cdot\delta x_4^{j} + \frac{\partial G_i^j}{\partial x_1^{j+1}}\cdot\delta x_1^{j+1} \\ &+ \frac{\partial G_i^j}{\partial x_2^{j+1}}\cdot\delta x_2^{j+1} + \frac{\partial G_i^j}{\partial x_3^{j+1}}\cdot\delta x_3^{j+1} + \frac{\partial G_i^j}{\partial x_4^{j+1}}\cdot\delta x_4^{j+1} = 0. \end{aligned} \tag{5.61}$$

Note that each individual equation depends on only two of the three variables: x_i^{j-1}, x_i^j, x_i^{j+1}. The goal is, given an initial guess for the x_i^j, to solve the set of linear equations for the 4J corrections δx_i^j, add the corrections to the variables themselves, and iterate until δx_i^j and G_i^j approach zero.

The full set of equations is solved for j running from 2 to $J - 1$. At the center ($j = 1$), the equations are somewhat simplified because the quantities labeled $j - 1$ are set to zero. The boundary conditions at the surface connect the mesh points M_J and M_{J-1}. The partial derivatives in Equation (5.61) at the surface must be obtained numerically, by incrementation of the parameters in the atmosphere integration. The net result is the solution of $4(J-1)$ difference equations plus four boundary conditions.

At the beginning of each time step, a trial model must be provided; generally the model at the previous time step is adequate as long as Δt is small. Thus, a first guess for all x_i^j and, therefore, for all G_i^j is available. These G_i^j in general will not be zero. The partial derivatives $\frac{\partial G_i^j}{\partial x_i^j}$ are also available from the equations themselves;

however in some cases they must be obtained numerically. For example, E_{nuc} is in general a complicated function of (ρ, T). To get $\partial E_{\mathrm{nuc}}/\partial T$, one increments T by, say, 0.1% and recalculates E_{nuc}. Then we simply have to solve this set of $4J$ *linear* equations for the $4J$ unknowns δx_i^j, which are the corrections to the first guess for x_i^j. Once the equations are solved, the corrections (or a fraction of them) are added to the variables themselves to provide a second guess to the solution. The functions G_i^j and their derivatives are then reevaluated and the system of linear equations is solved again for a new set of corrections δx_i^j. The iterative process is continued until all fractional corrections are smaller than a given tolerance, e.g., 10^{-4}, in which case the model is declared converged.

Exercise

Calculate $\frac{\partial G_2^j}{\partial x_2^j}$ and $\frac{\partial G_4^j}{\partial x_4^{j+1}}$. Note, in the latter derivative, ∇ is a function of T. (The answer is well hidden in the program STELLAR).

Exercise

Write out the full Taylor expansion (Equation 5.61) for the mass equation (5.40). Note that ρ_j is a function of T_j and P_j through the equation of state so that the derivatives $(\partial\rho/\partial T)_P$ and $(\partial\rho/\partial P)_T$ are needed. There are only four nonzero derivatives in the expansion.

The system of linear equations can be expressed in matrix form. Let us define

$$C_{ik}^j = \frac{\partial G_i^j}{\partial x_k^{j-1}}; \quad C_{ik}^1 = 0 \tag{5.62}$$

$$D_{ik}^j = \frac{\partial G_i^j}{\partial x_k^j} \tag{5.63}$$

$$E_{ik}^j = \frac{\partial G_i^j}{\partial x_k^{j+1}} \quad E_{ik}^J = 0. \tag{5.64}$$

Thus, D_{ik}^j is the derivative of the ith equation with respect to the kth variable at the mesh point j. For example, at the atmosphere zone the only nonzero Cs are C_{22}^J, involving r_{J-1}, and C_{33}^J, involving L_{J-1}. However, the Ds involve the quantities P_{atm}, T_{atm}, and Δr_{atm}, which are functions of R_J and L_J. Thus, derivatives of the functions ϕ, ψ, and ζ with respect to r_J and L_J are required. These are obtained numerically by incrementation of r_J at constant L_J and L_J at constant r_J, recalculating the atmosphere, and finding the resulting changes in P_{atm}, T_{atm}, and Δr_{atm}.

Equations (5.61) can be written as follows

$$-G_i^j = C_{ik}^j \delta x_k^{j-1} + D_{il}^j \delta x_l^j + E_{im}^j \delta x_m^{j+1} \tag{5.65}$$

where summation over k, l, and m from 1 to 4 is implied, but there is no summation over j. The matrix can be shown schematically as follows:

$$\begin{bmatrix} D^1 & E^1 & 0 & 0 & \cdot & \cdot & \cdot & 0 \\ C^2 & D^2 & E^2 & 0 & \cdot & \cdot & \cdot & 0 \\ 0 & C^3 & D^3 & E^3 & \cdot & \cdot & \cdot & 0 \\ 0 & 0 & C^4 & D^4 & E^4 & \cdot & \cdot & 0 \\ \cdot & \cdot & \cdot & \cdot & \cdot & \cdot & \cdot & \cdot \\ \cdot & \cdot & \cdot & \cdot & \cdot & \cdot & \cdot & \cdot \\ 0 & \cdot & \cdot & 0 & 0 & C^{J-1} & D^{J-1} & E^{J-1} \\ 0 & \cdot & \cdot & \cdot & 0 & 0 & C^J & D^J \end{bmatrix} \begin{bmatrix} \delta\mathbf{x}^1 \\ \delta\mathbf{x}^2 \\ \delta\mathbf{x}^3 \\ \cdot \\ \cdot \\ \cdot \\ \delta\mathbf{x}^{J-1} \\ \delta\mathbf{x}^J \end{bmatrix} = \begin{bmatrix} -\mathbf{G}^1 \\ -\mathbf{G}^2 \\ -\mathbf{G}^3 \\ \cdot \\ \cdot \\ \cdot \\ -\mathbf{G}^{J-1} \\ -\mathbf{G}^J \end{bmatrix}, \tag{5.66}$$

where each of the Cs, Ds and Es is a 4×4 matrix

$$C^j = \begin{bmatrix} C_{11}^j & C_{12}^j & C_{13}^j & C_{14}^j \\ C_{21}^j & C_{22}^j & C_{23}^j & C_{24}^j \\ C_{31}^j & C_{32}^j & C_{33}^j & C_{34}^j \\ C_{41}^j & C_{42}^j & C_{43}^j & C_{44}^j \end{bmatrix}, \tag{5.67}$$

with entirely analogous matrices for D^j and E^j. The vector $\mathbf{x}^j = (x_1^j, x_2^j, x_3^j, x_4^j)$, and the vector $\mathbf{G}^j = (G_1^j, G_2^j, G_3^j, G_4^j)$.

It is clear that the matrix is very sparse, with 4×12 blocks along the diagonal and with a large number of zeros. This form allows the solution of the overall matrix to be reduced to the solution of a series of much smaller submatrices. The procedure involves a process of elimination followed by back substitution, starting with the inner boundary.

We assume the following relation between the corrections δx at the grid points j and $j-1$

$$\delta x_k^{j-1} = A_k^{j-1} + B_{kl}^{j-1} \delta x_l^j, \tag{5.68}$$

where summation over l is implied. Insert this expression into Equation (5.65) to obtain, for $j = 2, \ldots, J-1$

$$-G_i^j = C_{ik}^j A_k^{j-1} + \left(C_{ik}^j B_{kl}^{j-1} + D_{il}^j\right) \delta x_l^j + E_{im}^j \delta x_m^{j+1}. \tag{5.69}$$

Now, we define the quantity S^j_{il}, which is a 4×4 matrix

$$S^j_{il} = C^j_{ik} B^{j-1}_{kl} + D^j_{il} \tag{5.70}$$

from which we obtain

$$\delta x^j_l = A^j_l + B^j_{lm} \delta x^{j+1}_m \tag{5.71}$$

where

$$A^j_l = -\left(S^j_{il}\right)^{-1} \left(G^j_i + C^j_{ik} A^{j-1}_k\right) \tag{5.72}$$

$$B^j_{lm} = -\left(S^j_{il}\right)^{-1} E^j_{im}. \tag{5.73}$$

For the special case $j = 1$, where the $C^j = 0$

$$-G^1_i = D^1_{il} \delta x^1_l + E^1_{im} \delta x^2_m, \tag{5.74}$$

with

$$A^1_l = -\left(D^1_{il}\right)^{-1} G^1_i \tag{5.75}$$

$$B^1_{lm} = -\left(D^1_{il}\right)^{-1} E^1_{im}. \tag{5.76}$$

Since the vector A^1 and the matrix B^1 are known, Equation (5.72) and Equation (5.73) allow us to calculate A^2 and B^2, and so, recursively, all As and Bs are obtained up to A^{J-1} and B^{J-1}.

The back-substitution step now begins. At $j = J$ the Es are zero so Equation (5.69) and Equation (5.68) can be combined to give

$$-G^J_i = C^J_{ik} \delta x^{J-1}_k + D^J_{il} \delta x^J_l \tag{5.77}$$

so that

$$\delta x^J_l = -\left(C^J_{ik} B^{J-1}_{kl} + D^J_{il}\right)^{-1} \left(G^J_i + C^J_{ik} A^{J-1}_k\right). \tag{5.78}$$

All quantities in this expression on the right-hand side can be evaluated at the surface, so the δxs at J can be obtained. Then all remaining δxs from $J - 1$ to $j = 1$ are obtained recursively through the use of Equation (5.68) and the known values of A^{j-1}_k and B^{j-1}_{kl}.

The iteration is completed by replacing all x^j_i by $x^j_i + \delta x^j_i$. The only matrix inversions required are those of the 4×4 matrices S^j_{il}. If some of the corrections are larger than, say, 10%, it may be necessary to limit the size of the corrections, scaling all of them such that the largest correction δx is only 10% of the corresponding x. After the corrections are added to the corresponding variables, the next iteration is started. In a typical stellar evolution calculation, three to five iterations are required to reach a convergence, which generally requires that all $|\delta x^j_i / x^j_i|$ be less than 10^{-4}.

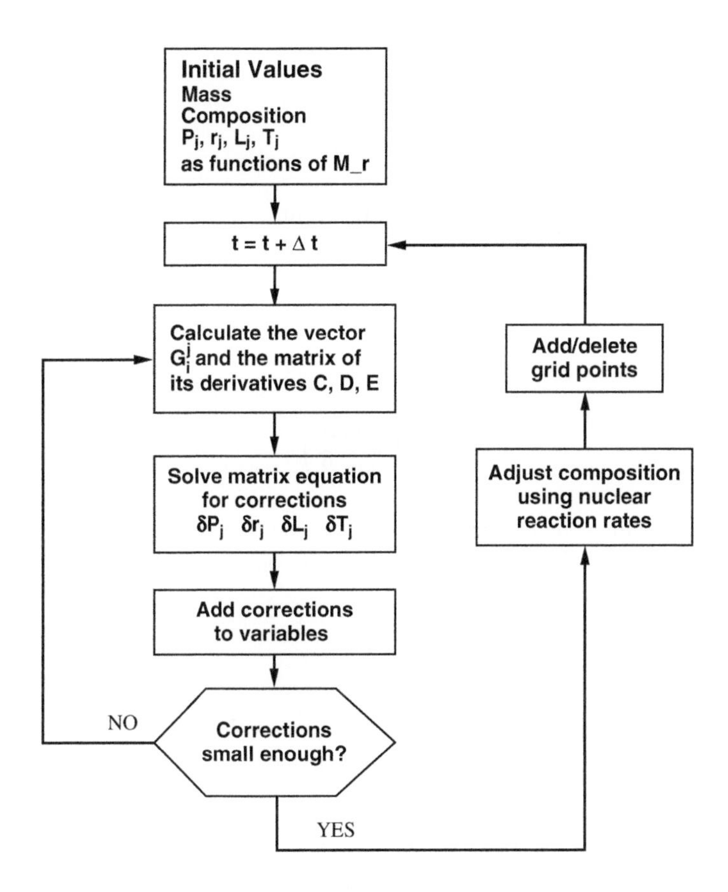

FIGURE 5.2 Flowchart for the calculation of stellar evolution.

5.6 PHYSICS PACKAGES

As discussed above, the basic dependent variables, which are the solutions to Equation (5.39) through Equation (5.42) are the pressure P, the radius r, the luminosity L, and the temperature T. However, additional unknowns appear in these equations, including the density ρ; the rate of nuclear energy generation E_{nuc}; the adiabatic gradient ∇_{ad}; the opacity κ_{tot}; the specific heat c_P, which is a function of the internal energy ϵ (Equation 5.9); and the thermodynamic derivative δ. Normally in the astrophysical literature, the quantities P, κ_{tot}, ϵ, and E_{nuc} are expressed in terms of ρ, T, and the chemical composition X_i. The procedure in STELLAR is to invert the equation of state so it gives $\rho(P, T, X_i)$ and $\epsilon(P, T, X_i)$. Then, $E_{\text{nuc}}(\rho, T, X_i)$ and $\kappa_{\text{tot}}(\rho, T, X_i)$ can be obtained. Once c_P and δ have been determined from the equation of state, the adiabatic gradient follows from

$$\nabla_{\text{ad}} = \frac{P\delta}{T\rho c_P}. \tag{5.79}$$

In the program STELLAR, the quantities α, δ, and

$$c_V = \left(\frac{\partial \epsilon}{\partial T}\right)_V \tag{5.80}$$

are obtained from tables, then c_P is obtained from

$$c_P - c_V = \frac{P\delta^2}{\rho T \alpha}, \tag{5.81}$$

where δ is obtained from Equation (5.10) and

$$\alpha = \left(\frac{\partial \ln \rho}{\partial \ln P}\right)_T. \tag{5.82}$$

This section outlines the physical relations that apply in most phases of stellar evolution, which are needed to obtain the pressure, opacity, and $E_{\rm nuc}$ as functions of density, temperature, and composition.

5.6.1 Equation of State

In the pre-main- and main-sequence stages of the evolution of most stars, the ideal gas equation holds. The pressure of the gas is given by

$$P = nk_BT, \tag{5.83}$$

where k_B is the Boltzmann constant and n the number of free particles per unit volume. If X, Y, and Z are the mass fractions of H, He, and heavy elements, respectively, and the gas is fully ionized, then

$$n = (2X + 0.75Y + 0.5Z)\rho/m_u, \tag{5.84}$$

where m_u is the mass corresponding to one atomic mass unit (1.66053×10^{-24} g). It has been assumed that the number of particles contributed per nucleus is 2 for H, 3 for He, and $A/2$ (an approximation) for the heavy elements, where A is the atomic weight. In the outer layers of the star, n must be adjusted to take into account partial ionization. The internal energy for an ideal gas is $1.5k_BT$ per particle or $1.5k_BTn/\rho = 1.5P/\rho$ per unit mass. The equation of state can also be written $P = R_g\rho T/\mu$, where μ, the mean atomic weight per free particle, is given by $\mu^{-1} = 2X + 0.75Y + 0.5Z$ for a fully ionized gas, and the gas constant $R_g = k_B/m_u = 8.31 \times 10^7$ erg $^\circ$K^{-1} g^{-1}.

Under conditions of high temperature and low density, the pressure of the radiation must be added. According to quantum theory, a photon has energy $h\nu$, where ν is the frequency and h Planck's constant, and momentum $h\nu/c$. The radiation pressure is the net rate of transfer of momentum per unit area, normal to an arbitrarily oriented surface. Under conditions in stellar interiors where the radiation is nearly isotropic and the system is near thermodynamic equilibrium, it can be shown (Clayton, 1968,

Section 2.1) that the radiation pressure is given by

$$P_R = \frac{1}{3} a T^4 \tag{5.85}$$

where a, the radiation density constant, equals 7.56×10^{-15} erg cm^{-3} °K^{-4}.

At high densities, an additional physical effect must be considered — the phenomenon of degeneracy. The effect is a consequence of the Pauli exclusion principle, which does not permit more than one particle to occupy one quantum state at the same time; it applies to elementary particles with half-integral spin, such as electrons or neutrons. In the case of electrons bound in an atom, the principle governs the distribution of electrons in various energy states. If the lowest energy states are filled, any electrons added subsequently must go into states of higher energy; only a discrete set of states is available. The same principle applies to free electrons. In a momentum interval $dp_x\, dp_y\, dp_z$ and in a volume element $dx\, dy\, dz$, the number of states is $(\frac{2}{h^3})\, dp_x\, dp_y\, dp_z\, dx\, dy\, dz$, where the factor of 2 arises from the two possible directions of the electron spin. Degeneracy occurs when most of these states, up to some limiting momentum, are occupied by particles. Complete degeneracy is defined as the situation in which all states are occupied up to a limiting momentum p_0 and all higher states are empty.

Under the assumption of complete electron degeneracy, a situation that holds in the interiors of most white dwarfs and in the dense core of highly evolved red giants, the electron pressure P_e is obtained as a function of the gas density from the equation (Chandrasekhar, 1939)

$$P_e = 6.003 \times 10^{22} f(x) \quad \text{dyne cm}^{-2}, \tag{5.86}$$

where

$$f(x) = x(2x^2 - 3)(x^2 + 1)^{1/2} + 3 \sinh^{-1} x \tag{5.87}$$

and

$$x = \frac{p_0}{m_e c} = 1.009 \times 10^{-2} \left(\frac{\rho}{\mu_e}\right)^{1/3} \tag{5.88}$$

where μ_e, the mean atomic weight per free electron, is $2/(1 + X)$ for a fully ionized gas. Given the density and μ_e, these expressions allow one to calculate the electron pressure, which is very close to the total pressure. In the white dwarf density regime, most of the electrons are forced into such high momentum states that their pressure is much higher than the ideal gas pressure, and the ions, which remain nondegenerate in this regime, contribute negligible pressure. The corresponding internal energy per unit mass is:

$$\epsilon_e = \frac{\pi m_e^4 c^5}{3h^3 \rho} g(x) \tag{5.89}$$

where

$$g(x) = 8x^3[(x^2 + 1)^{1/2} - 1] - f(x). \tag{5.90}$$

The following criterion indicates where electron degeneracy is becoming important in the equation of state

$$\frac{\rho}{\mu_e} > 2.4 \times 10^{-8} T^{3/2} \quad \text{g cm}^{-3}. \tag{5.91}$$

For much lower densities the ideal gas equation applies. For densities in the vicinity of the boundary, the degeneracy is partial, and Equation (5.86) does not apply. Assuming that the electrons do not move with relativistic velocities, the electron pressure is a function of density and temperature and is given by

$$P_e = \frac{8\pi k_B T}{3h^3}(2m_e k_B T)^{3/2} F_{3/2}(\eta) \tag{5.92}$$

$$n_e = \frac{\rho}{\mu_e m_u} = \frac{4\pi}{h^3}(2m_e k_B T)^{3/2} F_{1/2}(\eta) \tag{5.93}$$

$$F_n(\eta) = \int_0^\infty \frac{u^n}{e^{(u-\eta)}+1} du \tag{5.94}$$

$$\epsilon_e = \frac{3P_e}{2\rho}, \tag{5.95}$$

where the integral is over all momenta p; the variable u is a dimensionless energy $u = p^2/(2m_e k_B T)$. The degeneracy parameter η describes the importance of degeneracy. For large negative values, degeneracy is negligible, and for large positive values the expressions reduce to those for complete degeneracy. The Fermi–Dirac integrals (Equation 5.94) have been evaluated numerically and are available in tabular form (McDougall and Stoner, 1939; Clayton, 1968; Kippenhahn and Weigert, 1990). The most straightforward numerical procedure is, given ρ and T, to use Equation (5.93) to find $F_{1/2}$, interpolate in the table to find η, and then interpolate again to find $F_{3/2}$, which, from Equation (5.92), gives P_e.

5.6.2 Opacity

The energy transport in a radiative star is controlled by the opacity κ_R. Numerous atomic processes contribute to this quantity and, in general, the structure of a star can be calculated only with the aid of detailed tables of the opacity, calculated as a function of ρ, T, and the chemical composition. Starting at the highest temperatures characteristic of the stellar interior and proceeding to lower temperatures, the main processes are

1. Electron scattering, also known as Thomson scattering, in which a proton undergoes a change in direction, but no change in frequency during an encounter with a free electron.
2. Free-free absorption, in which a photon is absorbed by a free electron in the vicinity of a nucleus, with the result that the photon is lost and the electron increases its kinetic energy.
3. Bound-free absorption on metals, also known as photoionization, in which the photon is absorbed by an atom of a heavy element (e.g., iron) and one of the bound electrons is removed.
4. Bound-bound absorption of a heavy element, in which the photon induces an upward transition of an electron from a lower quantum state to a higher quantum state in the atom.
5. Bound-free absorption on H and He, which generally occurs near stellar surfaces where these elements are being ionized,

6. Bound-free and free-free absorption by the negative hydrogen ion H^-, which forms in stellar atmospheres in layers where H is just beginning to be ionized (e.g., the surface of the Sun).
7. Bound-bound absorptions by molecules, which can occur only in the atmospheres of the cooler stars ($T_{eff} < 4000$ K, although even the Sun shows a few molecular features in its spectrum).
8. Absorption by dust grains, which can occur in the early stages of protostellar evolution and possibly in the atmospheres of brown dwarfs, at temperatures below the evaporation temperature of grains (1400–1800 K).

In general, these processes are frequency dependent and the average over frequency must be obtained. The expression for the Rosseland mean opacity, defined in Equation (1.115) for the case of thermodynamic equilibrium, must be modified if scattering processes are significant

$$\frac{1}{\kappa_R} = \frac{\int_0^\infty \frac{[dB_\nu(T)/dT]\,d\nu}{\kappa_\nu[1-\exp(-h\nu/kT)]+\sigma_\nu}}{\int_0^\infty [dB_\nu(T)/dT]\,d\nu}. \tag{5.96}$$

Here, κ_ν refers to processes of true absorption, which have to be corrected for induced emission, and σ_ν refers to scattering processes. Detailed calculations of opacities add up the contributions from all of the relevant physical processes at a given frequency, then after the total opacity at each frequency is obtained, the Rosseland mean is performed. During calculation of stellar evolution, several different tables of opacities, each as a function of (ρ, T), must be available at the same time. For example, on the main sequence, it may be necessary to have tables for X = .75, Y = .23, Z = .02, as well as for X = .375, Y = .605, Z = .02 and X = 0, Y = .98, Z = .02, to take into account the conversion of hydrogen to helium in the central regions. A three-dimensional interpolation is needed, generally done in log space for (ρ, T) and linearly for composition.

The Thomson scattering from free electrons, in units of cm^2 g^{-1}, is given by $\sigma = 0.2(1 + X)$. This process is important generally in massive stars. The free-free absorptions and the bound-free absorptions on heavy elements have the approximate dependence $\kappa \propto \rho T^{-3.5}$. As one moves outward in a star from higher to lower T, the number of bound electrons per atom increases and, correspondingly, κ increases. A maximum in κ occurs in the range 10^4 K $< T <$ 10^5 K depending on density; this range corresponds to the zones where H and He, the most abundant elements, undergo ionization. Below this range in T, the above dependence is no longer valid and κ drops rapidly with decreasing T, reaching a minimum of about 10^{-2} at $T = 2000$ K. At lower T, where grains exist, the opacities are higher, on the order of 1 cm^2 g^{-1}.

Detailed opacity calculations are described by Seaton et al. (1994) and Iglesias and Rogers (1996). For temperatures $\log T \geq 3.75$, tables for a variety of compositions are available for download from the Web site: www-phys.llnl.gov/Research/OPAL. Also available are Opacity Project tables down to log T = 3.5 (http://astro.u-strasbg.fr/-OP.html). At lower temperatures molecules must be included. Ferguson et al. (2005) provide tables for $2.7 \leq \log T \leq 4.5$ with a number of different compositions. Conductive opacities are available from Hubbard and Lampe (1969).

5.6.3 Nuclear Reactions

Nuclear energy in moderate-mass stars is produced by conversion of hydrogen to helium or by conversion of helium to carbon and oxygen. Only in the most massive stars do nuclear reactions proceed farther to the synthesis of the heavier elements up to iron and nickel. At temperatures of 1 to 3×10^7 K, which characterize the interiors of most main sequence stars, only reactions involving hydrogen burning need to be considered. Helium burning starts at about 10^8 K. In this subsection, we discuss formulas for obtaining E_{nuc} for hydrogen-burning and helium-burning reactions only.

Two reaction sequences have been identified that result in the conversion of four protons into one helium nucleus, the proton–proton (pp) chains and the CNO cycle. The reactions of the main branch of the pp chains (PP I) are:

$$ {}^{1}\text{H} + {}^{1}\text{H} \rightarrow {}^{2}\text{D} + \text{e}^{+} + \nu \tag{5.97} $$

$$ {}^{2}\text{D} + {}^{1}\text{H} \rightarrow {}^{3}\text{He} + \gamma \tag{5.98} $$

$$ {}^{3}\text{He} + {}^{3}\text{He} \rightarrow {}^{4}\text{He} + {}^{1}\text{H} + {}^{1}\text{H} \tag{5.99} $$

The symbol e^+ denotes a positron, ν the neutrino, and ^{2}D the deuterium nucleus, composed of one proton and one neutron. The positron immediately reacts with an electron, resulting in the annihilation of both. The total energy production of the sequence is 26.7 MeV or 4.27×10^{-5} erg, corresponding to the mass difference between four protons and one ^{4}He nucleus, times c^2. Each sequence produces two neutrinos, of average energy 0.26 MeV each, which immediately escape from the star, as they interact only very weakly with matter, carrying with them a certain fraction of the energy. The remainder of the energy is deposited locally in the star, typically in the form of γ rays.

The second branch of the pp chains (PP II) occurs when the ^{3}He nucleus produced in Equation (5.98) reacts with ^{4}He. The sequence of reactions is then

$$ {}^{3}\text{He} + {}^{4}\text{He} \rightarrow {}^{7}\text{Be} + \gamma \tag{5.100} $$

$$ {}^{7}\text{Be} + \text{e}^{-} \rightarrow {}^{7}\text{Li} + \nu \tag{5.101} $$

$$ {}^{7}\text{Li} + {}^{1}\text{H} \rightarrow {}^{4}\text{He} + {}^{4}\text{He} \tag{5.102} $$

Under conditions of the solar interior, the reaction given by Equation (5.100) proceeds at approximately one-sixth of the rate of the reaction given by Equation (5.99), and as the temperature increases it becomes relatively more important. Equation (5.101) involves an electron capture on the ^{7}Be and the production of a neutrino with energy 0.86 MeV (90% of the time) or 0.38 MeV (10% of the time). The ^{7}Li nucleus immediately captures a proton and produces two nuclei of ^{4}He. This reaction also results in the destruction of any ^{7}Li that may be present at the time of the star's formation, since it becomes effective at temperatures of 2.8×10^6 K or above. Note that the second branch of the pp chains has an average neutrino loss of about 1.1 MeV, but that otherwise the net result is the same — the conversion of four protons to a ^{4}He nucleus.

The third branch of the pp chains (PP III) occurs if a proton, rather than an electron, is captured by ^{7}Be

$$^7\text{Be} + {}^1\text{H} \rightarrow {}^8\text{B} + \gamma \tag{5.103}$$

$$^8\text{B} \rightarrow \text{e}^+ + \nu + {}^8\text{Be} \tag{5.104}$$

$$\rightarrow {}^4\text{He} + {}^4\text{He} \tag{5.105}$$

Less than 0.1% of pp-chain completions in the Sun occur through Equation (5.103) to Equation (5.105); however, they are very important because the neutrino produced has an average energy of 6.7 MeV, high enough to be detectable in all current solar neutrino experiments.

The total rate of energy generation by the pp chains involves a complicated calculation of the rates of all the reactions given above, which are functions of the temperature, density, and concentration of the species involved (Clayton, 1968; Bahcall, 1989). A simplified expression for the energy generation can be obtained under the assumption of equilibrium, that is, the abundances of the intermediate species ^{2}D, ^{3}He, ^{7}Be, and ^{7}Li have reached steady state and the rate for the cycle is controlled by the slowest reaction, which is given by Equation (5.97). If T_6 is the temperature in units of 10^6 K

$$E_{\text{nuc,pp}} = 2.38 \times 10^6 \psi f_{11} g_{11} \rho X^2 T_6^{-2/3} \exp(-33.80 T_6^{-1/3}) \text{ erg g}^{-1} \text{ sec}^{-1}, \tag{5.106}$$

where ρ is the density, X is the fraction of hydrogen by mass, and

$$g_{11} = 1 + 0.0123\, T_6^{1/3} + 0.0109\, T_6^{2/3} + 0.0009\, T_6. \tag{5.107}$$

Neutrino losses have been subtracted. The general expression for the electron screening factor is:

$$f \approx \exp(0.188 Z_1 Z_2 \rho^{1/2} \zeta^{1/2} T_6^{-3/2}), \tag{5.108}$$

where

$$\zeta = \sum_i \frac{Z_i(Z_i + 1)X_i}{A_i} \tag{5.109}$$

in the weak-screening approximation that is appropriate to stars around 1 $M_\odot$ on the main sequence. Here Z_1 and Z_2 are the charges on the two nuclei which are reacting and, in the case of ζ, the sum is over all nuclei with charges Z_i, atomic weights A_i, and abundances by mass X_i. For the proton-plus-proton reaction, $f_{11} \approx \exp(0.27\rho^{1/2} T_6^{-3/2})$ for primordial composition, but, of course, it changes with time as hydrogen is depleted.

Equation (5.106) is based on the reaction rate for the $p+p$ reaction alone (Equation 5.97), which is by far the slowest in the entire chain. The function ψ is a correction factor, between 1 and 2, that accounts for (1) the increase by a factor 2 in energy production rate, with respect to that for PP I, that occurs when the reactions go through PP II or PP III, and (2) the different neutrino energies in the three chains. Thus, $\psi = 1$ corresponds to energy production by PP I alone. For the center of the present Sun, $\psi \approx 1.5$. To avoid the detailed calculation of reaction rates that, in general, is required to obtain ψ, we refer here to a figure in Kippenhahn and Weigert (1990; p. 164) where

ψ is plotted as a function of the temperature and helium abundance. Unless one wishes to calculate a detailed solar model, an approximation to ψ should be adequate. For stars of less than 0.5 $M_\odot$ on the main sequence, where internal temperatures are less than 9×10^6 K, one may set $\psi = 1$. However, in the regime of very-low-mass stars ($< 0.2M_\odot$), intermediate electron screening (Graboske et al., 1973) must be taken into account and the assumption of equilibrium in the pp chains breaks down. The program STELLAR takes the nonequilibrium into account by calculating the reaction rates for Equation (5.97) and Equation (5.99) separately.

Protons can also interact with the CNO nuclei, but because the Coulomb barrier is higher, these reactions become more important than the pp chains only at temperatures above 2×10^7 K. These reactions dominate for masses above about 1.5 $M_\odot$. The reactions are:

$$^{12}\mathrm{C} + {}^{1}\mathrm{H} \rightarrow {}^{13}\mathrm{N} + \gamma \tag{5.110}$$
$$^{13}\mathrm{N} \rightarrow {}^{13}\mathrm{C} + \mathrm{e}^{+} + \nu \tag{5.111}$$
$$^{13}\mathrm{C} + {}^{1}\mathrm{H} \rightarrow {}^{14}\mathrm{N} + \gamma \tag{5.112}$$
$$^{14}\mathrm{N} + {}^{1}\mathrm{H} \rightarrow {}^{15}\mathrm{O} + \gamma \tag{5.113}$$
$$^{15}\mathrm{O} \rightarrow {}^{15}\mathrm{N} + \mathrm{e}^{+} + \nu \tag{5.114}$$
$$^{15}\mathrm{N} + {}^{1}\mathrm{H} \rightarrow {}^{12}\mathrm{C} + {}^{4}\mathrm{He}. \tag{5.115}$$

The net effect is the conversion of four protons into a helium nucleus and the production of two positrons (which annihilate), and two neutrinos, which carry away energies of 0.71 and 1.0 MeV, respectively. The CNO nuclei simply act as catalysts; however their relative abundances change as a result of the operation of the cycle. The reaction rate of Equation (5.110), which uses up ^{12}C, under stellar conditions is about 100 times faster than that of Equation (5.113), which uses up ^{14}N. All other reactions are much faster than these two. The reaction chain tends to reach equilibrium, in which each CNO nucleus is produced as fast as it is destroyed.

The equilibrium is obtained only when most of the ^{12}C and other participating nuclei are converted to ^{14}N. This change in relative abundances could be observed if the layers in which the reactions occur could later be mixed to the surface of the star. To compare with observations, it is important to calculate the abundances of ^{12}C, ^{14}N, and ^{16}O in all layers of the star. A secondary branch of the CNO cycle affects the abundance of ^{16}O. Starting at ^{15}N, the reactions are:

$$^{15}\mathrm{N} + {}^{1}\mathrm{H} \rightarrow {}^{16}\mathrm{O} + \gamma \tag{5.116}$$
$$^{16}\mathrm{O} + {}^{1}\mathrm{H} \rightarrow {}^{17}\mathrm{F} + \gamma \tag{5.117}$$
$$^{17}\mathrm{F} \rightarrow {}^{17}\mathrm{O} + \mathrm{e}^{+} + \nu \tag{5.118}$$
$$^{17}\mathrm{O} + {}^{1}\mathrm{H} \rightarrow {}^{14}\mathrm{N} + {}^{4}\mathrm{He}. \tag{5.119}$$

This branch accounts for only 0.1% of the completions, but the ^{16}O is generated less rapidly than it is converted into ^{14}N by reactions given by Equation (5.117) to Equation (5.119), so that when the branch comes into equilibrium, the abundance ratio O/N will be reduced. The overall effect of the CNO cycle, apart from its energy generation, is the conversion of 98% of the CNO isotopes into ^{14}N. Once the cycle reaches equilibrium, the energy generation can be calculated by the rate of the slowest reaction in the main chain (Equation [5.113]) multiplied by the energy released over the whole cycle (24.97 MeV, after neutrino losses)

$$E_{\mathrm{nuc,CNO}} = 8.67 \times 10^{27} g_{14,1} \rho X X_{\mathrm{CNO}} T_6^{-2/3} \exp(-152.28 T_6^{-1/3}) \text{ erg g}^{-1} \text{ sec}^{-1}, \tag{5.120}$$

where X_{CNO} is the total mass fraction of all CNO isotopes and

$$g_{14,1} = 1 + 0.0027\, T_6^{1/3} - 0.00778\, T_6^{2/3} - 0.000149\, T_6. \tag{5.121}$$

The electron screening factor is not important in this case. At $T_6 \approx 25$, the energy generation goes approximately as T^{17}.

The burning of helium generally takes place at temperatures in the range $1-2\times10^8$ K, well above those for hydrogen burning. Helium burning proceeds through a three-particle reaction known as the triple-α process, which can be represented as

$$^4\mathrm{He} + {}^4\mathrm{He} \leftrightarrow {}^8\mathrm{Be} \tag{5.122}$$

$$^8\mathrm{Be} + {}^4\mathrm{He} \leftrightarrow {}^{12}\mathrm{C} + \gamma + \gamma. \tag{5.123}$$

The amount of energy produced per reaction is 7.275 MeV, or 0.606 MeV per atomic mass unit, a factor 10 less than in hydrogen burning (per amu). An expression for the rate of energy production is:

$$E_{\mathrm{nuc},3\alpha} = 5.09 \times 10^{11} f_{3\alpha} \rho^2 Y^3 T_8^{-3} \exp\left(-44.027 T_8^{-1}\right) \text{erg g}^{-1} \text{ sec}^{-1}, \tag{5.124}$$

where $T_8 = T/10^8$K, Y is the mass fraction of ^{4}He, and the weak electron screening factor (Clayton, 1968) is given approximately by

$$f_{3\alpha} = \exp(2.76 \times 10^{-3} \rho^{1/2} T_8^{-3/2}). \tag{5.125}$$

This reaction has a very steep temperature sensitivity, with $E_{\mathrm{nuc}} \propto T^{40}$ at $T_8 = 1$. A further helium burning reaction, which occurs at slightly higher temperatures than the triple-α process, combines a helium nucleus with a carbon nucleus to produce oxygen

$$^{12}\mathrm{C} + {}^4\mathrm{He} \rightarrow {}^{16}\mathrm{O} + \gamma, \tag{5.126}$$

with an energy production of 7.162 MeV per reaction and an uncertain reaction rate. The energy generation is (Kippenhahn and Weigert, 1990, Section 18.5.2)

$$E_{\mathrm{nuc},12\alpha} \approx 1.3 \times 10^{27} f_{12\alpha} X_C Y h_{12\alpha} \rho T_8^{-2} \exp(-69.2/T_8^{1/3}), \tag{5.127}$$

where X_C is the mass fraction of ^{12}C

$$h_{12\alpha} = \left(\frac{1 + 0.134 T_8^{2/3}}{1 + 0.01 T_8^{2/3}} \right)^2 \tag{5.128}$$

and the electron screening factor

$$f_{12\alpha} = \exp(2.24 \times 10^{-3} \zeta^{1/2} \rho^{1/2} T_8^{-3/2}). \tag{5.129}$$

The program STELLAR includes hydrogen-burning and helium-burning reactions up through Equation (5.126).

5.7 EXAMPLES

5.7.1 EVOLUTION OF THE SUN

An excellent way to test a stellar evolution code is to calculate the evolution of the Sun from the age-zero main sequence to the present time and to compare results with standard models of the present solar interior and the present observed values of $R_\odot$ and $L_\odot$ after an evolution time of 4.56×10^9 yr, the known age of the Sun. For the results shown here, the starting point is a chemically homogeneous model with $X = .71$, $Y = .27$, $Z = .02$ in the pre-main-sequence stage with a radius of about 6 $R_\odot$ at which point there is negligible nuclear burning and the energy source is gravitational contraction. Since a calculation using the Henyey method requires a first guess for all variables x_j^i at the initial time, such a model must be provided. At 6 $\mathrm{R}_\odot$ the model Sun is fully convective and the equation of state is that for an ideal gas, so if the actual convective gradient is close to the adiabatic gradient (which holds in most of the mass), then the relation $P \propto \rho^{5/3}$ should apply. This corresponds to a polytrope of index 1.5, so initial values of P, r, and ρ as a function of mass fraction can be obtained from the numerical solution of such a polytrope. The temperature then follows from the ideal gas law, and the luminosity is assumed to satisfy the relation

$$dL_r/dM_r = E_{\mathrm{grav}} \quad \text{with} \quad E_{\mathrm{grav}} \propto T \tag{5.130}$$

where E_{grav} is the net energy release per unit mass per unit time from contraction. The proportionality follows from the assumption of homologous contraction in the initial model (Schwarzschild, 1958, Section 19). Given the radius and the known $T_{\mathrm{eff}} \approx 4000$ K during the fully convective phase of evolution, the internal luminosities are scaled to the total luminosity $L = 4\pi R^2 \sigma_B T_{\mathrm{eff}}^4$.

Once the main sequence is reached, the physics is relatively simple. The time-dependent terms in the energy equation (5.8) become unimportant, and evolutionary changes occur entirely from the change in composition due to nuclear burning in the central regions. The timescale for such changes is long, and time steps on the order of 10^8 years can be taken. As long as the time step is small enough so that the change in composition in every zone is at most a few percent, then the change in composition can be calculated explicitly after each Henyey time step from Equation (5.23). The nuclear burning takes place in radiative zones, so the convective mixing effects are unimportant. Equation (5.106) can be used for the energy production; nonequilibrium effects, as well as the contribution of the CNO cycle, are not important.

The equation of state is that of an ideal gas; effects of electron degeneracy are small. Opacity tables are available for $Z = 0.02$ and several different (X, Y) combinations. Some complications occur in the very outer layers, where the assumption of complete ionization no longer holds and where the convective temperature gradient is superadiabatic. The layers in which these effects are important are thin enough so that they can be dealt with in an outer-layer atmosphere integration (see Section 5.4). If the atmospheric layers are not correctly treated, the radius of the model will not match that of the Sun after an evolution time of 4.56×10^9 yr. The key parameter in determining this match is α, the ratio of the mixing length to the pressure scale height in the superadiabatic layers of the atmosphere. Current solar models give a value $\alpha \approx 1.75$. In most cases, this parameter is assumed to be constant as a function of time.

Current state-of-the-art solar models include very detailed physics for the equation of state, nuclear reaction rates, and opacity; they even include the very long-term effect of the diffusion of helium downward in the Sun. This detail is necessary to provide an accurate calculation of the expected rate of production of neutrinos from the three branches of the pp chains, for the purpose of comparing with detection rates of neutrinos in experiments on Earth. These experiments now indicate that the "standard solar model" does, in fact, predict the correct number of neutrinos, and it also agrees well with other observed properties of the Sun, including the sound speed as a function of depth that is derived from helioseismology (involving the precise measurements of solar oscillations). However, a rather good solar model can be obtained even with the approximations outlined above.

Results for a solar model are shown in Figure 5.3 and Figure 5.4. During the evolution the central regions contract, reaching a central density of 1.48×10^2 g cm^{-3}, and heat to a central temperature of 1.56×10^7 K. The hydrogen mass fraction in the center has been reduced to 0.34, and most of the nuclear energy is produced in the inner 25% of the mass. The outer radius and luminosity increase slowly with time. The Sun has an outer convection zone that extends down to a radius of 0.72

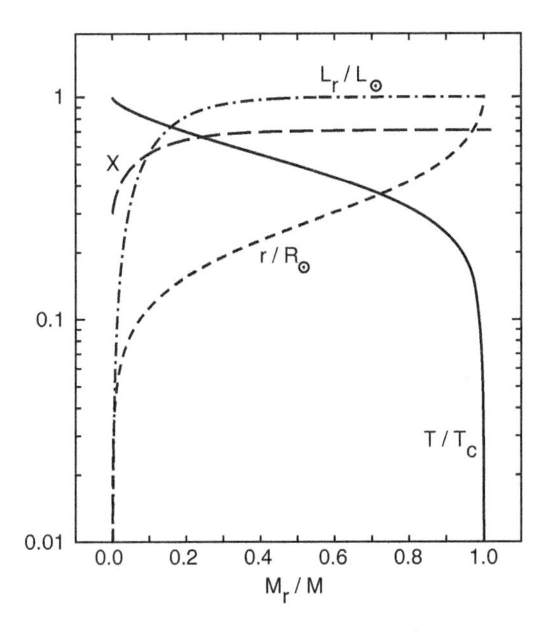

FIGURE 5.3 Solar model: various quantities are plotted as a function of the fractional mass M_r/M. The solid line gives the temperature, normalized to the central value of 1.56×10^7 K. The short dashed line gives the radius, normalized to the total radius of 7×10^{10} cm. The long dashed line gives the hydrogen mass fraction. The dashed dotted line gives the luminosity, normalized to the surface value of 3.85×10^{33} erg s^{-1}.

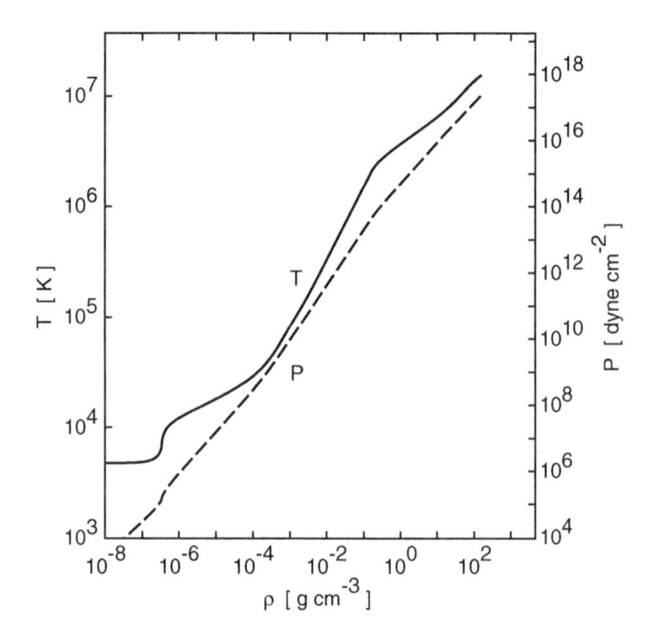

FIGURE 5.4 Solar model: the pressure and temperature distributions from surface to center are plotted as a function of density.

$R_\odot$; its lower boundary is evident in Figure 5.4 where the temperature curve changes slope at about $\log \rho = -1$.

5.7.2 Age Determination for a Star Cluster

Although the age of the solar system (and, therefore, presumably the Sun) can be determined accurately from radioactive dating of the oldest moon rocks and meteorites; the ages of other stars cannot be determined directly. One important indirect method, based on the theory of stellar evolution, compares the theoretical and observed positions of stars in the Hertzsprung–Russell (H–R) diagram of a star cluster. The method can be used, for example, to estimate the ages of the oldest objects in our galaxy, the globular clusters.

The stars in a cluster are assumed to be all of the same age, to have the same composition, and, except for the very nearest clusters, to all lie at the same distance from the Earth. The observed cluster diagram is compared with a theoretical line of constant age, known as an *isochrone*, obtained by calculating the evolution in the H–R diagram of a set of stars with different masses, but the same composition, and connecting points on their evolutionary tracks that correspond to the same elapsed times since formation. This procedure is illustrated in Figure 5.5, which shows how the age is determined for the Hyades, a nearby galactic cluster with a distance of 46.3 pc. The plot shows the positions of a number of single stars, which are known to be members of the cluster, in terms of the absolute visual magnitude and color index $B - V$. The distance and, therefore, the actual luminosities of the stars are well

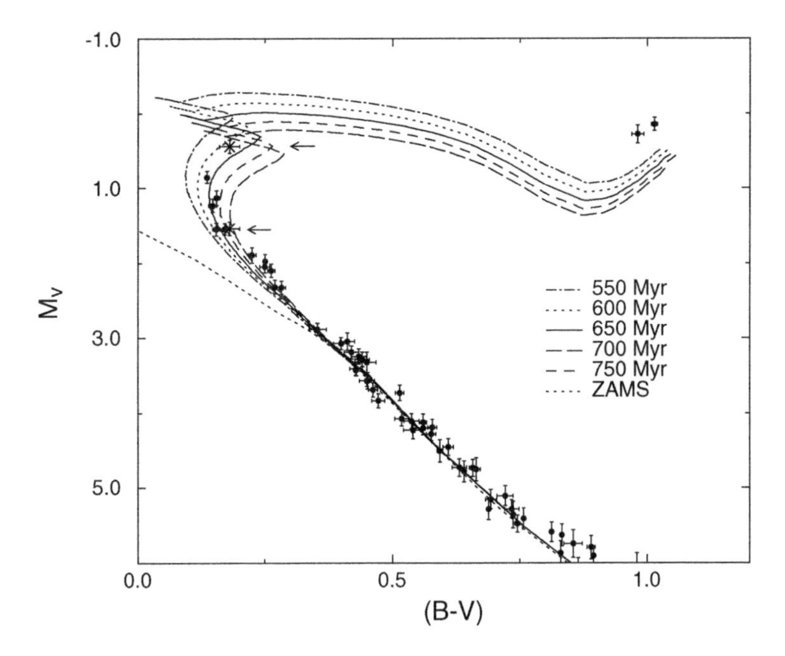

FIGURE 5.5 A Hertzsprung–Russell (H–R) diagram for the Hyades cluster. The luminosity is given in terms of the absolute visual magnitude M_v, and the observed color $(B - V)$ is an indicator of surface temperature (increasing to the left). The filled circles with error bars correspond to the observations of the individual single stars, whose distances have been determined from Hipparcos parallaxes. The symbols with arrows correspond to the two components of a spectroscopic binary. The dotted curve corresponds to the theoretical zero-age main sequence for the composition of this cluster. "Zero-age" refers to the time for a given mass when hydrogen burning first supplies the entire energy of the star. Other curves are theoretical isochrones, that is, predictors of how the H–R diagram should look at the indicated ages. (From Perryman, M. et al. (1998). *Astron. Astrophys.* **331**: 81. With permission.)

determined in this case from observations by the Hipparcos satellite. The composition is also relatively well measured; the cluster is slightly more metal-rich than the Sun. The theoretical evolutionary tracks are determined from models with the Hyades composition, and they produce H–R diagrams in terms of log $T_{\rm eff}$ and log $L/L_\odot$. These quantities must be converted to $B - V$ and M_v, respectively, to carry out the comparison.

The conversion can be done fairly accurately for the Hyades, but for metal-poor objects, such as globular clusters, it introduces uncertainties. The age is determined from the positions in the H–R diagram of the stars near the "turnoff" point, usually identified as the point of highest effective temperature on the main sequence. The time spent by a star on the main sequence decreases with increasing mass; thus, stars with masses above that corresponding to the turnoff mass have already left the main sequence and have evolved to the red giant region. In this case, the age is determined to be 625 Myr (million years), with an error of $\pm$ 50 Myr. The same procedure, of fitting isochrones from stellar evolutionary tracks to the observed

Hertzsprung–Russell diagram near the turnoff, can be used to age-date globular clusters and, thereby, to determine the approximate age of the galaxy. The reliability of such ages is not so much dependent on the numerical procedure, but rather on the quality of the physics that is incorporated in the calculations, including opacities and convection theory, and on the accuracy of the transformation from observed colors and magnitudes to luminosity and effective temperature.

REFERENCES

Bahcall, J. N. (1989) *Neutrino Astrophysics* (Cambridge: Cambridge University Press).

Baraffe, I., Chabrier, G., Allard, F. and Hauschildt, P. H. (1998) *Astron. Astrophys.* **337**: 403.

Böhm-Vitense, E. (1958) *Zeitschr. f. Astrophys.* **46**: 108.

Chandrasekhar, S. (1939) *An Introduction to the Study of Stellar Structure* (Chicago: University of Chicago Press).

Clayton, D. D. (1968) *Principles of Stellar Evolution and Nucleosynthesis* (New York: McGraw-Hill).

Eggleton, P. P. (1971) *Mon. Not. R. Astron. Soc.* **151**: 351.

Ferguson, J. W., Alexander, D. R., Allard, F., Barman, T., Bodnarik, J. G., Hauschildt, P. H., Heffner-Wong, A. and Tamanai, A. (2005) *Astrophys. J.* **623**: 585.

Graboske, H. C., DeWitt, H. E., Grossman, A. S. and Cooper, M. S. (1973) *Astrophys. J.* **181**: 457.

Henyey, L. G., Forbes, J. E. and Gould, N. L. (1964) *Astrophys. J.* **139**: 306.

Hubbard, W. B. and Lampe, M. (1969) *Astrophys. J. Suppl.* **18**: 297.

Iglesias, C. and Rogers, F. (1996) *Astrophys. J.* **464**: 943.

Kippenhahn, R. and Weigert, A. (1990) *Stellar Structure and Evolution* (Berlin: Springer).

Kippenhahn, R., Weigert, A. and Hofmeister, E. (1967) *Methods in Computational Physics* vol 7, B. Alder, S. Fernbach, and M. Rotenberg, Eds. (New York: Academic Press) p. 129.

Ludwig, H.-G., Freytag, B. and Steffen, M. (1999) *Astron. Astrophys.* **346**: 111.

McDougall, J. and Stoner, E. C. (1939) *Phil. Trans. Roy. Soc.* **237**: 67.

Mihalas, D. (1978) *Stellar Atmospheres,* 2nd ed. (San Francisco: W. H. Freeman).

Perryman, M., Brown, A., Lebreton, Y., Gómez, A., Turon, C. et al. (1998) *Astron. Astrophys.* **331**: 81.

Schwarzschild, M. (1958) *Structure and Evolution of the Stars* (Princeton: Princeton University Press).

Seaton, M. J., Yan, Y., Mihalas, D., and Pradhan, A. K. (1994) *Mon. Not. R. Astron. Soc.* **266**: 805.

Stein, R. F. and Nordlund, A. (1998) *Astrophys. J.* **499**: 914.

6 Grid-Based Hydrodynamics

In contrast to the basic stellar structure problem outlined in the previous chapter, in many problems in astrophysics the assumption of spherical symmetry is not adequate to describe the situation nor is the assumption of hydrostatic equilibrium. For example, the expansion of a supernova remnant into a nonuniform interstellar medium, the collapse of an interstellar cloud and its breakup into a binary system, the evolution of a disk that develops spiral waves as a result of gravitational instability, or the development of structure in the early universe after the Big Bang — all these problems require a three-dimensional hydrodynamic treatment. For some problems the situation can be reduced down to two space dimensions and the numerical techniques are the same, so in order to simplify matters this chapter concentrates on two-dimensional solutions to hydrodynamic problems on a grid. The methods discussed here can be generalized to three-dimensional problems in a straightforward manner.

Grid methods and particle methods, such as SPH (see Chapter 4), each have their own advantages and disadvantages, and it is often useful, in a complicated problem, to attempt a solution by both methods to compare results. In principle, since the equations solved are the same, a particle method should give exactly the same result as a grid method, but in practice the two methods unavoidably have different resolutions and different types of numerical inaccuracies as a function of position and time in a simulation, so the detailed results can be somewhat different. Whether the primary emphasis for a given problem should be on a grid-based method or a particle-based method depends on a number of considerations, including computer time requirements, the numerical resolution that is necessary, how the intrinsic numerical effects in the code would affect the solution, and whether or not shocks play an important role. Grid methods are considered preferable to particle methods for the treatment of shocks.

The core of this chapter is Sections 6.3 and 6.5, which discuss general grid-based techniques for Eulerian hydrodynamics and how they are applied in one particular code, the ZEUS 2-D code. The 3-D version of this code is provided on the CD that accompanies this book. However, an important aspect of any hydrodynamic calculation is the possible development of shocks. Section 6.1 introduces the basic properties of shocks and the Rankine–Hugoniot relations that describe the conservation of mass, momentum, and energy across a shock. The numerical treatment of shocks, on the one hand by a Riemann solver and on the other hand by the artificial viscosity technique, is then discussed. Section 6.2 illustrates the application of the artificial viscosity technique in the context of a one-dimensional Lagrangian code. However, in two and three space dimensions on a grid, Eulerian techniques rather than Lagrangian are highly preferred. It usually happens, however, that small regions of the flow require much higher numerical resolution than average. The method for treating such problems is discussed in Section 6.4. Section 6.6 treats problems in which a pseudo-three-dimensional solution can be obtained, under the assumption of certain symmetries, by the use of a two-dimensional code. Finally, in Chapter 6, Section 6.7, examples are presented on the results of two- and three-dimensional Eulerian grid-based simulations.

6.1 FLOW DISCONTINUITIES AND HOW TO HANDLE THEM

The solutions of the equations of hydrodynamics and magnetohydrodynamics are often discontinuous. As we have shown in Chapter 2, Section 2.7, such solutions cannot be adequately represented on a grid. Therefore, whenever they appear, a "special treatment" must be applied. The treatment of discontinuities is the key ingredient of every numerical scheme employed to simulate fluid flow (hereafter, such schemes will be referred to as *hydrocodes*). There are two basic kinds of noncontinuous features encountered in astrophysical flows: *shock fronts* (also called *shock waves* or *shocks*) and *contact discontinuities*. To define them precisely, let us first split the velocity vector into two components, v_n and v_t, oriented, respectively, perpendicularly and tangentially to the surface of discontinuity. A shock front is a surface across which P, ρ, v_n, and T change abruptly, whereas v_t remains continuous. A contact discontinuity is marked by abrupt changes in ρ, T, or v_t, but it preserves the smoothness of v_n and P. An example of a shock wave is a supernova envelope propagating supersonically into the interstellar gas; an example of a contact discontinuity is a sharp interface between a cold, dense region in the interstellar gas and an adjacent warm, lower density region at the same gas pressure.

Contact discontinuities are always related to initial and/or boundary conditions. As they do not generate jumps in the basic dynamical variables v_n and P, from the practical point of view, they are less challenging than the shocks. Problems they generate are usually limited to excessive numerical diffusion and, as such, they can be cured by increasing grid resolution and/or employing a higher-order scheme. As opposed to contact discontinuities, shocks may develop even if the initial and boundary conditions are perfectly smooth, as a result of a particular property of sound waves, which we shall discuss in the next section. Moreover, unless preventive measures are taken, the shocks will trigger self-amplifying instabilities, which are able to kill the simulation within just a few time steps.

6.1.1 STEEPENING OF SOUND WAVES

Consider a one-dimensional flow along the x-axis, assuming that the medium is originally not uniform in density and that the pressure p is a power-law function of density ρ. The evolution of the medium is governed by Equations (2.59) supplemented by the *polytropic relation*

$$P = K\rho^{\gamma}, \tag{6.1}$$

where K is the *polytropic constant*. In the standard notation

$$\gamma = \frac{n+1}{n} \geq 1, \tag{6.2}$$

where n is the *polytropic index*. Note that the ideal monoatomic gas with ϵ and P given, respectively, by Equation (1.14) and Equation (1.43) behaves like a polytropic medium with $\gamma = 1$ when the change it undergoes is isothermal, and like a polytropic medium with $\gamma = 5/3$ when the change it undergoes is adiabatic.

Exercise

Show that when Equation (6.1) is valid, then

1. Low amplitude sound waves propagate with a velocity

$$c_s = \sqrt{\gamma P/\rho}. \tag{6.3}$$

   Hint: follow the derivation of Equation (2.63).
2. The internal energy density and pressure of a nonradiating and nonconducting medium are related by a simple formula

$$P = (\gamma - 1)e. \tag{6.4}$$

   Hint: make use of the first law of thermodynamics written for 1 g of the medium.

Assume that ρ and v are single-valued functions of x and t. We may write

$$\begin{aligned} \frac{\partial \rho}{\partial t} + \frac{\partial(\rho v)}{\partial \rho}\frac{\partial \rho}{\partial x} &= 0 \\ \frac{\partial v}{\partial t} + \left(v + \frac{1}{\rho}\frac{\partial P}{\partial v}\right)\frac{\partial v}{\partial x} &= 0. \end{aligned} \tag{6.5}$$

The standard relation

$$d\rho = \frac{\partial \rho}{\partial x}dx + \frac{\partial \rho}{\partial t}dt \tag{6.6}$$

yields

$$\left.\frac{\partial x}{\partial t}\right|_\rho = -\frac{\frac{\partial \rho}{\partial t}}{\frac{\partial \rho}{\partial x}}. \tag{6.7}$$

From this, and an analogous relation for v, we get

$$\begin{aligned} \left.\frac{\partial x}{\partial t}\right|_\rho &= \frac{\partial(\rho v)}{\partial \rho} = v + \rho\frac{\partial v}{\partial \rho} \\ \left.\frac{\partial x}{\partial t}\right|_v &= v + \frac{1}{\rho}\frac{\partial P}{\partial v}. \end{aligned} \tag{6.8}$$

Assume further that there is a range of x and t in which ρ is a single-valued function of v (such an assumption is reasonable as long as the flow is smooth). Therefore,

$$\left.\frac{\partial x}{\partial t}\right|_\rho \equiv \left.\frac{\partial x}{\partial t}\right|_v \tag{6.9}$$

and

$$\rho\frac{dv}{d\rho} = \frac{1}{\rho}\frac{dP}{dv} = \frac{c_s^2}{\rho}\frac{d\rho}{dv}, \tag{6.10}$$

where the derivative of the polytropic relation (6.1) with respect to v and the formula for the sound speed from the last exercise were employed to obtain the final equality. Equation (6.10) is equivalent to

$$\frac{dv}{d\rho} = \pm\frac{c_s}{\rho}, \tag{6.11}$$

which, substituted into the first line of Equation (6.8), results in

$$\left.\frac{\partial x}{\partial t}\right|_\rho = v \pm c_s \tag{6.12}$$

or

$$\left.\frac{\partial x}{\partial t}\right|_v = v \pm c_s \tag{6.13}$$

Equation (6.13) has a general solution

$$x(v) = t \cdot (v \pm c_s(v)) + f(v), \tag{6.14}$$

where f is an arbitrary function of v. Consider the point $x(v)$, at which the velocity of the medium is equal to v. From Equation (6.14) it follows that its location on the x-axis changes at a velocity $v \pm c_s$. These changes may be thought of as resulting from the flow at a velocity v with respect to the reference frame with a perturbation propagating at c_s with respect to the medium. Because ρ is a single-valued function of v, exactly the same equation describes the motion of the point where the density of the medium is $\rho(v)$. Points with different densities move at different velocities, and the density profile $\rho(x)$ changes its shape. Differentiation of $v \pm c_s$, with respect to ρ, yields

$$\frac{d}{d\rho}(v \pm c_s) = \pm\frac{c_s}{\rho} \pm \frac{dc_s}{d\rho} = \pm\frac{c_s}{\rho}\left(1 + \frac{d\ln c_s}{d\ln\rho}\right) = \pm\frac{c_s}{2\rho}(1+\gamma), \tag{6.15}$$

leading to the very important conclusion: Propagation velocity *increases* with increasing ρ if the perturbation propagates *downstream* (i.e., in the direction of the flow) and *decreases* with increasing ρ if the perturbation propagates *upstream* (i.e., in the direction opposite to the flow). As a result, the density profile tends to steepen downstream and flatten upstream (Figure 6.1). Finally, the downstream-facing density profile evolves into a discontinuity (i.e., a shock front is formed). This behavior has a convincing, although not entirely precise, interpretation. In a wave composed of density maxima and minima, the velocity of sound is higher in the maxima, and the maxima propagate faster than the minima and eventually catch up with the minima. From the formal point of view, the steepening of the profile is inevitable, no matter how small the density amplitude and how tiny the associated difference in sound velocity. In the real world, the steepening may be entirely quenched by dissipative processes.

According to Equation (6.12) to Equation (6.15), the maximum velocity with which information may propagate across the grid is $|v| + c_s$. Therefore, as we already

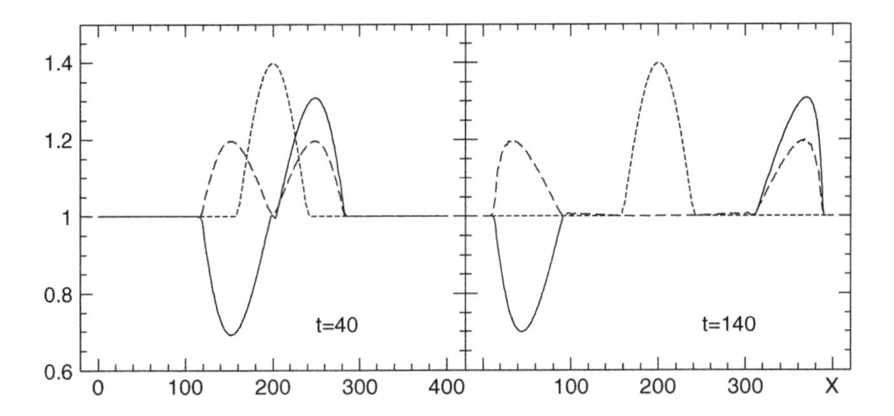

FIGURE 6.1 Propagation of large-amplitude sound waves followed with the help of Equation (2.59) and Scheme (6.41) adapted to Cartesian coordinates. An initial 40% pressure perturbation (*dotted line*) produces two waves, propagating toward the left and the right. In each wave the leading (*downstream*) part of the profile steepens with time, while the trailing (*upstream*) part flattens. Units of pressure and velocity on the vertical scale are arbitrary. *Dashed curves* and *solid curves:* pressure and velocity distributions, respectively, at two later times.

indicated at the end of Chapter 2, Section 2.5, explicit schemes dealing with the full set of hydrodynamical equations will have the time step limited by the condition

$$\Delta t \leq \frac{\Delta x}{|v| + c_s} . \tag{6.16}$$

A von Neumann-type analysis, which we shall not repeat here, shows that this is indeed the case (Potter, 1973).

6.1.2 Rankine--Hugoniot Conditions

Conservation laws imply useful relations between preshock and postshock parameters of the medium, known as the *Rankine–Hugoniot conditions*. Consider a steady one-dimensional flow with a shock front, in a reference frame co-moving with the shock (Figure 6.2b). Assume that all external forces vanish, and neglect any effects of radiation transfer or conduction. Let u be the velocity of the fluid with respect to the shock. From the first two lines of Equation (1.42) and from Equation (1.36), it follows (see Landau and Lifschitz, 1959) that on both sides of the shock the fluxes of mass ρu, momentum $\rho u^2 + P$, and total energy $u(\frac{1}{2}\rho u^2 + e + P)$ must have the same values

$$\begin{aligned} \rho_2 u_2 &= \rho_1 u_1 \\ \rho_2 u_2^2 + P_2 &= \rho_1 u_1^2 + P_1 \\ u_2\left(\frac{1}{2}\rho_2 u_2^2 + e_2 + P_2\right) &= u_1\left(\frac{1}{2}\rho_1 u_1^2 + e_1 + P_1\right). \end{aligned} \tag{6.17}$$

Obviously, the preshock flow must be supersonic: $u_1 > c_{s,1}$, where $c_{s,1}$ is the preshock sound velocity (otherwise the discontinuity would diffuse due to sound

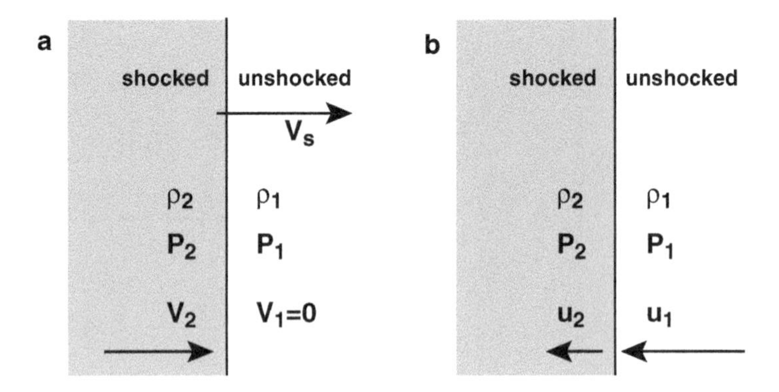

FIGURE 6.2 Schematic view of a one-dimensional shock front. Because of the pressure difference $P_2 - P_1 > 0$, the shock propagates to the right at a velocity v_s in the laboratory frame (*a*), sweeping through the unshocked medium at rest. In the reference frame co-moving with the shock (*b*), the low-pressure and high-velocity preshock medium enters the shock from the right to be processed into the high-pressure and low-velocity postshock medium. The velocities v_1, v_2 refer to gas velocities in the fixed (lab) frame, while u_1, u_2 refer to gas velocities in the frame of reference that moves with the shock front. Note that $u_1 = -v_s$.

waves propagating into the preshock medium). Given the preshock parameters ρ_1, u_1, e_1, P_1 and the equation of state, then with the help of Equation (6.17) one can compute ρ_2, u_2, e_2, and P_2. It is important to remember that, upon crossing the shock, the fluid dissipates part of its kinetic energy into heat (even when the flow is nonviscous), i.e., there is an entropy jump across a shock. The frequently used term *adiabatic shock*, therefore, is a misnomer, since the shock itself may be viewed as a source of heat. When talking about adiabatic shocks, one usually means shocks in a medium that does not radiate or conduct heat.

Exercise

Based on the Rankine–Hugoniot conditions, show that for the polytropic medium the following relation holds

$$\frac{\rho_2}{\rho_1} = \frac{P_2 + m^2 P_1}{P_1 + m^2 P_2}, \tag{6.18}$$

where

$$m^2 = \frac{\gamma - 1}{\gamma + 1}.$$

From the exercise, we see that for $\gamma = 5/3$ the compression ratio ρ_2/ρ_1 cannot exceed 4 even if the shock is extremely strong (i.e., if $P_2 \gg P_1$). On the other hand, for $\gamma = 1$ the compression ratio may become arbitrarily high. The first case corresponds to an "adiabatic" shock in an ideal, nonradiating, and nonconducting monoatomic gas, while the second corresponds to a shock in a monoatomic gas able to instantaneously radiate away all the dissipated energy. Such a shock is usually referred to as *isothermal*.

A shock is commonly characterized by its *Mach number*

$$\mathcal{M} = \frac{|u_1|}{c_{s,1}} = \frac{v_s}{c_{s,1}}, \tag{6.19}$$

where, again, v_s is the shock propagation velocity in the fixed frame. When $\mathcal{M}$ and preshock parameters are known, all postshock parameters can be calculated from the following relations

$$\frac{\rho_2}{\rho_1} = \frac{(\gamma+1)\mathcal{M}^2}{(\gamma-1)\mathcal{M}^2+2}$$
$$\frac{P_2}{P_1} = \frac{2\gamma\mathcal{M}^2-(\gamma-1)}{\gamma+1}. \tag{6.20}$$

Note that v_s is *different* from the velocity of the postshock medium, v_2: since $u_1 = -v_s$ and $u_2 = v_2 - v_s$, we have

$$v_s = v_2 \frac{\rho_2}{\rho_2 - \rho_1}. \tag{6.21}$$

From Equation (6.21) it follows that only in very strong, nearly isothermal shocks, in which $\rho_2 \gg \rho_1$, does v_s approach v_2. In general, it is very important to note the distinction between these two velocities.

Exercise

Show that, in the reference frame of the shock, the postshock medium has a Mach number

$$\mathcal{M}_2^2 = \frac{(\gamma-1)\mathcal{M}^2+2}{2\gamma\mathcal{M}^2-(\gamma-1)}. \tag{6.22}$$

The important conclusion of the exercise is that in the reference frame of the shock the postshock flow is subsonic.

6.1.3 Shock Tube and Riemann Problem

We now introduce the *Riemann problem* and show how it is related to the Rankine–Hugoniot conditions. A procedure for solving the Riemann problem (often referred to as the *Riemann solver*) is the backbone of many modern hydrocodes (see Section 6.3.3). The practical illustration of the Riemann problem is the *shock tube*, which, at the same time, nicely illustrates the meaning and behavior of a contact discontinuity.

To define the shock tube, imagine two large volumes of fluid, initally at rest, separated by a diaphragm located at x_0. Let the pressure and density of the fluid to the left and to the right of the diaphragm be, respectively, (P_1, ρ_1) and (P_5, ρ_5), where $P_1 > P_5$. Let v be the velocity measured in the reference frame in which the fluid

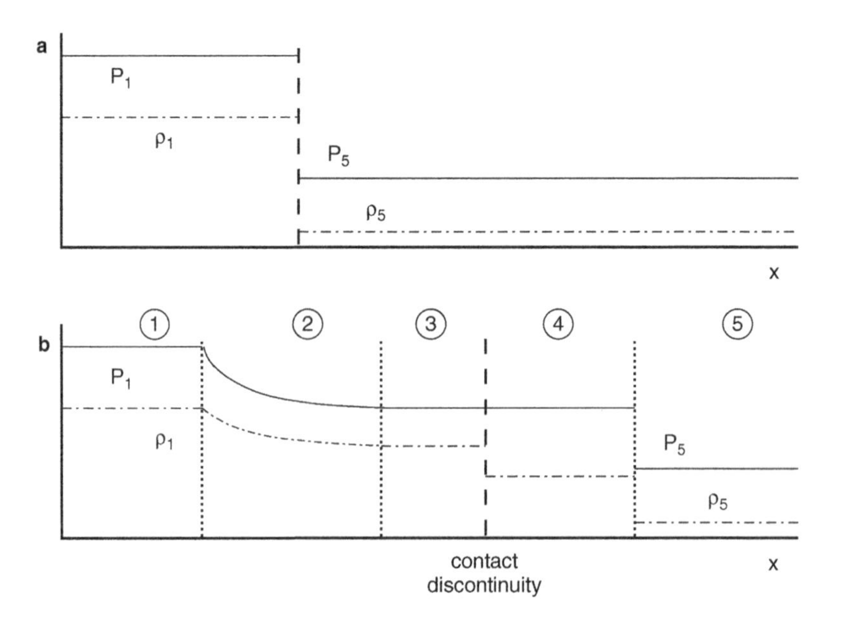

FIGURE 6.3 A shock tube. When the diaphragm (*vertical dashed line in upper panel*) is removed, the initial state with two fluids at rest (a) evolves into a flow pattern with five regions (b). The quantities P, ρ, and v in regions 2 to 4 can be calculated by solving the Riemann problem.

is originally at rest. At $t = 0$ the diaphragm is removed, and the "left" fluid begins to expand to the right, setting the "right" fluid into motion. It can be experimentally verified that a flow pattern quickly emerges, in which five distinct regions may be identified. From left to right, they are (1) undisturbed left fluid, (2) expanding left fluid, (3) decompressed left fluid, (4) compressed right fluid, and (5) undisturbed right fluid (Figure 6.3). Regions (3) and (4) are separated by the contact discontinuity marking the position at which the two fluids meet. Regions (4) and (5) are separated by a shock if the flow is supersonic.

The Riemann problem is defined as follows: Given P_1, ρ_1, P_5, and ρ_5 in regions (1) and (5), find the P, ρ, and v distributions in the remaining three regions. The practical purpose of such calculations will be explained in Section 6.3.3 and Section 6.3.4. Here we shall only note that the Riemann problem is admittedly one-dimensional, but a multidimensional flow may be split into one-dimensional flows along rows of grid cells. Once such splitting is performed, in every one-dimensional flow the solution of the Riemann problem provides an accurate description of the interactions between fluids from every pair of adjacent cells. The physical picture is that in each zone, the density and pressure are constant, and there is a discontinuity (which may be small) between each pair of zones. For the length of the time step, every pair may be approximated by a shock tube with the diaphragm located at the boundary between the cells. The diaphragm is assumed to be removed at the beginning of the time step. Note that even if the flow is smooth, with no shocks, the Rankine–Hugoniot conditions apply.

Exercise

Based on the Rankine–Hugoniot conditions, show that P_4 and v_4 obey the equation

$$v_4 = (P_4 - P_5)\sqrt{\frac{1 - m^2}{\rho_5(P_4 + m^2 P_5)}}. \tag{6.23}$$

After somewhat lengthy calculations, it may also be shown that in region (2) the following relation holds (see, e.g., Hawley et al., 1984):

$$v_2 = \sqrt{\frac{(1 - m^4)P_1^{1/\gamma}}{m^4 \rho_1}} \left(P_1^{\frac{\gamma-1}{2\gamma}} - P_2^{\frac{\gamma-1}{2\gamma}} \right). \tag{6.24}$$

On a (P, v) plane Equation (6.23) and Equation (6.24) define two curves representing, respectively, all possible states of the fluid in regions (2) and (4). There is no flow across the contact discontinuity, i.e., $v_3 = v_4$, and the pressure in regions (3) and (4) is uniform, i.e., $P_3 = P_4$. Therefore, the curves intersect at a point where the expansion flow of the "left" fluid in (2) and (3) smoothly matches the postshock flow of the "right" fluid in (4), so that P_3 may be calculated from

$$(P_3 - P_5)\sqrt{\frac{1 - m^2}{\rho_5 \left(P_3 + m^2 P_5\right)}} = \sqrt{\frac{(1 - m^4)P_1^{1/\gamma}}{m^4 \rho_1}} \left(P_1^{\frac{\gamma-1}{2\gamma}} - P_3^{\frac{\gamma-1}{2\gamma}} \right). \tag{6.25}$$

Once P_3 is known, either Equation (6.23) or Equation (6.24) yields v_3. Next, ρ_4 and ρ_3 can be calculated from Equation (6.18) and Equation (6.1), respectively. Note that in regions (3) and (4) pressures are equal, but densities are different, although the equation of state (Equation 6.1) suggests that they should be equal, too. The reason is that upon crossing the shock front the fluid heats up, and as a result the values of the polytropic constant K in regions (3) and (4) become different. To complete the solution of the Riemann problem, we have to find the functions $v(x, t)$ and $\rho(x, t)$ describing the flow in region (2). Again, we shall only give the final formulae here, referring the reader interested in their derivation to the paper by Hawley et al. (1984). The formulae read

$$v(x, t) = (1 - m^2)\left(\frac{x}{t} - c_{s,1}\right)$$

$$\rho(x, t) = \rho_3 \left(1 + \frac{\gamma - 1}{2}\frac{v(x, t) - v_3}{c_{s,3}}\right)^{\frac{2}{\gamma-1}}, \tag{6.26}$$

where $c_{s,1}$ and $c_{s,3}$ stand for the velocity of sound in regions (1) and (3), respectively. Note that the contact discontinuity advances at a velocity v_4, which is lower than the shock propagation velocity v_s (Equation 6.21). Therefore, the distance between the contact discontinuity and the shock increases in time as the postshock medium is "piled up" between the two. Note also that on the one hand, the very nature of Riemann solvers makes them ideal tools for dealing with discrete (i.e., noncontinuous) distributions of the fluid on a numerical grid. In particular, they allow the simulation to precisely trace shock fronts, which in astrophysical flows are the rule rather than the exception. However, on the other hand, the exact solution of the Riemann problem is so complicated that many hydrocodes employ approximate Riemann solvers (see, e.g., Toro, 1999). A still simpler, but also less accurate, approach to the problem of shock propagation across the grid will be discussed in the following section.

6.1.4 Artificial Viscosity

Physically a shock is not infinitely sharp, but is spread out over a few particle mean free paths. However, this distance is generally far too small to be resolved on a typical grid. Numerically, for reasons explained in Chapter 2, Section 2.7, a shock has to be smeared over a few grid cells. If the shock is too steep, wavelengths too short to be properly propagated across the grid will dominate the Fourier expansion of the solution, and the scheme will break down or it will excite unwanted postshock oscillations. To prevent those possibilities, one introduces *artificial viscosity*; even Riemann solvers may need it, although at a relatively low level. It may appear either explicitly as a viscous term in the momentum equation, or as a procedure increasing the internal viscosity of the code (see Section 6.3.2). In both cases, its sole purpose is the same — to make the flow more diffusive wherever and whenever it proves to be necessary. In the following, we shall concentrate on the explicit approach. A well-implemented artificial viscosity:

- Should be operational in the vicinity of shock fronts only (we do not want to make the whole flow artificially viscous).
- Should not broaden shock fronts too strongly (they should remain as similar to discontinuities as possible).
- Should preserve the continuity of mass, momentum and energy fluxes across the shock, i.e., fulfill the shock jump conditions (Equation 6.17).

A simple and yet very effective prescription for artificial viscosity was given by von Neumann and Richtmyer (1950), who proposed to supplement the standard hydrodynamic equations with an *artificial pressure* term

$$Q = \begin{cases} q^2\rho(\Delta x)^2 \left|\frac{\partial v}{\partial x}\right|^2 & \text{if } \frac{\partial v}{\partial x} < 0 \\ 0 & \text{if } \frac{\partial v}{\partial x} > 0 \end{cases}, \tag{6.27}$$

where q is the dimensionless *artificial viscosity parameter* whose value has to be determined for every scheme separately by numerical experiments (in most cases $0.05 \leq q \leq 2$). The discretized version of Equation (6.27) reads

$$Q = q^2\rho(\Delta x)^2 \left|\frac{\Delta v}{\Delta x}\right|^2 = q^2 \Delta x |\Delta v| \rho \left|\frac{\Delta v}{\Delta x}\right|, \tag{6.28}$$

where Δv is the velocity change across Δx.

The general strategy then is, in the Eulerian hydrodynamic equations (1.42) or the Lagrangian equations (1.62), to replace $\partial P/\partial x_i$ by $\partial(P+Q)/\partial x_i$ in the momentum equation and to replace P in the energy equation by $P + Q$. Thus, we effectively introduce a diffusive term into the momentum equation and convert kinetic energy into heat, as must physically occur in a shock front. As a result, steep velocity gradients are smeared into gentler "ramps." A specific example is given in the next section.

The quantity

$$\nu_{av} = q^2 \Delta x |\Delta v| \tag{6.29}$$

is called the *artificial viscosity coefficient* (note that its physical unit is cm^2 s^{-1}, the same as that of the kinematic viscosity coefficient). It becomes large wherever the velocity abruptly jumps across a grid cell; it also grows larger on a coarser grid (i.e., if Δx increases). As defined by Equation (6.27), Q differs from 0 only when the medium contracts ($\frac{\partial v}{\partial x} < 0$), and expanding regions of the flow are not affected.

In Chapter 2, Section 2.7, we demonstrated how harmful the internal (numerical) viscosity of a scheme can be. Now we introduce artificial viscosity, and a skeptical reader might wonder whether that makes sense. The answer is simple: Internal viscosity is not only harmful, but also uncontrollable. Artificial viscosity is useful and controllable; it effectively operates in shock fronts only, leaving the rest of the flow practically unaffected. It simulates the actual dissipation that occurs in shock fronts, where kinetic energy of ordered motion is converted into internal energy. To appreciate its usefulness, let us see how it influences the flow on various length scales.

A quantitative indicator of the importance of viscosity on a length scale l is the *Reynolds number*

$$\mathcal{R} = \frac{lv}{\nu}, \tag{6.30}$$

which also may be thought of as the ratio of viscous timescale $\tau_\nu = l^2/\nu$ and dynamical timescale $\tau_{dyn} = l/v$. Large $\mathcal{R}$ means that on a given l many dynamical timescales are needed before any effects of diffusion become visible, i.e., that the flow is practically nonviscous on l.

Consider a wave of length λ. Its propagation across the grid is described by the associated Reynolds number

$$\mathcal{R}_\lambda = \frac{\lambda |\Delta v|_\lambda}{q^2 \Delta x |\Delta v|} = \frac{n_\lambda |\Delta v|_\lambda}{q^2 |\Delta v|}, \tag{6.31}$$

where $n_\lambda \equiv \lambda/\Delta x$. $|\Delta v|_\lambda$ is the average amplitude of velocity variations on scale λ, and $|\Delta v|$ is the same amplitude on scale Δx. At short wavelengths

$$n_\lambda \longrightarrow 1, \quad |\Delta v|_\lambda \longrightarrow |\Delta v|, \quad \text{and} \quad \mathcal{R}_\lambda \longrightarrow \frac{1}{q^2}, \tag{6.32}$$

which means that short waves with λ comparable to the grid resolution limit are strongly damped. On the other hand, if λ is large ($n_\lambda \gg 1$) and the flow is smooth ($|\Delta v|_\lambda \sim n_\lambda |\Delta v|$),

$$\mathcal{R}_\lambda \longrightarrow \frac{n_\lambda^2}{q^2} \gg 1, \tag{6.33}$$

and we see that long waves are only very weakly damped. Thus, the artificial viscosity has the desirable property of only affecting the very shortest length scales.

The original prescription of von Neumann and Richtmyer (1950) was found to work fine in Cartesian coordinates, but in a spherical collapse where the infall velocity $|v| \propto r$, it presents a problem, producing significant artificial viscosity where there is no shock. To see how the problem arises, consider the simple case of collapse of a sphere of uniform density under the influence of gravity, without pressure, starting at rest. The Lagrangian equation of motion then is

$$\frac{dv_r}{dt} = -\frac{GM_r}{r^2}, \tag{6.34}$$

where v_r is the radial component of velocity and M_r is the mass within a sphere of radius r. This equation can be integrated, with the result

$$v_r^2 = 2GM_r\left(\frac{1}{r} - \frac{1}{r_0}\right), \tag{6.35}$$

where r_0 is the initial position of mass element M_r. The general property of the solution is that if the initial density is uniform, the collapse proceeds with density increasing with time, but remaining uniform in space, such that all mass elements reach the center at the same finite time (the *free-fall* time). If we pick a time such that $r << r_0$ when the density is ρ, it is clear that

$$v_r^2 = \frac{8}{3}\pi G\rho r^2. \tag{6.36}$$

Exercise

Show that, for the free-fall problem (Equation 6.36), the acceleration caused by the artificial viscosity (Equation 6.28) divided by the acceleration of gravity is $2\,\Delta x/r$, where Δx is the zone size and $q = 1$. Thus, near the center where Δx is a large fraction of r, the effect can be significant because of geometrical effects, even though the flow is smooth. Note that in the equation of motion (Equation 1.62) the artificial viscosity Q is treated the same way as the pressure P; also v_r and its derivative are negative.

The problem was solved by defining the artificial viscosity as

$$\begin{aligned} Q &\equiv -q^2\rho(\Delta r)^2|\nabla\cdot\mathbf{v}|\left(\frac{\partial v_r}{\partial r} - \frac{1}{3}\nabla\cdot\mathbf{v}\right) \\ &= -q^2\rho(\Delta r)^2|\nabla\cdot\mathbf{v}|\frac{2}{3}\left(\frac{\partial v_r}{\partial r} - \frac{v_r}{r}\right), \end{aligned} \tag{6.37}$$

where v_r stands for the radial component of the velocity. Note that, in the free-fall example above, $Q = 0$. The latter formula was subsequently generalized onto multidimensional flows without any specific symmetry, yielding the *artificial pressure tensor*

$$Q_{ij} \equiv -0.5q^2\rho(\Delta x)^2|\nabla\cdot\mathbf{v}|\tilde{\sigma}_{ij}, \tag{6.38}$$

where $\tilde{\sigma}_{ij}$ is given by Equation (1.69), and Δx is a characteristic size of a grid cell (Tscharnuter and Winkler, 1979).

A final point regarding artificial viscosity is that it introduces further limitations on the time step, as discussed in Section 6.5.4.

6.2 A SIMPLE LAGRANGIAN HYDROCODE

The technique of artificial viscosity we discussed in the preceding section is implemented in a simple Lagrangian hydrocode LH1, described below and provided on the CD-ROM accompanying the book. LH1 is a one-dimensional code working in spherical coordinates, with which useful exercises can be performed related to supernova explosions and stellar winds. Similar codes are used in advanced research to simulate stellar pulsations.

In all of these problems, a natural independent variable is the mass M_r contained within a radius r, and grid points are attached to mass elements. The standard relation

$$dM_r = 4\pi r^2 \rho dr \tag{6.39}$$

holds, with which the general Lagrangian set of equations (Equation 1.62), together with Equation (1.58) may be transformed into

$$\begin{aligned}
\frac{1}{\rho} &= 4\pi r^2 \frac{dr}{dM_r} \\
\frac{dv}{dt} &= -4\pi r^2 \frac{dP}{dM_r} \\
\frac{d\epsilon}{dt} &= -4\pi P \frac{d(r^2 v)}{dM_r} \\
\frac{dr}{dt} &= v.
\end{aligned} \tag{6.40}$$

Again, we have four equations with five unknown functions $(\rho, r, v, P, \epsilon)$, and a closing equation — the equation of state — must be supplied. (Note that as before $\epsilon = e/\rho$ is the internal energy per unit mass.) For our purpose, the simple Equation (6.4) will be entirely sufficient.

We set up a staggered grid and refer to the simple space–time grid in Figure 2.1. The radii r_j^n are defined at cell corners, the velocities $v_j^{n+1/2}$ are defined at the centers of vertical cell edges, and $\rho_{j+1/2}^n$, $P_{j+1/2}^n$, and $\epsilon_{j+1/2}^n$ are defined at the centers of horizontal cell edges. The following explicit difference representation of Equation (6.40) results (Benz, 1991):

$$\begin{aligned}
v_j^{n+1/2} &= v_j^{n-1/2} - A_j^n \left(P_{j+1/2}^n - P_{j-1/2}^n\right) \frac{\Delta t^n}{\Delta m_j} \\
r_j^{n+1} &= r_j^n + v_j^{n+1/2} \Delta t^{n+1/2} \\
\rho_{j+1/2}^{n+1} &= \frac{3}{4\pi} \frac{\Delta m_{j+1/2}}{\left(r_{j+1}^{n+1}\right)^3 - \left(r_j^{n+1}\right)^3} \\
\epsilon_{j+1/2}^{n+1} &= \epsilon_{j+1/2}^n - P_{j+1/2}^{n+1/2} \left(A_{j+1}^{n+1/2} v_{j+1}^{n+1/2} - A_j^{n+1/2} v_j^{n+1/2}\right) \frac{\Delta t^{n+1/2}}{\Delta m_{j+1/2}},
\end{aligned} \tag{6.41}$$

where

$$\Delta m_{j+1/2} = \frac{4\pi}{3}\rho^0_{j+1/2}\left[\left(r^0_{j+1}\right)^3 - \left(r^0_j\right)^3\right]$$
$$\Delta m_j = 0.5(\Delta m_{j+1/2} + \Delta m_{j-1/2})$$
$$A^n_j = 4\pi\left(r^n_j\right)^2$$
$$\text{and}\quad \Delta t^n = 0.5\left(\Delta t^{n-1/2} + \Delta t^{n+1/2}\right)$$

and indices $_0$ refer to the initial state at $t = 0$. Note that $P^{n+1/2}_{j+1/2}$ in the fourth equation is not known in advance. In our case, with a γ-law equation of state,

$$P^{n+1/2}_{j+1/2} = (\gamma - 1)\rho^{n+1/2}_{j+1/2}\epsilon^{n+1/2}_{j+1/2} = 0.5(\gamma - 1)\rho^{n+1/2}_{j+1/2}\left(\epsilon^n_{j+1/2} + \epsilon^{n+1}_{j+1/2}\right),$$

where $\rho^{n+1/2}_{j+1/2}$ is a suitable average, and the equation for ϵ^{n+1} may be easily solved. For a general equation of state, this procedure is not possible, and the solution must be found iteratively. The simplest iterative procedure consists of just two steps:

- Approximate $P^{n+1/2}_{j+1/2}$ with $P^n_{j+1/2}$; solve the energy equation to get the first approximation for $\epsilon^{n+1}_{j+1/2}$; from the equation of state, calculate $P^{n+1}_{j+1/2}\big|_1$.
- Approximate $P^{n+1/2}_{j+1/2}$ with $0.5(P^n_{j+1/2}+P^{n+1}_{j+1/2}|_1)$; solve the energy equation again to get the second approximation of $\epsilon^{n+1}_{j+1/2}$.

Scheme (6.41) defines the sequence of operations to be performed within one time step. Given $v^{n+1/2}_j$ and r^{n+1}_j, $\Delta t^{n+3/2}$ can be found from the CFL (Courant-Friedrichs-Lewy) condition (Equation 2.72), and the calculations may be advanced through the next time step.

Scheme (6.41) works only as as long as there are no shock waves in the flow. (The exercise at the end of this section demonstrates what happens when a shock wave develops.) If we want to employ it for flows with shocks, we have to additionally implement artificial viscosity according to Equation (6.37). There are various ways of implementing the artificial pressure in spherical coordinates; we give one example in the following. One would like to define the artificial viscous pressure, just like the "normal" pressure, at the center of the horizontal cell edge (Figure 2.1). Unfortunately, the most natural centering with Q proportional to

$$\frac{v_{j+1} - v_j}{r_{j+1} - r_j} - \frac{v_{j+1} + v_j}{r_{j+1} + r_j}$$

causes Q to vanish in the first grid cell, where $r_1 = v_1 = 0$. This problem is avoided if we introduce volume-averaged values of the divergence of the velocity and the (radial) velocity gradient

$$\overline{\nabla\cdot\mathbf{v}} \equiv \frac{1}{\Delta V_{j+1/2}}\int_{\Delta V_{j+1/2}} \nabla\cdot\mathbf{v}\, d\mathcal{V} = \frac{4\pi}{\Delta V_{j+1/2}}\int \frac{\partial}{\partial r}(r^2 v)dr \approx \frac{A_{j+1}v_{j+1} - A_j v_j}{\Delta V_{j+1/2}}$$
$$\overline{\frac{\partial v}{\partial r}} \equiv \frac{1}{\Delta V_{j+1/2}}\int_{\Delta V_{j+1/2}} \frac{\partial v}{\partial r}\, d\mathcal{V} = \frac{4\pi}{\Delta V_{j+1/2}}\int \frac{\partial v}{\partial r} r^2 dr \approx \frac{A_{j+1/2}(v_{j+1} - v_j)}{\Delta V_{j+1/2}},$$

where $\Delta V_{j+1/2}$ is the volume of a spherical shell and $A_{j+1/2} \equiv 0.5(A_{j+1} + A_j)$. An integration by parts is required for the second relation. Then using

$$\overline{\frac{\partial v}{\partial r}} - \frac{1}{3}\overline{\nabla \cdot \mathbf{v}}$$

and noting that $\Delta V_{j+1/2} = 4\pi r_{j+1/2}^2 \Delta r$, one obtains from Equation (6.37) the following expression for the artificial viscosity

$$Q_{j+1/2} \approx -q^2 \rho_{j+1/2}^{n+1/2} \left| v_{j+1}^{n+1/2} - v_j^{n+1/2} \right| \times \left[v_{j+1} \left(1 - \frac{A_{j+1}}{3A_{j+1/2}} \right) - v_j \left(1 - \frac{A_j}{3A_{j+1/2}} \right) \right].$$

Additionally, we require that $Q = 0$ wherever $\nabla \cdot \mathbf{v} > 0$ (i.e., where the flow diverges). Let us note that because v is centered at $t^{n+1/2}$ (i.e., in the middle of the time step), so is Q.

According to Equation (1.65) and Equation (1.71), terms $\frac{dQ}{M_r}$ and $Q\nabla \cdot \mathbf{v}$ must now be added to the right-hand sides of the momentum and energy equations, respectively. The derivation involves taking the volume average of the (dv/dt) term in the second of Equations (6.40), and it is rather complicated, so it will not be given here. The details are discussed in Benz (1991) and Bowers and Wilson (1991, Section 4.4). The revised momentum equation (first of Scheme 6.41) becomes

$$v_j^{n+1/2} = v_j^{n-1/2} - A_j^n \left(P_{j+1/2}^n - P_{j-1/2}^n \right) \frac{\Delta t^n}{\Delta m_j} - \frac{1}{2} \left[Q_{j+1/2}^{n-1/2} \left(3A_{j+1/2}^n - A_j^n \right) - Q_{j-1/2}^{n-1/2} \left(3A_{j-1/2}^n - A_j^n \right) \right] \frac{\Delta t^n}{\Delta m_j} \tag{6.42}$$

and the energy equation (fourth of Scheme 6.41) becomes

$$\epsilon_{j+1/2}^{n+1} = \epsilon_{j+1/2}^n - P_{j+1/2}^{n+1/2} \left(A_{j+1}^{n+1/2} v_{j+1}^{n+1/2} - A_j^{n+1/2} v_j^{n+1/2} \right) \frac{\Delta t^{n+1/2}}{\Delta m_{j+1/2}} - \frac{1}{2} Q_{j+1/2}^{n+1/2} [v_{j+1}(3A_{j+1/2} - A_{j+1}) - v_j(3A_{j+1/2} - A_j)]^{n+1/2} \frac{\Delta t^{n+1/2}}{\Delta m_{j+1/2}}, \tag{6.43}$$

where the square bracket in the artificial viscosity term is evaluated at $t^{n+1/2}$. Note again that the effect of these terms is to remove kinetic energy from the flow into the shock front and convert it into internal energy.

Exercise

Supernova explosions can be simulated with a spherically symmetric equivalent of the shock tube. At $t = 0$, a small sphere centered at the origin of the coordinate system (we shall refer to it as a "fireball" containing "ejecta") is loaded with internal energy, and it begins to expand into the static, uniform, ambient medium. After a number of time steps, a characteristic structure develops, consisting of (1) freely expanding ejecta, (2) reverse shock, (3) shocked ejecta, (4) contact discontinuity (a boundary between the ejecta and the ambient medium), (5) shocked ambient medium, (6) main shock, and (7) undisturbed ambient medium. The evolution of the fireball can be followed with the help of the Lagrangian code described above and provided on the CD-ROM (file lh1.f).

- Compile the code and run it with input data provided in file "lh1.dat." From the output file, plot distributions of $v(r)$, $\rho(r)$, $P(r)$, and $Q(r)$ (the suitable instruction set for Supermongo can be found in file "pltmod"). Note that all distributions are normalized.
- Identify all flow regions and features listed above. Observe the formation of the reverse shock and its propagation toward the center of the ejecta. Note that the original internal energy of the ejecta is first transformed into kinetic energy and then, upon passing through the reverse shock, into thermal energy again. Explain the origin of the reverse shock.
- Identify regions in which the artificial viscosity Q is effectively working, and observe their relation to shock fronts.
- Perform one run with artificial viscosity parameter q set to 0. Observe the differences in the evolution of the fireball (in particular, watch the total energy). How far can you proceed before the solution breaks down? Repeat the same calculation with various values of q.
- Perform several runs with progressively smaller numbers of grid points (adjust the value of nx in "commons.f" and recompile the code before each run). Observe how the features of the flow become more diffused, and how the total energy conservation becomes progressively poorer.
- Adapt the code to stellar wind modeling (add a source of mass at $r = 0$). For wind and supernova models, plot the location of the main shock as a function of time. Which dependence is steeper?

To end with, let us say that, although in one-dimensional hydrocodes the Lagrangian formulation is often useful, in two or three dimensions on a finite-difference grid it can prove to be problematical. Even if the grid is initially set so that the spacing is uniform, unless the flow is very well-ordered, the mass elements will quickly become highly distorted geometrically with the result that the finite difference approximation will soon become very inaccurate. Codes have been developed that overcome this difficulty by mapping the existing (distorted) Lagrangian grid back on to a uniform grid every few time steps, but this procedure introduces interpolation errors. It seems preferable to use the SPH method for multidimensional astrophysical problems in which a Lagrangian code is needed rather than an Eulerian one.

6.3 BASIC EULERIAN TECHNIQUES

6.3.1 Conservation of Physical Quantities

By its very nature, a Lagrangian scheme conserves mass; unless mass is explicitly added or removed, the total mass of the flow does not change in time. However, in general, this is not true for Eulerian schemes. A *conservative* Eulerian scheme must be based on the following *conservative form* of the Navier–Stokes Equations (1.73)

$$\frac{\partial \rho}{\partial t} + \frac{\partial}{\partial x_j}(\rho v_j) = 0$$

$$\frac{\partial}{\partial t}(\rho v_i) + \frac{\partial}{\partial x_j}\left(\rho v_i v_j + P\delta_{ij} - \sigma_{ij}\right) = \rho F_i, \quad (i = 1, 2, 3)$$

$$\frac{\partial \mathcal{E}}{\partial t} + \frac{\partial}{\partial x_j}[(\mathcal{E} + P)v_j - \sigma_{jk}v_k] = \rho v_j F_j, \tag{6.44}$$

in which the total energy density $\mathcal{E}$ replaces the internal energy density e. In order to illustrate the idea underlying this scheme, consider a staggered grid with ρ defined at cell centers and $\mathbf{v}$ defined at cell faces. Assume that the mass flow per unit area per unit time ρv_j through all faces of all cells is known. Integrating the continuity equation over the volume of a cell and applying the Gauss theorem, we get

$$\frac{\partial}{\partial t}\Delta m = -\int_S \rho v_j \hat{n}_j dA,$$

where Δm is the mass contained within the cell, S is the surface of the cell, and $\hat{\boldsymbol{n}}$ is the unit vector normal to the surface of the cell. In practice, the surface integral turns into a sum of mass fluxes through all faces of the cell

$$\int_S \rho v_j \hat{n}_j dA = \sum_j \mathcal{F}_j,$$

where $\mathcal{F}_j$ is the flux, defined here as the total mass per second flowing through the jth face. When the summation over all cells is performed, every flux is counted twice, but with an opposite sign (whatever flows out of a given cell flows into adjacent ones). As a result, the total mass contained in the grid stays constant, provided, of course, that there is no flow through the grid boundary. Analogous procedures of integrating the second and third of Equations (6.44) over the volume of a cell indicate that total momentum and total energy of the flow will also be conserved provided that no external forces act on the fluid. Note that Scheme (2.57), which we used to integrate the continuity equation, does not conserve the total mass when the velocity of the flow is not uniform. Note also that a scheme which is based directly on Equations (1.73), as is the ZEUS code, does not guarantee that the total energy of the flow will be conserved. When Equations (1.73) are used, the total energy of the flow should be monitored during the simulation in order to estimate the quality and accuracy of the solution.

6.3.2 Advection

The first two terms on the left-hand sides of Equations (6.44) describe the rate of change of mass density, momentum density, and energy density as a result of "pure transport" of mass, momentum, and total energy across the grid, commonly referred to as *advection*. Numerically, advection should result in no net change to the total mass, energy, and momentum. The procedure that performs advection in a hydrocode, thus, should calculate the fluxes not only accurately, but also *consistently*: All physical variables should be advected in the same way as is the mass. In other words, when a mass element moves from one grid cell to another, its momentum and energy should move along with it. In the following, we shall concentrate on a *consistent transport* algorithm implemented in the popular, publically available hydrocode ZEUS (Stone and Norman, 1992). Further concepts and techniques implemented in ZEUS are described in Section 6.5. In that implementation, mass fluxes are calculated first. Subsequently, multiplied by velocity components, they yield momentum fluxes. (Appropriate averaging is necessary here; see Section 6.5 for details.) Internal energy fluxes are given by mass fluxes multiplied by specific internal energy.

In order to illustrate the procedure of flux calculation, consider the advection of mass on a one-dimensional staggered grid with ρ defined at cell centers, and v defined at cell faces. The general scheme for advection may be written as

$$\rho_{i+1/2}^{n+1} = \rho_{i+1/2}^{n} - \frac{\Delta t^{n+1/2}}{\Delta V} \left(\mathcal{F}_{i+1} - \mathcal{F}_i\right), \tag{6.45}$$

where ΔV is the volume of the grid cell, $\mathcal{F}_{i+1}$ and $\mathcal{F}_i$ are the mass fluxes through the right and left faces of the cell, respectively, and $\Delta x_{i+1/2} = x_{i+1} - x_i$. In Cartesian coordinates $\Delta V = \Delta x_{i+1/2}\Delta y_{j+1/2}\Delta z_{k+1/2}$, but in a one-dimensional simulation we may set $\Delta y_{j+1/2} = \Delta z_{k+1/2} = 1$. We have

$$\mathcal{F}_i = A_i \bar{\rho} v_i, \tag{6.46}$$

where A_i is the area of the face between cells $i-1$ and i, and $\bar{\rho}$ is the *interpolated* density defined so that it represents as accurately as possible the flow of mass across A_i during $\Delta t^{n+1/2}$. Both the value and exact location of $\bar{\rho}$ depend on the method employed to calculate the fluxes.

A stable advection scheme is obtained if $\bar{\rho}$ (in general, the interpolated value of the variable being advected) is taken *upstream* (upwind) of the face. Note that the advection scheme (2.57), whose stability was proved in Chapter 2, Section 2.5, is just an upwind scheme. In the simplest case $\bar{\rho} = \rho_{in}$, where ρ_{in} is the density in the cell from which the flow is coming. Consider a face located at x_i, which separates cells with densities $\rho_{i-1/2}$ and $\rho_{i+1/2}$. If $v_i > 0$, then

$$\bar{\rho} = \rho_{i-1/2},$$

while, if $v_i < 0$, then

$$\bar{\rho} = \rho_{i+1/2}.$$

This procedure is known as the *donor cell* method. Since it is a first-order scheme, it turns out to be quite diffusive (see Figure 2.10).

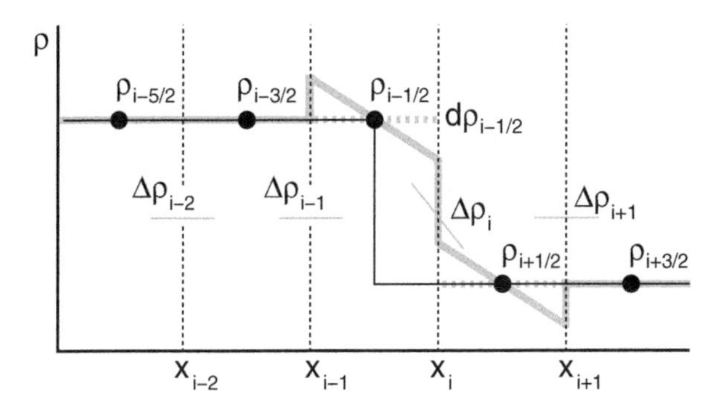

FIGURE 6.4 The van Leer monotonization. *Thin black line*: the original, piecewise constant distribution of ρ. The slopes of the *thin grey lines* indicate the face-centered slopes $\Delta\rho_i$. *Thick grey line*: piecewise linear distribution of ρ with nonmonotonized slopes. The slope at $i + 1/2$ is defined as $0.5(\Delta\rho_i + \Delta\rho_{i+1})$; note the spurious maximum at x_{i-1} and the spurious minimum at x_{i+1}. *Dotted thick grey line*: piecewise linear distribution with slopes monotonized according to Equation (6.48).

An improved method for calculating $\bar{\rho}$ and other interpolated quantities (namely ρv_i and e in the second and third of Equations (1.73), respectively), taking into account the variation in density between a cell's center and its boundary, was proposed by van Leer (1977). The variation in density across a cell is assumed to be linear, and in contrast to the donor-cell technique, the method is accurate to the second order in space. However, one might expect it to be dispersively unstable and, indeed, experience has shown that problems occur, especially when in the piecewise linear distribution of the advected quantity spurious maxima emerge, which were not present in the initial data. To cope with these problems, van Leer proposed the "monotonization" procedure, the net result of which is a reduction in the slope of the linear distribution and the removal of any spurious maxima (Figure 6.4). Let x_i be the grid line across which the flow is going. We define the face-centered slope

$$\Delta\rho_i = (\rho_{i+1/2} - \rho_{i-1/2})/\Delta x_i,$$

where $\Delta x_i = 0.5(\Delta x_{i-1/2} + \Delta x_{i+1/2})$. If $v_i > 0$, then

$$\bar{\rho} = \rho_{i-1/2} + (\Delta x_{i-1/2} - v_i \Delta t)\frac{d\rho_{i-1/2}}{2}.$$

Otherwise, if $v_i < 0$, then

$$\bar{\rho} = \rho_{i+1/2} - (\Delta x_{i+1/2} + v_i \Delta t)\frac{d\rho_{i+1/2}}{2}, \tag{6.47}$$

where $d\rho_{i+1/2}$ is the monotonized cell-centered slope defined as follows

$$d\rho_{i\pm 1/2} = \frac{2\Delta\rho_i \Delta\rho_{i\pm 1}}{\Delta\rho_i + \Delta\rho_{i\pm 1}} \quad \text{if} \quad \Delta\rho_i \Delta\rho_{i\pm 1} > 0$$
$$d\rho_{i\pm 1/2} = 0 \quad \text{if} \quad \Delta\rho_i \Delta\rho_{i\pm 1} \leq 0. \tag{6.48}$$

If the density variation is monotonic across the cell being considered, the scheme is accurate to the second order in space; if not, it reduces to the first-order donor-cell method. In other words, the method is stabilized at the expense of its accuracy by a controlled local increase of its diffusivity. Note that the harmonic mean is always closer to the smaller of the two original slopes. Note also that $\bar{\rho}$ is not defined at the face of the cell, but upstream from the face at a distance corresponding to half the distance covered by the flow in one time step. Thus, in a sense, $\bar{\rho}$ is centered at $t^{n+1/2}$. Even higher accuracy can be obtained if, instead of a linear fit, a piecewise parabolic method (PPM) is used to approximate the density distribution, as originally described by Colella and Woodward (1984). With such an approximation, the advection scheme is third-order accurate in space.

6.3.3 Godunov Method for Calculating Fluxes

Entirely different methods for calculating fluxes are based on Riemann solvers. Since this type of method was first proposed by Godunov (1959), they are often referred to as *Godunov schemes*. To illustrate the workings of a Godunov scheme consider a nonviscous one-dimensional flow along the x-axis assuming $\boldsymbol{F} = 0$. Equations (6.44) may be written in a compact form

$$\frac{\partial U}{\partial t} + \frac{\partial \mathcal{F}(U)}{\partial x} = 0, \tag{6.49}$$

where $U \equiv (\rho, \rho v, \mathcal{E})$ is the *vector of state*, with all components defined at cell centers, and $\mathcal{F}(U) \equiv (\rho v, \rho v^2 + P, (\mathcal{E} + P)v)$ is the *vector of fluxes*. At t^n, two Riemann problems may be defined for a grid cell centered at $x_{i+1/2}$, the first of them describing interactions between states $U^n_{i-1/2}$ and $U^n_{i+1/2}$ at x_i (the left face of the cell), and the second one describing interactions between states $U^n_{i+1/2}$ and $U^n_{i+3/2}$ at x_{i+1} (the right face of the cell) (see Figure 6.5). Let the solutions of those problems be $\mathcal{U}_i(x, t)$ and $\mathcal{U}_{i+1}(x, t)$, respectively. We demand that the solutions do not interfere, i.e., that

$$v^{max}\,\Delta t^{n+1/2} \leq \frac{1}{2}\Delta x_{i+1/2},$$

where v^{max} is the maximum velocity at which the waves resulting from $(U^n_{i-1/2}, U^n_{i+1/2})$ and $(U^n_{i+1/2}, U^n_{i+3/2})$ interactions propagate into the cell between x_i and x_{i+1}. Then, at the end of the time step the vector of state is given by

$$\begin{aligned} U^{n+1}_{i+1/2} = \frac{1}{\Delta x_{i+1/2}} \Bigg[&\int_{x_i}^{x_{i+1/2}} \mathcal{U}_i(x, t^{n+1})dx \\ &+ \int_{x_{i+1/2}}^{x_{i+1}} \mathcal{U}_{i+1}(x, t^{n+1})dx \Bigg]. \end{aligned} \tag{6.50}$$

Alternatively, Equation (6.50) may be written as

$$\begin{aligned} U^{n+1}_{i+1/2} &= U^n_{i+1/2} + \frac{1}{\Delta x_{i+1/2}} \int_{t^n}^{t^{n+1}} [\mathcal{F}(\mathcal{U}_i(x_i, t)) - \mathcal{F}(\mathcal{U}_{i+1}(x_{i+1}, t))]\,dt \\ &= U^n_{i+1/2} + \frac{\Delta t^{n+1/2}}{\Delta x_{i+1/2}} [\bar{\mathcal{F}}(\mathcal{U}_i(x_i)) - \bar{\mathcal{F}}(\mathcal{U}_{i+1}(x_{i+1}))], \end{aligned} \tag{6.51}$$

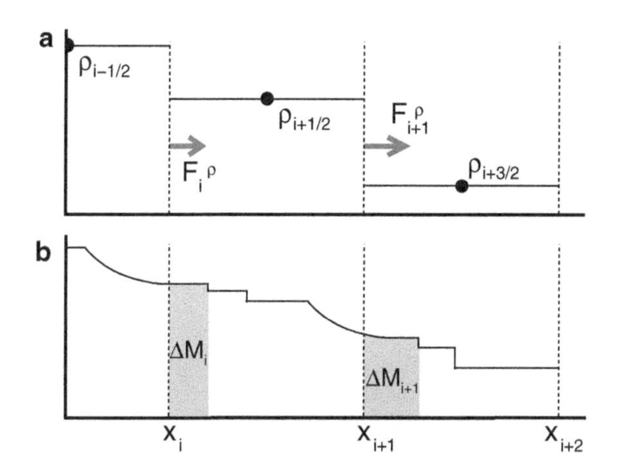

FIGURE 6.5 Advection of mass in Godunov schemes. Based on initial states (a), the Riemann problem is solved at each x_i (b). Once the solution is obtained, the amounts of mass ΔM_i and ΔM_{i+1} can be found that, within the time step, have flowed, respectively, into and out of the grid cell located between x_i and x_{i+1}. Alternatively, fluxes $\mathcal{F}_i^{\rho}$ and $\mathcal{F}_{i+1}^{\rho}$ can be calculated, where $\Delta M_i = \Delta t\, \mathcal{F}_i^{\rho}$. The same scheme is applied to each advected quantity.

where $\bar{\mathcal{F}}$ is the flux averaged over the time step $\Delta t^{n+1/2}$. Equation (6.51) is strictly conservative. The fluxes $\bar{\mathcal{F}}$ are calculated accurately and consistently; however, the scheme turns out to be almost as diffusive as the donor-cell scheme. As before, its quality improves dramatically when a piecewise linear or piecewise parabolic fit to the distribution of U is made, allowing for variation of U within grid cells. Then, however, an additional complication of the scheme is necessary: Since the Riemann problem, by definition, involves states that are spatially uniform, the linear or parabolic fits must be appropriately averaged before they enter the Riemann solver. Readers interested in details of a parabolic Godunov scheme are referred to the paper of Colella and Woodward (1984). In general, Godunov-type hydrocodes are much more accurate than ZEUS-type hydrocodes; however, they also are much more complex and more difficult to adapt to particular problems. As such, they are not recommended for beginners in the field of hydrosimulations.

6.3.4 Operator Splitting

Consider a two-dimensional flow in Cartesian coordinates (x, y). The general (not necessarily conservative) form of Eulerian hydrodynamic equations is:

$$\frac{\partial \mathbf{U}(x, y, t)}{\partial t} = \mathbf{G}(\mathbf{U}, \nabla \mathbf{U}, \nabla^2 \mathbf{U}, x, y, t), \tag{6.52}$$

where $\mathbf{U} = (\rho, \rho v_x, \rho v_y, e)$ is the vector of state introduced at the end of the previous section (v_x, v_y, and e are the velocity in the x-direction, the velocity in the y-direction, and the internal energy density, respectively). The operator $\mathbf{G}$ can be written as a sum

$$\mathbf{G} = \mathbf{G}_1 + \mathbf{G}_2 + \cdots + \mathbf{G}_n$$

and the equation can be solved by successive operations — first from all processes described by $\mathbf{G}_1$, then from all processes described by $\mathbf{G}_2$, and so on. $\mathbf{G}_1$ might be advection terms, $\mathbf{G}_2$ gravity or other external forces, and $\mathbf{G}_3$ heat diffusion.

When the advection procedure is explicitly extracted out of $\mathbf{G}$, Equation (6.52) takes the form

$$\frac{\partial \mathbf{U}}{\partial t} + \nabla \mathcal{F} = \mathbf{S}, \tag{6.53}$$

where

$$\mathcal{F} = [\rho \mathbf{v}, \rho v_x \mathbf{v}, \rho v_y \mathbf{v}, e\mathbf{v}]$$

and $\mathbf{S}$ is the *vector of sources*, which in this case contains pressure gradients and $P\nabla \cdot \mathbf{v}$ work as well as contributions from all external forces and interactions. Equation (6.53) may be solved with an explicit method

$$\frac{\mathbf{U}^{n+1} - \mathbf{U}^n}{\Delta t} + \nabla \mathcal{F}^{n+1/2} = \mathbf{S}^{n+1/2},$$

which is second-order accurate in time. An operator split solution would involve the following substeps:

1. The source terms are applied for half the time step
$$^1\mathbf{U} = \mathbf{U}^n + \mathbf{S}^n \frac{\Delta t}{2} \tag{6.54}$$
2. The fluxes $\mathcal{F}$ are calculated based on $^1\mathbf{U}$, then the advection terms are applied
$$^2\mathbf{U} = {^1\mathbf{U}} - \Delta t \nabla \mathcal{F} \tag{6.55}$$
3. The source terms are recalculated based on $^2\mathbf{U}$, producing $^2\mathbf{S}$, then they are included for the second half of the time step
$$\mathbf{U}^{n+1} = {^2\mathbf{U}} + {^2\mathbf{S}} \frac{\Delta t}{2}. \tag{6.56}$$
The resulting scheme is only quasi-second order accurate in time; however, it offers an improvement in accuracy over the procedure in which the calculations of all updates to the variables are based on the values $\mathbf{U}^n$.

The splitting of operators may proceed even farther. The multidimensional advection substep may be decomposed into a sequence of one-dimensional advection steps, commonly referred to as *sweeps*. Let A_{xy} be the two-dimensional advection operator, acting as described in substep 2 of the operator-split scheme (Equation 6.54 to Equation 6.56). Define A_x and A_y as analogous operators performing advection in directions x and y, respectively. Let $^x\mathbf{U}$ be the result of A_x acting on a state $\mathbf{U}$ for the full length of the time step, and $^{x/2}\mathbf{U}$ be the result of A_x acting on $\mathbf{U}$ for half the length of the time step. With $^y\mathbf{U}$ and $^{y/2}\mathbf{U}$ defined analogously for the operator A_y, the action of A_{xy} may be performed in the following sweeps

$$\begin{aligned}
^{x/2}\mathbf{U} &= A_{x/2}(\mathbf{U}) \\
^{x}\mathbf{U} &= A_{x/2}(^{x/2}\mathbf{U}) \\
^{xy/2}\mathbf{U} &= A_{y/2}(^{x}\mathbf{U}) \\
^{xy}\mathbf{U} &= A_{y/2}(^{xy/2}\mathbf{U}),
\end{aligned}$$

i.e.,

$$\mathbf{A}_{xy} = A_{y/2} \circ A_{y/2} \circ A_{x/2} \circ A_{x/2}. \tag{6.57}$$

An alternative ordering of sweeps

$$\mathbf{A}_{xy} = A_{x/2} \circ A_{y/2} \circ A_{y/2} \circ A_{x/2} \tag{6.58}$$

is also possible. Experience shows that in many cases entirely satisfactory results can be achieved with a procedure involving just two sweeps

$$\mathbf{A}_{xy} = A_y \circ A_x, \tag{6.59}$$

with the condition that at each time step the ordering of sweeps is reversed (if at t^n we begin with the x-sweep, then at t^{n+1} the y-sweep should be made first). Scheme (6.57) to Scheme (6.59), which effectively reduce the multidimensional problem to a collection of one-dimensional problems, may be easily generalized onto three dimensions.

It has to be stressed that there are not any strict rules on how to perform the splitting; numerical experiments and experience play a vital role. Also, there is considerable freedom in the choice of the order of the substeps. A practical rule is that one should avoid splitting terms that nearly cancel each other, such as $(1/\rho)\nabla P$ and $\nabla\Phi$ in a close-to-equilibrium configuration, or emission and absorption of photons in an optically thick medium. The advantages of operator splitting are obvious: Every operation $\mathbf{G}_i$ may be treated independently by either an explicit or an implicit method, whichever is appropriate, and the resulting hydrocode has a clear, modular design that is easy to modify. The main disadvantage is that there is no rigorous mathematical proof for the validity of the process. As remarked by Müller (1994), "It is by no means obvious that operator splitting yields solutions identical to those obtained by solving the original equations." Thus, one should always remember that the operator-split solution is only an approximation to the correct solution of the full nonlinear multidimensional equations. It is highly recommended to check an operator-split scheme against analytic solutions, where available, or at least against numerical solutions obtained with independent methods.

6.3.5 Accuracy, Convergence, and Efficiency

The question arises as to how much improvement in accuracy one obtains by going to higher-order methods. Figure 6.6 illustrates the results of advecting a one-dimensional "square pulse" involving a density discontinuity of a factor 10. The pulse is originally centered at $x = 30$ and is 50 cells wide. It is assumed to propagate in the $+x$-direction at a constant velocity, and the figure shows what it looks like when it has propagated five times its original width, so that the pulse center is at $x = 280$. This is a particularly stiff test because a discontinuity cannot be resolved no matter how many grid points are used. The diffusion in the donor cell case is clear: The original discontinuity has spread over $\sim$ 50 cells, and the pulse has completely lost its shape. In the van Leer treatment, the situation has improved considerably, but the discontinuity is still smeared into a ramp extending over 14 cells. In the third-order piecewise parabolic

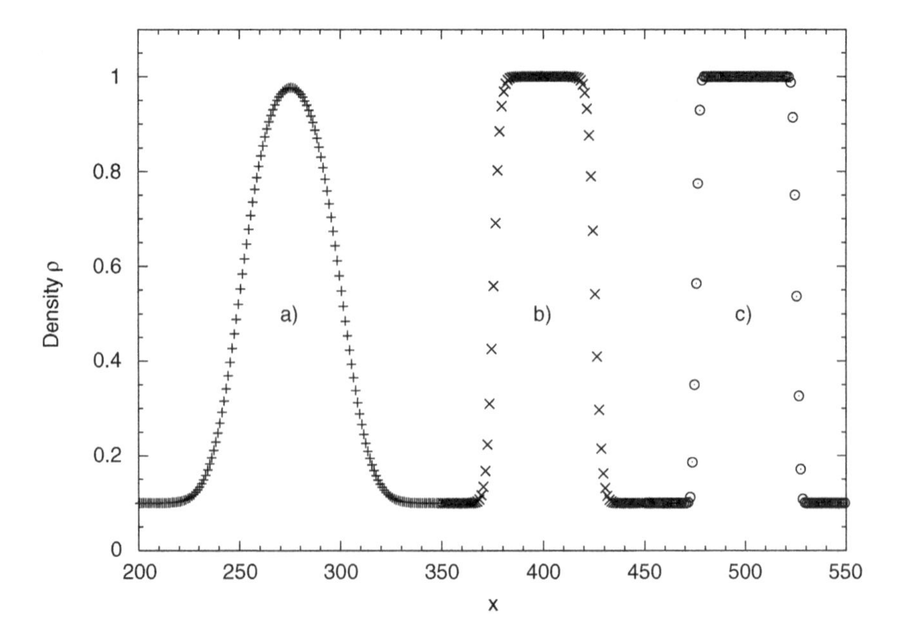

FIGURE 6.6 The advection of a square density pulse shown after it has travelled five times its original width, as calculated by (a) the donor-cell method, (b) the van Leer second-order scheme, and (c) the piecewise parabolic scheme. The centers of pulses (b) and (c) have been arbitrarily offset from that of pulse (a). Each cell along the x-axis has a width of 1. (Data from Stone, J. M. and Norman, M. L. (1992) *Astrophys. J. Suppl.* **80**: 753.)

advection scheme (often referred to as PPM), the width of the ramp is reduced to 6 cells.

If we define a measure of the global error

$$\epsilon_{nm} = \frac{1}{L}\sum_{j=1}^{L} |q_j - q_{an}(x_j)|, \tag{6.60}$$

where q_j is the numerical solution at grid point x_j and $q_{an}(x_j)$ is the corresponding analytical solution, we expect that ϵ_{nm} should decrease with increased numerical resolution in x as $(\Delta x)^r$, where r is the so-called convergence rate. In principle, r should be close to the order of accuracy of the scheme.

The square pulse test provides a good qualitative illustration of the accuracy of various schemes, but for a quantitative measurement of the accuracy, a density distribution without discontinuities is more appropriate. Stone and Norman (1992) reported the results of the advection test performed with a Gaussian density pulse. After the pulse has propagated to a point 10 times its original width, the errors calculated from Equation (6.60) are 10%, 1%, and 0.1 % for the donor cell, van Leer, and PPM schemes, respectively. The convergence rates, as more cells are added, are about $r = 0.5$, 2, and 3, for the same three cases. We see that for the higher-order schemes the error is indeed significantly smaller, and the convergence rate is significantly higher. However, this positive effect is partly offset by the increase in CPU time required per time step and by the complexity of programming. We encounter here the problem of the efficiency of the code, which, for purely practical reasons, is a property as important as the code's accuracy or convergence rate.

We shall proceed with a brief discussion of the efficiency of a Eulerian code. Assume a uniform grid whose physical extent X is the same in all dimensions. The CPU time needed to complete a simulation may be estimated from

$$t_{CPU} = \left(\frac{X}{\Delta x}\right)^n \frac{t}{\Delta t} t_{cell}, \tag{6.61}$$

where n is the number of dimensions, Δx is the size of the grid cell, t is the physical time of the evolution of the modeled system, Δt is the average length of the time step, and t_{cell} is the actual CPU time required to perform all calculations necessary to advance one grid cell for one time step. If an explicit scheme, subject to the CFL condition, is used, then

$$\Delta t \approx \frac{\Delta x}{v_c}, \tag{6.62}$$

where v_c is the characteristic velocity of the medium, and we get

$$t_{CPU} = X^n t v_c \Delta x^{-(n+1)} t_{cell}. \tag{6.63}$$

Suppose that scheme $\mathcal{A}$ produces results as accurate as scheme $\mathcal{B}$, but per dimension uses m times more grid cells. Then, in terms of CPU time, $\mathcal{A}$ is m^{n+1} times less efficient.

Estimating m for pairs of different schemes is not simple and, in general, m is problem-dependent. As an example, consider again the square-pulse test described at the beginning of this section. To keep the physical width of the ramp constant while switching from the PPM to lower-order schemes, one would have to decrease Δx by a factor of 2.5 or 8 for the second- or first-order schemes, respectively, with a corresponding increase of the total number of grid points, $(X/\Delta x)^n$. As a result, for $n = 3$ the CPU time needed to complete the simulation would increase by a factor of 40 or 4100 if the second- or the first-order schemes, respectively, were used. Higher-order schemes are more complicated than the lower-order ones and, in terms of the CPU time per cell per time step, they may be two to three times slower. However, this example clearly demonstrates that they are the obvious choice for multidimensional simulations.

6.4 ADAPTIVE MESH REFINEMENT

As we have shown in Chapter 2, Section 2.7, a finite-difference scheme cannot reliably simulate the details of flow that are smaller than about 10 grid cells. Unfortunately, it often happens (e.g., during galaxy formation or protostellar collapse) that the expected outcome of a simulation is an object or structure many orders of magnitude smaller than the computational domain. In such cases, the desired spatial resolution cannot be achieved with a uniform Eulerian grid, as it would lead to unacceptable computational costs. If the small-scale structure is well localized (i.e., during the simulation it does not move across the computational domain), a technique of *nested grids* may be used (Yorke and Kaisig, 1995). In this technique several refined grids, nested within each other, are placed on the base grid in the region where the small-scale structure is expected to develop, and the whole grid structure remains static during the simulation. An example of a simulation performed with the help of nested grids is presented in Section 6.7.3. Another example of a case where the required grid structure is known in advance is that of the accretion of a giant planet in a disk (D'Angelo et al., 2002). If the planet is assumed to be in a circular orbit and if the calculations are performed in a frame that rotates with the planet's angular frequency, then the disk as a whole can be simulated with a relatively coarse grid. The detailed structure around the planet, where the flow from disk to planet occurs and where a subdisk is expected to form, can be highly resolved with a set of nested subgrids, on each of which the spatial resolution is better than that on the next outermost subgrid. For reasons of accuracy, it is not recommended for one to vary the resolution from one subgrid to the next by more than a factor of 2.

However it often happens that the region where a higher resolution is needed keeps on changing its location on the base grid. Such problems may be solved with the help of a technique known as *adaptive mesh refinement* (AMR) whose basic idea is to match the local resolution of the grid to the momentary requirements set by the configuration of the flow (Berger and Oliger, 1984; Berger and Colella, 1989; Klein and McKee, 1994). To that end, AMR automatically generates (or deletes) finer grids that follow small-scale structures as they form within the computational domain. Thus, as opposed to the nested grid technique, the AMR technique employs a *dynamical grid*. There are several different approaches to AMR. Here, we shall focus on a technique known as *block structured* AMR, in which the base grid is divided into fixed *blocks*. Since the subject is highly complex, we have to restrict ourselves to a brief presentation only, following the description of the method by Fryxell et al. (2000). Readers interested in details of block-structured AMR are referred to the original publication, where they will also find further references to papers presenting other AMR implementations.

The block-structured AMR builds the hierarchy of progressively finer grids (*subgrids*) by "halving" the blocks in each coordinate wherever and whenever refinement is required. Consider a two-dimensional case with each base block containing $(nx)_b \times (ny)_b$ cells. The process of halving a base block creates four *child* blocks, each containing the same number of grid cells as the original (*parent*) block. The child blocks are restricted to fit exactly within the borders of the parent block. Thus, with $nx_b \times ny_b$ cells, but with physical dimensions two times smaller, the resolution offered by the child block is two times finer than that of the parent block. As the simulation proceeds, each child block may be halved once more to produce next-generation child blocks (if even better resolution is locally needed), or it may be cancelled (if the local increase in resolution is no longer necessary). The criteria for refinement or derefinement of blocks may vary from one problem to another. Usually, they are based on the value of the modified second-derivative error norm, which in one dimension is given by

$$E_i^f \equiv \frac{|f_{i+1} - 2f_i + f_{i-1}|}{|f_{i+1} - f_i| + |f_i - f_{i-1}|}, \tag{6.64}$$

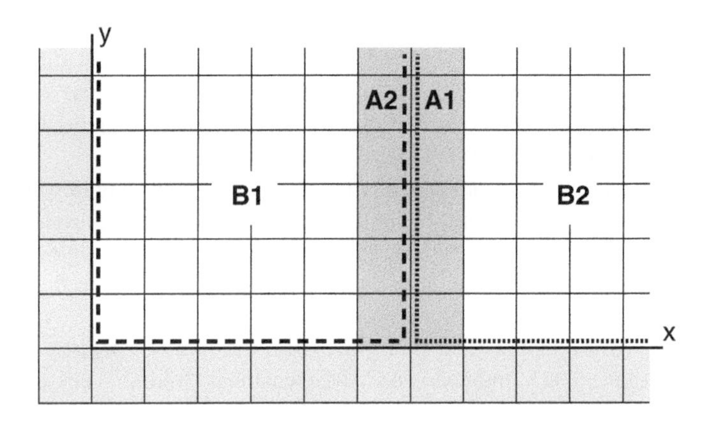

FIGURE 6.7 A schematic view of two neighboring AMR blocks with ghost zones and the boundary of the domain. *Solid line*: boundary of the computational domain. *Dashed line*: active area of block B1. *Dotted line*: active area of block B2. *Light grey:* ghost zones filled by boundary conditions. *Dark grey:* ghost zones enabling interblock communication. In blocks B1 and B2, they are filled with data from areas A1 and A2, respectively.

where f stands for any hydrodynamic variable chosen for refinement. When the maximum value of E_i^f in a block is larger than an adjustable constant C_1, that block is marked for refinement. On the other hand, if it is smaller than another constant, C_2, the block is marked for derefinement. The values of C_1 and C_2 may vary depending on the problem, but often $C_1 = 0.8$ and $C_2 = 0.2$ (Fryxell et al., 2000). Such a procedure works well, but it unnecessarily marks for refinement blocks in which only short wavelength "ripples" of a small amplitude are present. To disable refinement in such cases, the term $\hat{\epsilon}|f_{i+1} + 2f_i + f_{i-1}|$ is added to the denominator of Equation (6.64), with $\hat{\epsilon}$ of order 10^{-2}.

Each block is composed of an *active area* and *ghost zones*. The ghost zones consist of several layers of *ghost cells* (Figure 6.7), through which the block communicates with all neighboring blocks. The number of layers depends on the form of the equations and on the order of the method used to solve the equations. If the boundary of the block is a part of the boundary of the domain, the ghost cells along that boundary are filled by appropriate boundary conditions, otherwise they are filled with data from neighboring blocks. Once the ghost zones of a given block are filled, that block can have its variables updated to the next time step in parallel and independently of any other blocks.

The backbone of the block-structured AMR code consists of routines that:

- Manage refinement and derefinement of blocks.
- Distribute the blocks among the available processors.
- Handle the interblock and interprocessor communication.
- Build and update the database that contains information about location of blocks and relationships between blocks.

Within each block the equations of hydrodynamics are solved with the help of a conservative Eulerian scheme. The flux through the face of a cell that forms an interface with smaller cells of a more refined neighbor block is defined as the sum of fluxes through the appropriate faces of the smaller cells. This definition ensures the overall conservation of mass, momentum, and energy.

A significant variation in spatial resolution implies an equally significant variation in the length of the time step. For the evolution on various refinement levels to proceed coherently, at each level the time step is calculated as an integer multiple of the value used at the finest level. It is also required that all blocks on the same refinement level use the same time step.

A snapshot from a simulation performed with the block-structured AMR code is shown in Figure 6.8. Simulated is a two-dimensional "fireball," i.e., a supernova explosion in a uniform density interstellar medium. The calculation is analogous to the one discussed in Section 6.2, but here the intrinsically spherically symmetric

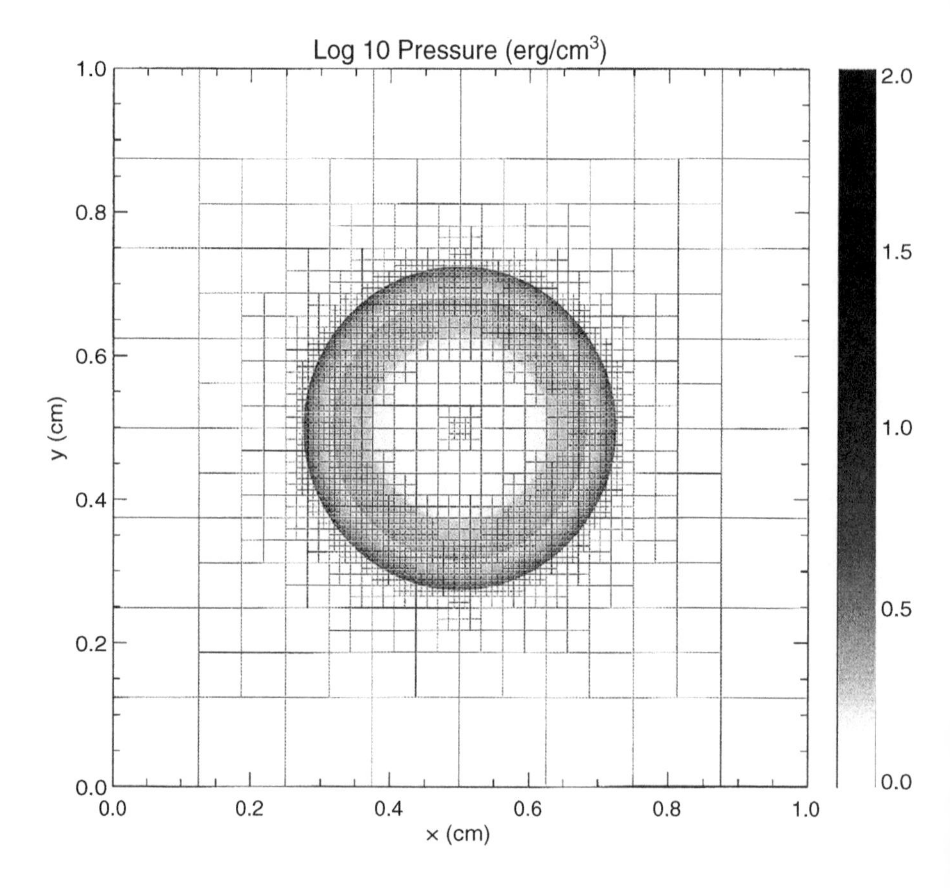

FIGURE 6.8 Pressure field in the two-dimensional simulation of a fireball with eight levels of refinement. Overlaid on the pressure greyscale map are the outlines of the AMR blocks. (From Fryxell, B. et al. (2000) *Astrophys. J. Suppl.* **131**: 273. With permission.)

explosion is taken to be cylindrically symmetric and is run on a two-dimensional Cartesian grid. The initial density ρ_0 and the explosion energy are taken to be 1; in the actual interstellar medium the density would be much lower (10^{-24} g cm^{-3}), the energy much larger (10^{51} erg), and the scale on the order of tens of parsecs. With eight levels of refinement, excellent agreement is obtained with the analytic solution (Sedov, 1959).

Compared to a run in which the resolution is uniform in the whole domain and equal to that of the highest refinement level, an almost 10-fold reduction of the effective simulation time may be achieved with AMR (this includes the time spent on block management as well as interblock and interprocessor communication). Unfortunately, such a significant speedup is only possible when the maximally refined fraction of the domain does not exceed $\sim$ 10%. It becomes less significant when that fraction increases, and ultimately it turns into a slowdown when more than $\sim 2/3$ of the domain is maximally refined. These numbers depend on the machine used to perform the simulation, and they also vary with the number of dimensions of the computational domain. However, the general trend is always the same: Appreciable speedup is achieved only when the maximally resolved fraction of the domain is small. Therefore, it does not make much sense to use many refinement levels when small-scale structures are expected to develop throughout the whole computational domain (a good example of such a case is stellar convection).

On the other hand, AMR codes are very well suited for parallel processing. Independently of the number of refinement levels, the computational workload can be efficiently distributed among many processors, resulting in dramatic speedup of the simulation when the code is run on a massively parallel machine. However, because of their complexity, AMR codes should be viewed as advanced tools to be used by experienced researchers rather than beginners.

In any case there is an important physical reason for using AMR in problems involving self-gravity, for example, calculations of gravitational collapse. As discussed in Chapter 4, Section 4.3 for the case of SPH, the mass and spatial resolution must be sufficiently fine so that the Jeans mass and the Jeans length are resolved. These quantities are given by Equation (4.32) and Equation (4.33), respectively. Truelove et al. (1997, 1998) showed that the Jeans number, defined by $J = (\Delta x/R_J)$, where Δx is the grid spacing, must be less than about 0.25 to avoid nonphysical effects such as artificial fragmentation. During a gravitational collapse in which the density increases rapidly but the temperature does not change appreciably in certain limited regions of the grid, the Jeans length can become quite small, in which case a natural method for satisfying the Jeans condition is AMR.

6.5 A MULTIDIMENSIONAL EULERIAN HYDROCODE

In this section, we shall introduce the hydrocode ZEUS, originally described by Stone and Norman (1992). While less accurate than modern Godunov-type codes employing AMR techniques, ZEUS is much easier to operate. It is also general enough to enable various kinds of numerical experiments without significant modifications.

ZEUS can be used for calculations in Cartesian, cylindrical, or spherical coordinates in one, two, or three dimensions. To simplify the presentation, while discussing all the basic points in the method, we here consider a two-dimensional Eulerian grid (fixed in space) in Cartesian coordinates (x, y), with uniform cell size. The time differencing is explicit, and the method can be generalized to three space dimensions, to nonuniform cell size and, in fact, to a moving grid. For the moment, we neglect radiation transport and magnetic fields, but we do include gravity. The basic equations to be solved, then, are the equations of continuity, momentum, and energy (Equations 1.73), with the addition of appropriate boundary conditions.

The *active grid*, on which the flow is evolved, is embedded in a ghost zone, composed of two layers of ghost cells, which are necessary for providing boundary conditions. Let the grid cells be labelled with the subscript j running in the x-direction from 0 at the inner boundary of the active grid to J at the outer boundary. In the y-direction, the cells are labelled with k, running from 0 to K. Then, ghost cells beyond the inner boundary of the active grid are labelled -2 and -1, and beyond the outer boundary they are labelled $J+1$, $K+1$, $J+2$, $K+2$. The edges of the cells on the active grid, correspondingly, are numbered starting at 0 and running to $J+1$ and $K+1$, while the ghost cell edges are numbered -2 and -1 at the inner edge, and $J+2$, $J+3$ and $K+2$, $K+3$ at the outer edges. A schematic diagram of the full grid setup is shown in Figure 6.9.

In ZEUS a staggered grid is used, as illustrated in Figure 6.10. As we already indicated in Chapter 2, Section 2.3, the advantage of such a grid is that it facilitates the setting up of centered difference equations, which are of second-order accuracy in space. Scalar quantities, such as density and pressure, are defined at the centers of the cells, while vector quantities, such as velocity, are defined at the edges of the cells.

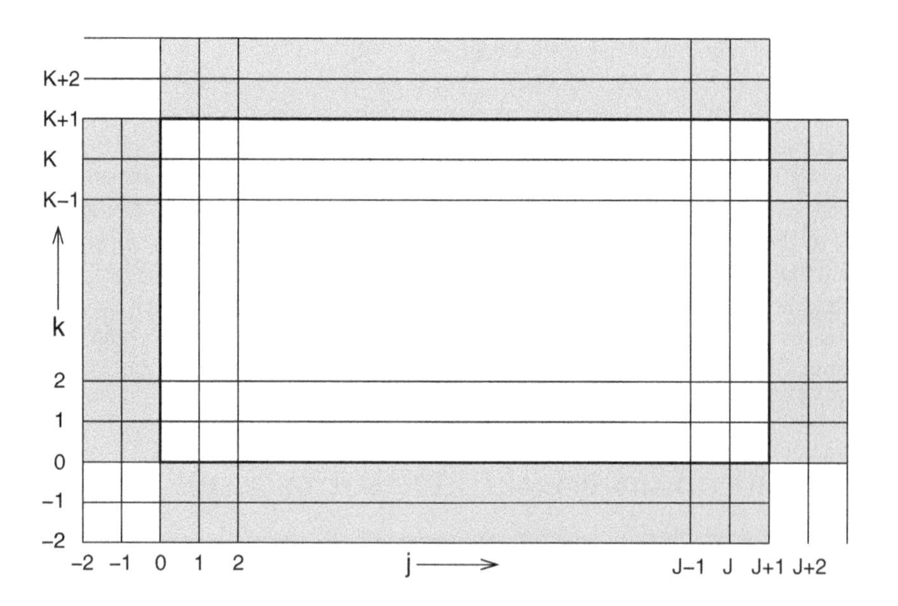

FIGURE 6.9 Schematic diagram of a full two-dimensional grid of the ZEUS code (*white*: active area, *grey*: ghost zones filled by boundary conditions). For clarity, only four rows and columns of grid cells in the active zone are shown.

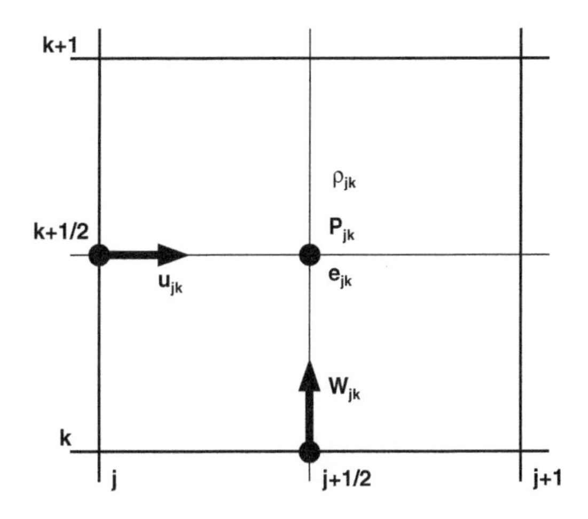

FIGURE 6.10 Schematic diagram of a two-dimensional grid of the ZEUS code indicating the centering of variables.

The coordinates at the cell edges are labeled x_j and y_k, while those at the cell centers are labeled $x_{j+1/2}$ and $y_{k+1/2}$, where

$$x_{j+1/2} = 0.5(x_{j+1} + x_j)$$
$$y_{k+1/2} = 0.5(y_{k+1} + y_k).$$

The velocity $\mathbf{v}$ has two components: $u_{j,k}$ in the x-direction and $w_{j,k}$ in the y-direction. The component $u_{j,k}$ is defined at the center of the left-hand boundary of a cell, so its coordinates are $(x_j, y_{k+1/2})$. The component $w_{j,k}$ is defined at the center of the lower boundary of a cell, so its coordinates are $(x_{j+1/2}, y_k)$.

For the purposes of this section, the continuity equation, the two components of the momentum equation, and the energy equation can be written as

$$\frac{\partial \rho}{\partial t} + \frac{\partial}{\partial x_j}(\rho v_j) = 0 \tag{6.65}$$

$$\frac{\partial}{\partial t}(\rho v_i) + \frac{\partial}{\partial x_j}(\rho v_i v_j) = -\frac{\partial P}{\partial x_i} - \rho \frac{\partial \Phi}{\partial x_i}; \quad i = 1, 2 \tag{6.66}$$

$$\frac{\partial e}{\partial t} + \frac{\partial}{\partial x_j}(e v_j) = -P \frac{\partial v_j}{\partial x_j}. \tag{6.67}$$

As before, we shall refer to the quantities on the right-hand side of the above equations as the source terms, while the second term on the left-hand side of each equation will be referred to as the advection term (cf. Section 6.3.4). A separate solution for the gravitational potential Φ, defined at the centers of cells, is obtained from the Poisson Equation (2.1). We assume that the pressure P and internal energy e obey the ideal gas law (Equation [1.14] and Equation [1.43]).

To solve Equation (6.65) to Equation (6.67), ZEUS employs operator splitting. The velocity and energy are first updated in time by application of the source terms.

Then, the updated values are further modified by application of the advection terms. In each such substep, the most recently obtained values of all variables are used. The details of the difference schemes are presented in the following two subsections. Boldface superscripts **n** and **n + 1** will be used in equations in which a physical variable is only partially updated between time levels n and $n+1$.

6.5.1 Source terms

The continuity equation has no source terms. The momentum equation, with source terms involving the gravitational force and the pressure gradient, is updated in its nonconservative form, which reads

$$\frac{\partial u}{\partial t} = -\frac{1}{\rho}\frac{\partial P}{\partial x} - \frac{\partial \Phi}{\partial x} \tag{6.68}$$

$$\frac{\partial w}{\partial t} = -\frac{1}{\rho}\frac{\partial P}{\partial y} - \frac{\partial \Phi}{\partial y}, \tag{6.69}$$

where the advection terms have been left out because they are inactive in this substep. The difference form of Equation (6.68) and Equation (6.69) is

$$\begin{aligned}
\frac{u_{j,k}^{\mathbf{n+1}} - u_{j,k}^{n}}{\Delta t} &= -\frac{2\left(P_{j,k}^{n} - P_{j-1,k}^{n}\right)}{\Delta x_j\left(\rho_{j,k}^{n} + \rho_{j-1,k}^{n}\right)} - \frac{\Phi_{j,k}^{n} - \Phi_{j-1,k}^{n}}{\Delta x_j} \\
\frac{w_{j,k}^{\mathbf{n+1}} - w_{j,k}^{n}}{\Delta t} &= -\frac{2\left(P_{j,k}^{n} - P_{j,k-1}^{n}\right)}{\Delta y_k\left(\rho_{j,k}^{n} + \rho_{j,k-1}^{n}\right)} - \frac{\Phi_{j,k}^{n} - \Phi_{j,k-1}^{n}}{\Delta y_k},
\end{aligned} \tag{6.70}$$

where $\Delta x_j = x_{j+1/2} - x_{j-1/2}$ and $\Delta y_k = y_{k+1/2} - y_{k-1/2}$. The velocity components u and w are defined on the left-hand edge and on the lower edge of a cell, respectively, so both difference equations are spatially centered. The gravitational potential Φ is obtained from the solution to the Poisson equation using densities $\rho_{j,k}^{n}$, and the pressure is obtained from $e_{j,k}^{n}$ and the equation of state. With this explicit time differencing, the contribution of source terms to Equation (6.68) and Equation (6.69) is only first-order accurate in time.

The final source term is the compressional heating term on the right-hand side of the energy equation (6.67). It is preferable here to use a time-centered scheme to update the internal energy in order to get better conservation of overall energy. A simple scheme involving one iteration can be used if the equation of state is more complicated than that of a polytrope; otherwise an analytical formula can be obtained to relate $e^{\mathbf{n+1}}$ to e^{n}. (Note that the same procedure was applied to solve the Lagrangian energy equation in Section 6.2.) The first step is:

$$\frac{e_{j,k}^{\mathbf{n+1,p}} - e_{j,k}^{n}}{\Delta t} = -P_{j,k}^{n}\left(\frac{u_{j+1,k}^{n} - u_{j,k}^{n}}{x_{j+1} - x_j} + \frac{w_{j,k+1}^{n} - w_{j,k}^{n}}{y_{k+1} - y_k}\right), \tag{6.71}$$

where the superscript $\mathbf{n+1,p}$ indicates a provisional value. The value of $P_{j,k}^{\mathbf{n+1,p}}$ is then obtained from $e_{j,k}^{\mathbf{n+1,p}}$ and the equation of state. We then define

$P^{\mathbf{n}+1/2} = 0.5(P^n + P^{\mathbf{n}+1,\mathbf{p}})$. The final step is:

$$\frac{e^{\mathbf{n}+1}_{j,k} - e^n_{j,k}}{\Delta t} = -P^{\mathbf{n}+1/2}_{j,k} \nabla \cdot \mathbf{v}_{j,k}, \tag{6.72}$$

where the divergence of velocity is calculated just as in Equation (6.71).

An additional update to u, w, and e is required if the simulation includes artificial viscosity as described in Section 6.1. In its standard version, ZEUS employs the simplest possible artificial viscosity, which is the classical von Neumann viscosity (Equation 6.27), acting independently in directions x and y. The x-component of the artificial pressure is given by

$$\begin{aligned} Q_{x(j,k)} &= q^2 \rho_{j,k} (u_{j+1,k} - u_{j,k})^2 \quad \text{if } (u_{j+1,k} - u_{j,k}) < 0 \\ &= 0 \quad \text{otherwise.} \end{aligned}$$

Similarly, the y-component is:

$$\begin{aligned} Q_{y(j,k)} &= q^2 \rho_{j,k} (w_{j,k+1} - w_{j,k})^2 \quad \text{if } (w_{j,k+1} - w_{j,k}) < 0 \\ &= 0 \quad \text{otherwise.} \end{aligned}$$

Then the updates to the momentum equation become

$$\frac{u^{\mathbf{n}+1}_{j,k} - u^{\mathbf{n}}_{j,k}}{\Delta t} = -\frac{2(Q_{x(j,k)} - Q_{x(j-1,k)})}{\Delta x_j \left(\rho^n_{j,k} + \rho^n_{j-1,k}\right)} \tag{6.73}$$

$$\frac{w^{\mathbf{n}+1}_{j,k} - w^{\mathbf{n}}_{j,k}}{\Delta t} = -\frac{2(Q_{y(j,k)} - Q_{y(j,k-1)})}{\Delta y_k \left(\rho^n_{j,k} + \rho^n_{j,k-1}\right)}, \tag{6.74}$$

where the boldface superscripts **n** on the velocities indicate that the values updated for the gravitational acceleration are taken. Finally, the energy dissipated by artificial viscosity is accounted for in the energy equation

$$\frac{e^{\mathbf{n}+1}_{j,k} - e^{\mathbf{n}}_{j,k}}{\Delta t} = -Q_{x(j,k)} \left(\frac{u^{\mathbf{n}}_{j+1,k} - u^{\mathbf{n}}_{j,k}}{x_{j+1} - x_j} \right) - Q_{y(j,k)} \left(\frac{w^{\mathbf{n}}_{j,k+1} - w^{\mathbf{n}}_{j,k}}{y_{k+1} - y_k} \right),$$

where the boldface superscript **n** on the energy indicates that the value updated for compressional heating is taken. The velocities are those obtained before the application of Equation (6.73) and Equation (6.74), i.e., they are not updated for effects of artificial viscosity.

6.5.2 Advection Terms

As we mentioned in Section 6.3.2, the ZEUS code uses so-called "consistent transport" as its advection method, meaning that all physical variables are advected in the same way as is the mass ("together with the mass"). The process of advection follows the scheme given by Equation (6.45). A slight modification is needed only in the case of momentum densities that are not defined at the centers of cells. To calculate ρu and

ρw at the same locations as u and w, one must average the values of ρ across cell interfaces — one of the disadvantages of the staggered grid.

For the density update, the following two mass fluxes must be known for each grid cell

$$\mathcal{F}^{\rho}_{x(j,k)} = u_{j,k}\bar{\rho}A_{x(j,k)}$$
$$\mathcal{F}^{\rho}_{y(j,k)} = w_{j,k}\bar{\rho}A_{y(j,k)}.$$

(Note that there is no need to calculate $\mathcal{F}^{\rho}_{x(j+1,k)}$ and $\mathcal{F}^{\rho}_{y(j,k+1)}$ in the cell (j,k), since they are calculated, respectively, in the cells $(j+1,k)$ and $(j,k+1)$.) As before, $\bar{\rho}$ is the value of the density interpolated to the face of the cell through which the flow is going, and A is the area of that face. In two dimensions $A_x = \Delta y \Delta z$ and $A_y = \Delta x \Delta z$, where we may set $\Delta z = 1$. If u and w are both positive over a cell, then $\mathcal{F}^{\rho}_{x(j+1,k)}$ represents the mass flowing per second out of the cell through the right face, $\mathcal{F}^{\rho}_{x(j,k)}$ represents the mass flowing per second into the cell through the left face, and $\mathcal{F}^{\rho}_{y(j,k+1)}$ and $\mathcal{F}^{\rho}_{y(j,k)}$ are the corresponding expresssions for the mass flow through top and bottom faces, respectively.

In ZEUS, the advection is split into one-dimensional sweeps. To advect mass, the following operations are performed

$$^{1}\rho^{\mathbf{n+1}}_{j,k} = \rho^{n}_{j,k} - \frac{\Delta t}{\Delta V}\left(\mathcal{F}^{\rho}_{x(j+1,k)} - \mathcal{F}^{\rho}_{x(j,k)}\right) \quad (6.75)$$

$$\rho^{n+1}_{j,k} = {}^{1}\rho^{\mathbf{n+1}}_{j,k} - \frac{\Delta t}{\Delta V}\left(\mathcal{F}^{\rho}_{y(j,k+1)} - \mathcal{F}^{\rho}_{y(j,k)}\right), \quad (6.76)$$

where the *control volume* ΔV is equal to the volume of the cell (i.e., $\Delta V = \Delta x \Delta y \Delta z$). Note that the density was not affected by update due to source terms and that ρ^{n+1} is the completely updated value. For every mass flux the corresponding internal energy flux is calculated from

$$\mathcal{F}^{e} = \bar{\epsilon}\mathcal{F}^{\rho},$$

where the specific internal energy $\bar{\epsilon}$ is interpolated in the same way as is the density. Then, operations analogous to Equation (6.75) and Equation (6.76) follow

$$^{1}e^{\mathbf{n+1}}_{j,k} = e^{\mathbf{n+1}}_{j,k} - \frac{\Delta t}{\Delta V}\left(\mathcal{F}^{e}_{x(j+1,k)} - \mathcal{F}^{e}_{x(j,k)}\right)$$

$$e^{n+1}_{j,k} = {}^{1}e^{\mathbf{n+1}}_{j,k} - \frac{\Delta t}{\Delta V}\left(\mathcal{F}^{e}_{y(j,k+1)} - \mathcal{F}^{e}_{y(j,k)}\right),$$

where $e^{\mathbf{n+1}}$ is the internal energy density updated for effects of compressional work and artificial viscosity.

We now consider advection in the momentum equation. The velocities are located on cell edges, so the advection surfaces are offset from the actual faces of the cell; however, because of assumed uniformity of the grid, the control volume ΔV does not change. The x-momentum fluxes in directions x and y are given by

$$\mathcal{F}^{u}_{x(j,k)} = \bar{u}_{x}\frac{1}{2}\left(\mathcal{F}^{\rho}_{x(j-1,k)} + \mathcal{F}^{\rho}_{x(j,k)}\right) \quad (6.77)$$

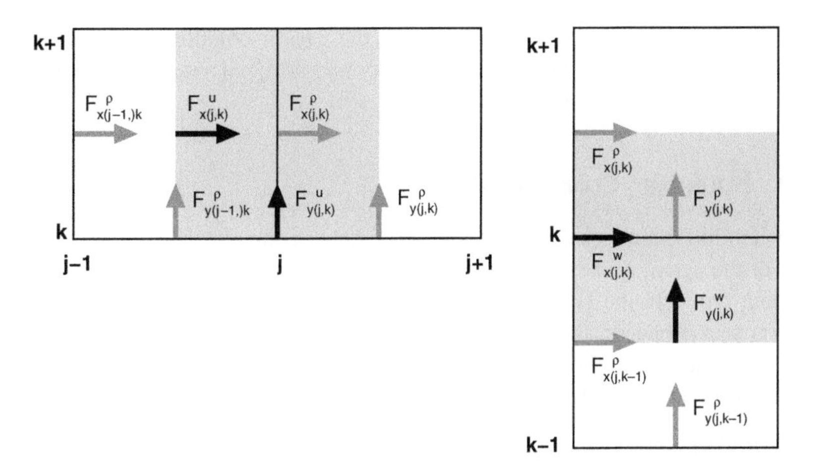

FIGURE 6.11 Momentum fluxes in ZEUS (*solid black lines*: boundaries of grid cells, *shaded areas*: control volumes for the advection of momentum, *grey arrows*: mass fluxes, *black arrows*: momentum fluxes) calculated according to Equation (6.77) to Equation (6.80).

and

$$\mathcal{F}^{u}_{y(j,k)} = \bar{u}_y \frac{1}{2}\left(\mathcal{F}^{\rho}_{y(j-1,k)} + \mathcal{F}^{\rho}_{y(j,k)}\right), \tag{6.78}$$

where $\bar{u}_x$ and $\bar{u}_y$ are the values of u interpolated in directions x and y, respectively. Similarly, the y-momentum fluxes are given by

$$\mathcal{F}^{w}_{x(j,k)} = \bar{w}_x \frac{1}{2}\left(\mathcal{F}^{\rho}_{x(j,k-1)} + \mathcal{F}^{\rho}_{x(j,k)}\right) \tag{6.79}$$

and

$$\mathcal{F}^{w}_{y(j,k)} = \bar{w}_y \frac{1}{2}\left(\mathcal{F}^{\rho}_{y(j,k-1)} + \mathcal{F}^{\rho}_{y(j,k)}\right), \tag{6.80}$$

where, as in the case of u, $\bar{w}_x$, and $\bar{w}_y$ are the values of w interpolated in directions x and y, respectively (Figure 6.11).

The advection step is completed with a momentum-density update performed separately for each component of momentum and each direction according to

$$\begin{aligned} {}^{1}(\rho u)^{\mathbf{n+1}}_{j,k} &= (\rho u)^{\mathbf{n+1}}_{j,k} - \frac{\Delta t}{\Delta V}\left(\mathcal{F}^{u}_{x(j+1,k)} - \mathcal{F}^{u}_{x(j,k)}\right) \\ (\rho u)^{n+1}_{j,k} &= {}^{1}(\rho u)^{\mathbf{n+1}}_{j,k} - \frac{\Delta t}{\Delta V}\left(\mathcal{F}^{u}_{y(j,k+1)} - \mathcal{F}^{u}_{y(j,k)}\right) \end{aligned}$$

and

$$\begin{aligned} {}^{1}(\rho w)^{\mathbf{n+1}}_{j,k} &= (\rho w)^{\mathbf{n+1}}_{j,k} - \frac{\Delta t}{\Delta V}\left(\mathcal{F}^{w}_{x(j+1,k)} - \mathcal{F}^{w}_{x(j,k)}\right) \\ (\rho w)^{n+1}_{j,k} &= {}^{1}(\rho w)^{\mathbf{n+1}}_{j,k} - \frac{\Delta t}{\Delta V}\left(\mathcal{F}^{w}_{y(j,k+1)} - \mathcal{F}^{w}_{y(j,k)}\right), \end{aligned}$$

where $(\rho u)^{n+1}_{j,k}$ and $(\rho w)^{n+1}_{j,k}$ are, respectively, the x and y components of the momentum density, updated for the effects of gravity and artificial viscosity. The densities are interpolated to the positions of u and w.

6.5.3 Boundary Conditions

ZEUS handles boundary conditions with the help of the ghost zones located at the edges of the active grid (see Figure 6.9). In general, boundary conditions can be of two main types. In the *Dirichlet* condition, the value of a function (which could be zero), is prescribed on the boundary. In the *Neumann* condition, the derivative of the function normal to the boundary surface is prescribed. Common combinations of these conditions are the following:

- *Reflecting Boundary Condition*: The cell-centered variables ρ, e in the ghost cells are set to the corresponding value in the last active cell, thus $\rho_{J+1,k} = \rho_{J,k}$ and $\rho_{J+2,k} = \rho_{J-1,k}$ with an analogous treatment at the K-boundary ($k = K$). The same treatment is given to the w-component of the velocity at the J-boundary and the u-component of the velocity at the K-boundary. The normal component of the velocity $u_{J+1,k}$ is set to zero, and $u_{J+2,k} = u_{J,k}$. Similarly, at the K-boundary, $w_{j,K+1} = 0$ and $w_{j,K+2} = w_{j,K}$.
- *Inflow Boundary Condition*: Here all of the variables in the ghost cells, ρ, e, u, and w, are set to prescribed values, which can vary in time. The component of the velocity normal to the boundary must be directed toward the interior of the active zone.
- *Outflow Boundary Condition*: Here all of the variables in the ghost cells, ρ, e, u, and w, are set to the corresponding values in the active cells. (Note this is a Neumann boundary condition with derivatives of all variables set to 0 at the boundary of the active zone.) Thus, at the outer J-boundary, $\rho_{J+1,k} = \rho_{J,k}$, $e_{J+1,k} = e_{J,k}$, $w_{J+1,k} = w_{J,k}$, and $u_{J+1,k} = u_{J,k}$. An analogous treatment holds at the K-boundary. If the outflow is subsonic, then reflected waves can be generated at the boundary causing numerical problems. If the outflow is supersonic, this problem does not arise.
- *Periodic Boundary Conditions*: The material that flows out of the grid on one side is assumed to reenter the grid on the opposite side with the same physical characteristics. Thus, $\rho_{-2,k} = \rho_{J-1,k}$, $\rho_{-1,k} = \rho_{J,k}$, $\rho_{J+1,k} = \rho_{0,k}$, and $\rho_{J+2,k} = \rho_{1,k}$. The same treatment applies to the internal energy and the tangential component of velocity. For the normal component of velocity, say, for the J-direction, $u_{-1,k} = u_{J-1,k}$ and $u_{J+2,k} = u_{1,k}$. At the boundaries themselves, $j = 0$ and $j = J$, the value of u is computed from the difference equations.

6.5.4 Time Step Control

As an explicit code, ZEUS is subject to limitations resulting from the CFL condition (Equation [6.16]). On a two-dimensional grid, two time steps for each cell can be determined

$$\Delta t_1 = \frac{\Delta x}{c_s + |u|}$$
$$\Delta t_2 = \frac{\Delta y}{c_s + |w|}. \tag{6.81}$$

The time step is then chosen to be the minimum of all values of Δt_1 and Δt_2. An important additional constraint on the time step must be considered because the artificial viscosity is calculated explicitly, and it introduces a diffusion term into the momentum equation. For explicit diffusion in two dimensions, the von Neumann stability analysis yields the condition

$$\Delta t \leq \frac{1}{2\nu_d}\left[\frac{1}{(\Delta x)^2} + \frac{1}{(\Delta y)^2}\right]^{-1}, \tag{6.82}$$

which, on a uniform grid with $\Delta x = \Delta y$, reduces to

$$\Delta t \leq \frac{(\Delta x)^2}{4\nu_d} \tag{6.83}$$

(Peaceman and Rachford, 1955). In Section 6.1.4 it was shown that the effective kinematic artificial viscosity coefficient is

$$\nu_{av} = q^2 \Delta x |\Delta v|,$$

where q is the artificial viscosity parameter, generally of order unity, and Δv is the velocity difference across a cell. Thus, we define for each cell two additional time steps

$$\Delta t_3 = \frac{\Delta x}{4q^2|\Delta u|} \qquad \Delta t_4 = \frac{\Delta y}{4q^2|\Delta w|}.$$

The actual time step may then be given by

$$\Delta t = C_0 \min[\Delta t_1, \Delta t_2, \Delta t_3, \Delta t_4], \tag{6.84}$$

where the minimum is taken over all cells and C_0 is the so-called *Courant number*, a safety factor often taken to be 0.5. In practice, instead of Conditions (6.81) to (6.84), ZEUS uses a slightly different algorithm. First, four auxiliary time steps are calculated for each cell

$$\Delta t_1 = \frac{1}{c_s}\min(\Delta x, \Delta y)$$
$$\Delta t_2 = \frac{\Delta x}{|u|}$$
$$\Delta t_3 = \frac{\Delta y}{|w|}$$
$$\text{and} \quad \Delta t_4 = \min\left(\frac{\Delta x}{4q^2|\Delta u|}, \frac{\Delta y}{4q^2|\Delta w|}\right). \tag{6.85}$$

The final time step is then found from

$$\Delta t = C_0 \left[\max\left(\Delta t_1^{-2} + \Delta t_2^{-2} + \Delta t_3^{-2} + \Delta t_4^{-2}\right)\right]^{-1/2}, \tag{6.86}$$

where the maximum of the sum in parentheses is taken over all cells.

6.6 $2\frac{1}{2}$-DIMENSIONAL SIMULATIONS

Although today's computers allow for complex three-dimensional simulations, for purely economical reasons it is not recommended to overload them with computational exercises. In this section, we shall discuss techniques that allow us to simulate certain three-dimensional objects in just two dimensions, thus reducing the CPU time by a factor of ~ 100.

6.6.1 AXIAL SYMMETRY

Many interesting astrophysical problems may be solved in cylindrical (R, Z, ϕ) or spherical (r, θ, ϕ) coordinates under the assumption of axial symmetry. Because of the symmetry constraint, all derivatives in the ϕ-direction are zero and no ϕ-dependence appears in the equations. However, unlike in Cartesian coordinates, the equations contain the third velocity component. In analogy to the linear momenta per unit volume ρu and ρw, it is usually introduced in the form of the angular momentum per unit volume

$$A = \rho R v_\phi, \tag{6.87}$$

where v_ϕ is the linear velocity component in the ϕ direction. The quantity A is actually treated as a scalar and is defined at cell centers.

A scheme of this sort is sometimes called "$2\frac{1}{2}$-D" (even though it is really only two-dimensional). A $2\frac{1}{2}$-D calculation is one of the options offered by ZEUS and it may be chosen by suitable arrangement of control files (no modification of the code is needed). The axisymmetric setup is described here separately only because of the significant role it has played (and often still plays) in astrophysical simulations.

Taking cylindrical coordinates, we assume that the configuration is mirror-symmetric about the equatorial plane. We fix the rotation axis as the Z-axis and the equatorial plane as the R-axis. Then, the computational domain that is required includes only the quadrant with positive Z and positive R, since the configuration is also symmetric about the Z-axis. We use here u for the velocity component in the R-direction and w as the velocity component in the Z-direction. The continuity equation, the equation of motion in all three coordinates, and the energy equation read now as

$$\frac{\partial \rho}{\partial t} + \nabla \cdot (\rho \mathbf{v}) = 0. \tag{6.88}$$

$$\frac{\partial (\rho u)}{\partial t} + \boldsymbol{\nabla} \cdot (\rho u \mathbf{v}) = -\rho \frac{\partial \Phi}{\partial R} - \frac{\partial P}{\partial R} + \frac{A^2}{\rho R^3} \tag{6.89}$$

$$\frac{\partial (\rho w)}{\partial t} + \boldsymbol{\nabla} \cdot (\rho w \mathbf{v}) = -\rho \frac{\partial \Phi}{\partial Z} - \frac{\partial P}{\partial Z} \tag{6.90}$$

$$\frac{\partial A}{\partial t} + \nabla \cdot (A \mathbf{v}) = 0. \tag{6.91}$$

$$\frac{\partial e}{\partial t} + \nabla \cdot (e \mathbf{v}) = -P \nabla \cdot \mathbf{v}. \tag{6.92}$$

The effect of angular momentum on the R component of the force is the last term of Equation (6.89); it is the standard centrifugal acceleration. The equation of motion in the ϕ-direction (Equation [6.91]) simply expresses conservation of angular momentum; if no external torques are present, the only way the angular momentum in a Eulerian cell can change is through advection.

At all points on the Z axis ($R = 0$), special boundary conditions must be satisfied

$$\frac{\partial \rho}{\partial R} = 0 \qquad \frac{\partial w}{\partial R} = 0$$
$$u = 0 \qquad A = 0.$$

Similarly, at all points on the R-axis ($Z = 0$)

$$\frac{\partial \rho}{\partial Z} = 0 \qquad \frac{\partial u}{\partial Z} = 0$$
$$w = 0 \qquad \frac{\partial A}{\partial Z} = 0.$$

The numerical solution to the difference equations in the cylindrical case proceeds in a very similar way to that for the Cartesian case described in Section 6.5. The main differences are:

1. In the R-component of the momentum equation, the update of the velocity in the R-direction for the effect of centrifugal acceleration is performed in a separate operation by

$$\frac{u^{\mathbf{n+1}}_{j,k} - u^{\mathbf{n}}_{j,k}}{\Delta t} = \frac{0.25\left(A^n_{j,k}/\rho^n_{j,k} + A^n_{j-1,k}/\rho^n_{j-1,k}\right)^2}{R^3_j}. \tag{6.93}$$

2. The treatment of the angular momentum per unit volume A in Equation (6.91) is the same as that for the density in Equation (6.88).
3. The divergence of the velocity that appears in Equation (6.92) looks like

$$\nabla \cdot \mathbf{v}(j,k) = \frac{R_{j+1}u_{j+1,k} - R_j u_{j,k}}{0.5(R_{j+1} + R_j)(R_{j+1} - R_j)} + \frac{w_{j,k+1} - w_{j,k}}{Z_{k+1} - Z_k}. \tag{6.94}$$

   Note that this expression is cell-centered as needed for the energy equation.
4. An actual cell in the cylindrical system is three-dimensional: it has a square (or, in general, rectangular) cross section of height ΔZ and width ΔR, and it extends through the full 360° in ϕ; thus it looks like a thin annulus. Because of the assumed symmetry with respect to the equatorial plane, a mirror cell appears at $(R, -Z)$. The control volume of one of these cells, for a variable defined at cell center, is $\Delta V = 2\pi R_{j+1/2}\Delta R \Delta Z$, where $R_{j+1/2} = 0.5(R_{j+1} + R_j)$.

The advection is carried out according to the principles of Section 6.3.2. The only differences are the area and volume factors. Thus, the density is updated according to

$$\rho^{\mathbf{n+1}}_{j,k} = \rho^{\mathbf{n}}_{j,k} - \frac{\Delta t}{\Delta V}\left(\mathcal{F}^{\rho}_{R(j+1,k)} - \mathcal{F}^{\rho}_{R(j,k)} + \mathcal{F}^{\rho}_{Z(j,k+1)} - \mathcal{F}^{\rho}_{Z(j,k)}\right), \tag{6.95}$$

where, as before, the operation may be split into one-dimensional sweeps. The $\mathcal{F}^{\rho}$s in Expression (6.95) are fluxes of mass across the four faces of the volume, calculated by

$$\begin{aligned}
\mathcal{F}^{\rho}_{R(j+1,k)} &= u_{j+1,k}\bar{\rho}A_{R(j+1,k)} \\
\mathcal{F}^{\rho}_{R(j,k)} &= u_{j,k}\bar{\rho}A_{R(j,k)} \\
\mathcal{F}^{\rho}_{Z(j,k+1)} &= w_{j,k+1}\bar{\rho}A_{Z(j,k+1)} \\
\mathcal{F}^{\rho}_{Z(j,k)} &= w_{j,k}\bar{\rho}A_{Z(j,k)},
\end{aligned}$$

where $\bar{\rho}$ is the value of the density interpolated as before (Equation 6.46 and Equation 6.48), and the As are the areas of cell faces. In cylindrical geometry, we have

$$\begin{aligned}
A_{R(j+1,k)} &= 2\pi R_{j+1}\Delta Z \\
A_{R(j,k)} &= 2\pi R_j \Delta Z \\
A_{Z(j,k)} &= \pi(R_{j+1} + R_j)(R_{j+1} - R_j) \\
A_{Z(j,k+1)} &= A_{Z(j,k)}.
\end{aligned} \tag{6.96}$$

For the case of the advection of the w-component of the velocity, which is centered at $(R_{j+1/2}, Z_k)$, the volume element ΔV is the same, the A_Zs are the same, and in the expressions for the A_Rs, ΔZ must be understood to mean $Z_{k+1/2} - Z_{k-1/2}$ rather than $Z_{k+1} - Z_k$.

For the case of the advection of the u-component of the velocity, which is centered at $(R_j, Z_{k+1/2})$, the volume element is changed slightly to

$$\Delta V = 2\pi R_j \Delta R \Delta Z,$$

where $\Delta R = (R_{j+1/2} - R_{j-1/2})$ and $\Delta Z = Z_{k+1} - Z_k$. In this case, the areas become

$$\begin{aligned}
A_{R(j+1,k)} &= 2\pi R_{j+1/2}\Delta Z \\
A_{R(j,k)} &= 2\pi R_{j-1/2}\Delta Z \\
A_{z(j,k+1)} &= \pi(R_{j+1/2} + R_{j-1/2})(R_{j+1/2} - R_{j-1/2}) \\
A_{z(j,k)} &= A_{z(j,k+1)}.
\end{aligned}$$

For the axisymmetric problem in spherical coordinates (r, θ), Equation (6.88), Equation (6.91), and Equation (6.92) still apply. If we define v_r, v_θ, and v_ϕ as the velocity components in the r, θ, and ϕ directions, respectively, then Equation (6.89) and Equation (6.90) are modified as

$$\begin{aligned}
\frac{\partial(\rho v_r)}{\partial t} + \nabla\cdot(\rho v_r \mathbf{v}) &= -\rho\frac{\partial\Phi}{\partial r} - \frac{\partial P}{\partial r} + \frac{\rho}{r}(v_\theta^2 + v_\phi^2) \\
\frac{\partial(\rho v_\theta)}{\partial t} + \nabla\cdot(\rho v_\theta \mathbf{v}) &= -\frac{1}{r}\left(\rho\frac{\partial\Phi}{\partial\theta} + \frac{\partial P}{\partial\theta}\right) - \frac{\rho}{r}(v_r v_\theta - v_\phi^2\cot\theta).
\end{aligned}$$

6.6.2 Radiation Transport

Most astrophysical problems involve radiative transport of energy. The simplest approach to the problem of two-dimensional radiation transfer, which will be discussed below, is based on the diffusion approximation (Equation 1.117). The resulting numerical scheme may be included as a separate routine in the $2\frac{1}{2}$-D code described in Section 6.6.1.

According to Equation (1.117), under the assumption of local thermodynamic equilibrium the radiative flux of energy is related to the gradient of radiation energy density. If, as is the case in local thermodynamic equilibrium, the gas temperature and the radiation temperature are closely coupled, then the change in internal energy per unit volume of the gas, by radiation effects alone, is given by

$$\left.\frac{\partial e}{\partial t}\right|_{rad} = -\nabla \cdot \mathbf{F}. \tag{6.97}$$

The only change in the hydrodynamic equations is in the energy equation (6.67), which is now given by Equation (1.128)

$$\frac{\partial e}{\partial t} + \frac{\partial}{\partial x_j}(ev_j) = -P\frac{\partial v_j}{\partial x_j} - \nabla \cdot \mathbf{F}. \tag{6.98}$$

The radiation transfer term is treated as a separate operation, in the spirit of operator splitting. Taking the case of cylindrical symmetry of the previous section as an example, there are no temperature gradients in the ϕ-direction. However, there may be optically thin regions in the computational domain, in which Equation (1.117) does not apply, so a flux limiter λ must be introduced, yielding

$$\mathbf{F} = -\frac{c\lambda}{\kappa_R \rho}\boldsymbol{\nabla}(aT^4),$$

where λ is defined by Equation (1.122) or Equation (1.123). By analogy to Equation (1.127), the effect of radiation on the internal energy distribution can be described by

$$\rho c_V \frac{\partial T}{\partial t} = \frac{1}{R}\frac{\partial}{\partial R}\left(\frac{4acT^3 R\lambda}{\rho\kappa_R}\frac{\partial T}{\partial R}\right) + \frac{\partial}{\partial Z}\left(\frac{4acT^3\lambda}{\rho\kappa_R}\frac{\partial T}{\partial Z}\right), \tag{6.99}$$

where the specific heat c_V is assumed to be a constant. Equation (6.99) is just the diffusion equation, with the diffusion coefficient given by

$$\nu_d = \frac{4acT^3\lambda}{\kappa_R \rho^2 c_V}. \tag{6.100}$$

(Note that λ usually varies across the grid.) If we integrate it with an explicit scheme, the stability condition (Equation [6.83]) restricts the time step to

$$\Delta t \leq \frac{1}{16}\frac{\rho c_V T}{aT^4 \lambda}(\kappa_R \rho \Delta x)\left(\frac{\Delta x}{c}\right), \tag{6.101}$$

where $\kappa_R \rho \Delta x$ is the optical depth of the layer, $\rho c_V T$ the thermal energy density, and aT^4 the radiation energy density. In the center of the Sun, the ratio $\rho c_V T / aT^4$ is about 1000, $\kappa_R \approx 1$ cm^2 g^{-1}, and $\rho \approx 100$ g cm^{-3}. The total radius of the sun is 7×10^{10} cm, and to make a model with adequate spatial resolution, one would need about 100 cells, with thickness $\Delta x \approx 10^9$ cm. The time step, therefore, would have to be less than about 10^4 years. However, the nuclear evolution time of the Sun is 10^{10} yr, so the calculation of its evolution with explicit radiation transfer would involve on the order of one million time steps, during each of which there is negligible physical change. Thus, in this and many other situations, an implicit solution is required, as discussed in Chapter 2, Section 2.5.

Another situation to consider is a protoplanetary disk, which is still accreting material from the surrounding cloud. At 10 AU from the star, the density may be 10^{-11} g cm^{-3} and the temperature 100 K. The ratio of the thermal energy density to the radiation energy density is 10^5, $\kappa_R \rho \Delta x \approx 10$, and $\Delta x/c = 33$ if the resolution element Δx is taken to be 10^{12} cm, about 0.1 AU. The maximum Δt then works out to be 6×10^6 seconds, much shorter than the viscous evolution time scale of the disk, $\approx 10^5$ yr. If, however, the disk is evolved on a hydrodynamical time scale, to take into account the details of the flow, the CFL time step limit $\Delta x/c$, where c is the sound speed, is about 10^7 sec, not much longer. Nevertheless it is still worthwhile to do the radiation transport implicitly. Note also that the limit on the radiation time step decreases as $(\Delta x)^2$, while that resulting from the CFL condition decreases only as Δx.

To do the problem implicitly, we shall use the difference equation (2.73) with α set to 0, that is, with backward time differencing

$$\frac{u_j^{n+1} - u_j^n}{\Delta t} = \nu_d \frac{u_{j+1}^{n+1} + u_{j-1}^{n+1} - 2u_j^{n+1}}{(\Delta x)^2}. \tag{6.102}$$

During the radiation transfer step, κ_R, ρ, and T^3 can be considered to be constant in time, evaluated at the previous time step, but still functions of position. Then Equation (6.99) can be written in the form

$$\frac{\partial T}{\partial t} = \frac{1}{R}\frac{\partial}{\partial R}\left(KR\frac{\partial T}{\partial R}\right) + \frac{\partial}{\partial Z}\left(K\frac{\partial T}{\partial Z}\right) \tag{6.103}$$

where

$$K = \frac{4acT^3\lambda}{\kappa_R \rho^2 c_V} \tag{6.104}$$

is the diffusion coefficient, denoted by K instead of ν_d in order to simplify the subscripts. The two-dimensional problem is solved by breaking it up into a series of one-dimensional problems. First, at a fixed value of $R = R_j$, we write Equation (6.103) implicitly for the strip in the Z-direction (second term on the right-hand side) but explicitly in the R-direction (first term on the right-hand side). If the Z-values are indexed by k and the grid is uniform (i.e., ΔZ is constant), we get the following equation for the first approximation of T^{n+1} (time step $\Delta t/2$ is used)

$$\frac{{}^1T_{j,k}^{n+1} - T_{j,k}^n}{\Delta t} = L_j + M_k, \tag{6.105}$$

where

$$L_j = 2\left[\frac{R_{j+1}K_{1,j+1,k}\left(T_{j+1,k}^n - T_{j,k}^n\right) - R_j K_{1,j,k}\left(T_{j,k}^n - T_{j-1,k}^n\right)}{(R_{j+1} + R_j)(\Delta R)^2}\right],$$

and

$$M_k = \frac{K_{2,j,k+1}{}^1T^{n+1}_{j,k+1} + K_{2,j,k}{}^1T^{n+1}_{j,k-1} - K_{2,j,k+1}{}^1T^{n+1}_{j,k} - K_{2,j,k}{}^1T^{n+1}_{j,k}}{(\Delta Z)^2}.$$

(Note that on the staggered grid the Ts are evaluated at the centers of cells, the K_1s at the vertical edges in the (R, Z) plane where the u velocities are defined, and the K_2s on the horizontal edges where the w velocities are defined.) We then write this equation in the form analogous to Equation (2.79)

$$^1T^{n+1}_{j,k-1}A_k +^1 T^{n+1}_{j,k}B_k +^1 T^{n+1}_{j,k+1}C_k = -E_k \tag{6.106}$$

where j is fixed and

$$A_k = -\frac{K_{2,j,k}}{2(\Delta Z)^2}$$
$$C_k = -\frac{K_{2,j,k+1}}{2(\Delta Z)^2}$$
$$B_k = \frac{(K_{2,j,k} + K_{2,j,k+1})}{2(\Delta Z)^2} + \frac{1}{\Delta t}$$
$$E_k = -\frac{T^n_{j,k}}{\Delta t} - L_j/2.$$

Equation (6.106) can be easily solved for $^1T(Z)$ using the method described in Chapter 2, Section 2.5. The temperatures are updated at all points on the grid by solving along all one-dimensional strips in Z, each at a constant value of R. Each strip is solved independently of the others, using the old temperature values $T^n_{j,k}$. The intermediate values of temperature thus obtained, $^1T^{n+1}_{j,k}$, then play the role of the old temperature values $T^n_{j,k}$ in a set of sweeps in the R-direction, each at constant Z, to solve for the final $T^{n+1}_{j,k}$. For R-sweeps the Ls in Equation (6.105) are written implicitly and the Ms explicitly. The same time step $\Delta t/2$ is used.

Exercise

From Equation (6.103) set up a difference equation for solving for the temperature $T_{j,k}$ at a fixed k (corresponding to a given value of Z) as a function of j (i.e., for all values of R). Calculate the coefficients A_j, B_j, C_j, and E_j needed for the solution. You should come up with

$$B_j = \frac{R_{j+1}K_{1,j+1,k} + R_jK_{1,j,k}}{(R_{j+1} + R_j)(\Delta R)^2} + \frac{1}{\Delta t}. \tag{6.107}$$

This basic procedure is known as an *alternating direction implicit* technique. However, as formulated so far, with one set of sweeps in the R-direction and one set of sweeps in the Z-direction, it is by no means guaranteed to provide the correct solution to the original two-dimensional radiative transfer equation, since there is insufficient coupling between various regions of the grid. An iterative procedure is necessary. Peaceman and Rachford (1955) invented a scheme in which a pseudo-time t^* is introduced, and the radiative transfer equation (6.97) is replaced by

$$\frac{\partial e}{\partial t^*} = -\frac{\partial e}{\partial t} - \nabla \cdot \mathbf{F}. \tag{6.108}$$

The difference equation (6.105) is modified to become

$$\frac{T_{j,k}^{n+1} - T_{j,k}^{n}}{\Delta t^*} = L_j + M_k - \frac{\left(T_{j,k}^{n+1} - T_{j,k}^{n}\right)}{\Delta t}, \tag{6.109}$$

where L_j and M_k are alternately implicitly and explicitly differenced as above. The coefficients A_k and C_k remain unchanged, while B_k and E_k are slightly modified.

The system is relaxed to "steady state" $\partial e/\partial t^* = 0$ with a set of different values of the pseudo-time step Δt^*. The result should be a solution of the original differential equation. During each pseudo-time step the solution described above, involving a set of Z-sweeps followed by a set of R-sweeps, is carried out. After the solution has been repeated for all values of Δt^*, the results are substituted back into the original differential equation (6.97) to make sure that it is satisfied to a given level of accuracy. If not, the steady-state relaxation is continued until convergence is achieved. Convergence may be improved by taking closer intervals in the pseudo-time step; the optimum set of intervals is best determined by trial and error for a given problem.

Here is an example of a choice of successive values of Δt^*. First take minimum and maximum values of a dimensionless time, τ_{min} and τ_{max}, say, 1 and 3000, respectively. Divide the interval between them into M equal logarithmic intervals

$$\tau_m = \tau_{min}\xi^m; \quad m = 0, 1, \ldots M$$

where

$$\log \xi = \frac{1}{M}\left(\log \tau_{max} - \log \tau_{min}\right). \tag{6.110}$$

To determine the actual time steps, note that they must be related to the physical diffusion time of photons across the grid, and that both neighboring cells and distant cells must be coupled. The photon diffusion time in an optically thick medium is:

$$t_{diff} \approx \frac{\kappa\rho(\Delta x)^2}{c}, \tag{6.111}$$

where Δx is the length scale involved (Shu, 1992). To obtain the actual Δt^*, set $\tau_{min} = t_{diff}/4$, taking $(\Delta x)^2 = (\Delta R)^2 + (\Delta Z)^2$ to be comparable to the size of the smallest zone. Then, to provide good coupling on all scales, τ_{min} and τ_{max} should be in the approximate ratio

$$\frac{(\Delta R)^2 + (\Delta Z)^2}{(R_J^2 + Z_K^2)},$$

where J and K refer to the outermost cells in the grid (see Chapter 7, Section 7.2.2). The smallest value of Δt^* will couple neighboring cells, while the larger values will couple cells across the whole grid.

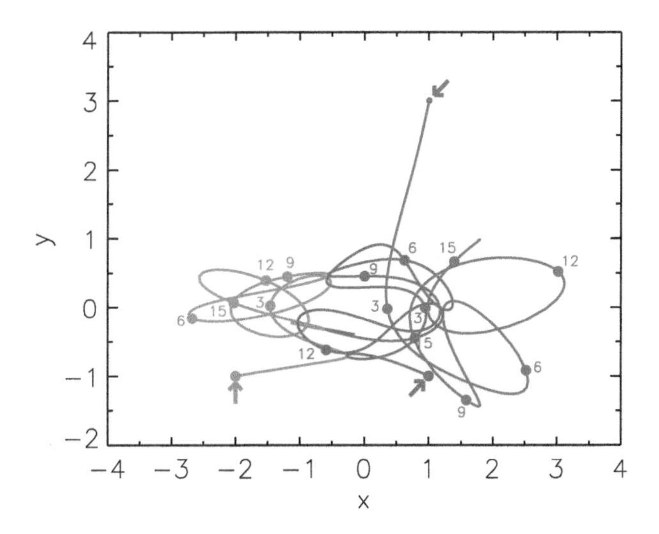

FIGURE 3.7 Integration of the three-body Pythagorean problem. The initial positions of the particles (indicated by arrows) are at the vertices of a right triangle. The particles have masses 3 (red line), 4 (green line) and 5 (blue line) grams. Positions in the (x, y) plane are expressed in cm. In these units, the unit of time is 3872 s. The equations are integrated for 16 time units, and the positions at selected times are marked on the curves. An initial close encounter between mass 4 and mass 5 occurs at about $x = -0.25$, $y = -0.75$. The end result of the simulation (not shown), which is determined at about $t = 60$, is the formation of a binary by masses 4 and 5 and the ejection of mass 3 from the system. (After Szebehely and Peters, 1967. Figure courtesy of Evan Kirby.)

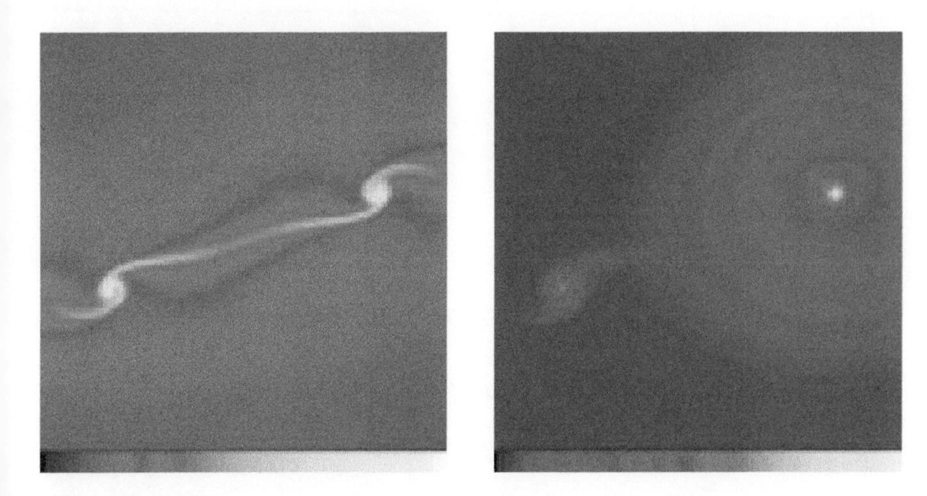

FIGURE 4.2 An SPH model, with 150,000 particles, including radiation transport, of the collapse of a protostellar cloud. *Left panel:* Column density integrated parallel to the rotation axis, on a logarithmic color scale with log N ranging from -1 to $+3.5$, where N is given in g cm^{-2}. *Right panel:* Gas temperature, weighted along each column according to the local mass, on a logarithmic color scale ranging from log $T = 1$ (black) to log $T = 3$ (white). Because of slight asymmetries in the SPH initial conditions, the protostar on the right has evolved slightly farther and has heated the surrounding gas more than the protostar on the left. (From Whitehouse, S. C. and Bate, M. R. 2006. *Monthly Notices, Royal Astronomical Society* **367**:32. With permission.)

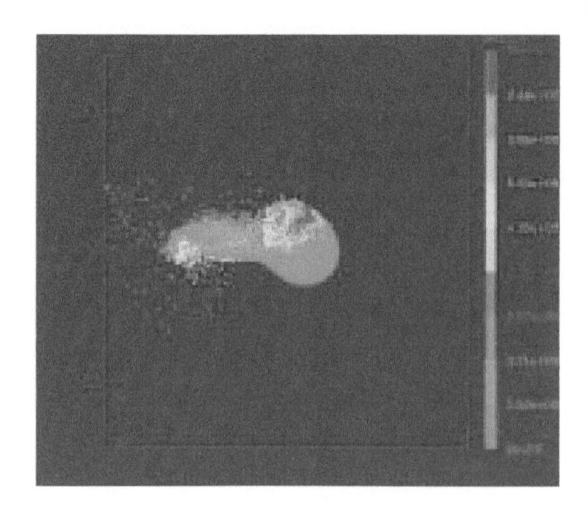

FIGURE 4.3 An SPH model of the collision between a Mars-sized object ($0.14\ M_\oplus$) with the protoearth ($0.81\ M_\oplus$) at a time of 0.86 hours. The color scale gives the temperature from 7000 K to 2000 K. The particle positions are projected onto the (x, y) plane. The linear scale is given in units of 1000 km. The total number of particles is 120,000. (From Canup, R. 2004 *Icarus* **168**:433. With permission.)

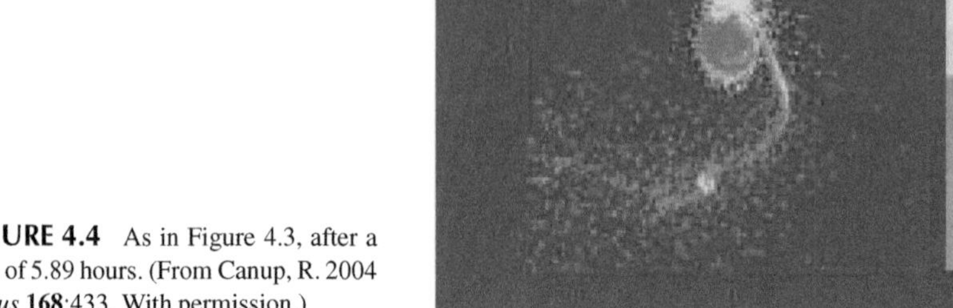

FIGURE 4.4 As in Figure 4.3, after a time of 5.89 hours. (From Canup, R. 2004 *Icarus* **168**:433. With permission.)

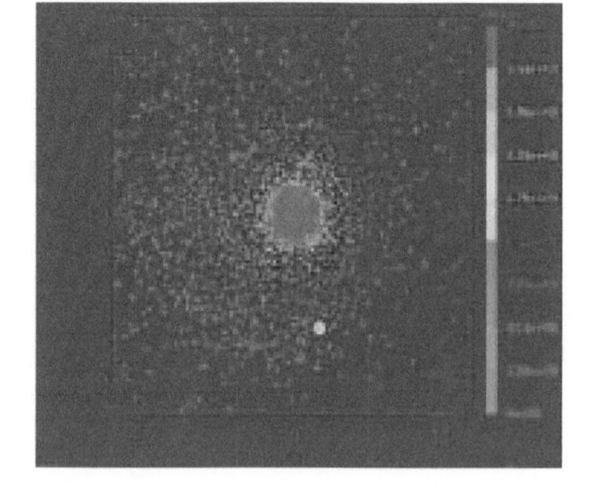

FIGURE 4.5 As in Figure 4.3 after a time of 21.9 hours. Retained material orbiting the Earth forms a disk with a clump with about 60% of a lunar mass. (From Canup, R. 2004 *Icarus* **168**:433. With permission.)

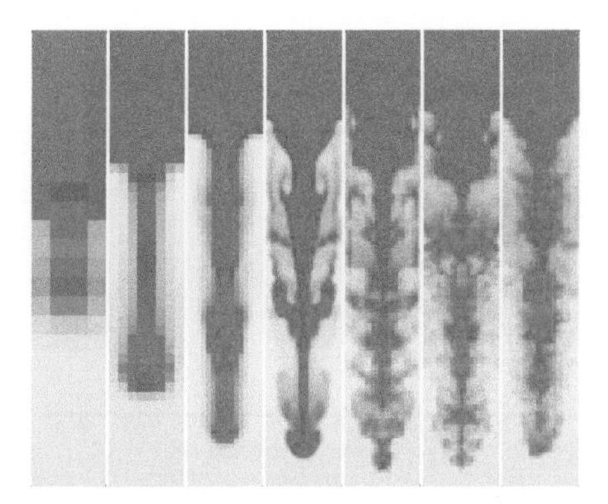

FIGURE 6.12 Three-dimensional Rayleigh–Taylor instability calculated with the piecewise-parabolic method and adaptive mesh refinement with seven different spatial resolutions. The calculations are done on a Cartesian (x, y, z) grid with gravity pointing in the downward $(-y)$ direction with a value of 1 cm s^{-2}. The color scale indicates density, in the (y, z) plane through the center of the grid in x, with very dark red corresponding to a density of 2.5 g cm^{-3} and very light yellow corresponding to a density of 0.5 g cm^{-3}. The time is 3.1 s after the start of the calculation in all cases. The grid has dimensions 0.25 cm $\times$ 1.5 cm $\times$ 0.25 cm, in the x-, y-, and z-directions, respectively. The number of grid points across the horizontal dimension (z) of the plane, from left to right, is 4, 8, 16, 32, 64, 128, and 256. (Images courtesy of Bruce Fryxell, based on calculations using the FLASH hydrodynamics code developed at the Center for Astrophysical Thermonuclear Flashes, University of Chicago, Chicago, IL.)

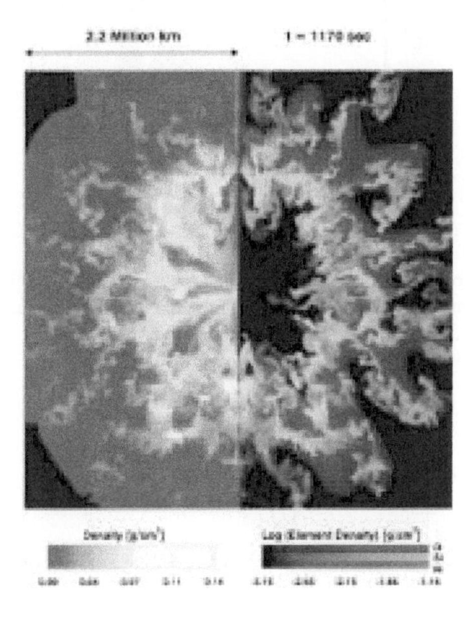

FIGURE 6.13 Hydrodynamic calculation showing Rayleigh–Taylor instability and mixing in a supernova explosion. The supernova shock has at this time expanded to a radius of 2.2×10^{11} cm, while the outer radius of the star (15 $M_\odot$) is at 3.9×10^{12} cm. The highest velocities in the outward-moving clumps are $\sim$ 4000 km s^{-1}. The left frame gives the distribution of the log of the total density on the two-dimensional axially symmetric grid. The right frame gives the partial densities of the elements oxygen, silicon, and nickel. (From Woosley, S. E. et al. 2002 *Rev. Mod. Phys.* **74**: 1015; Kifonidis, K. et al. 2003 *Astron. Astrophys.* **408**: 621. With permission.)

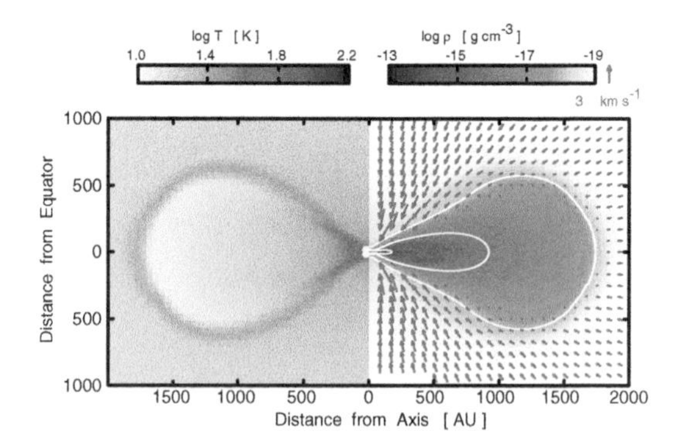

FIGURE 6.14 Two-dimensional hydrodynamic calculation with flux-limited radiation transport of the collapse of a protostar of 2 $M_{\odot}$. Temperature structure (*left*) and density structure (*right*) of the resulting disk in the (R, Z) plane. Arrows indicate the velocity, with length proportional to speed.

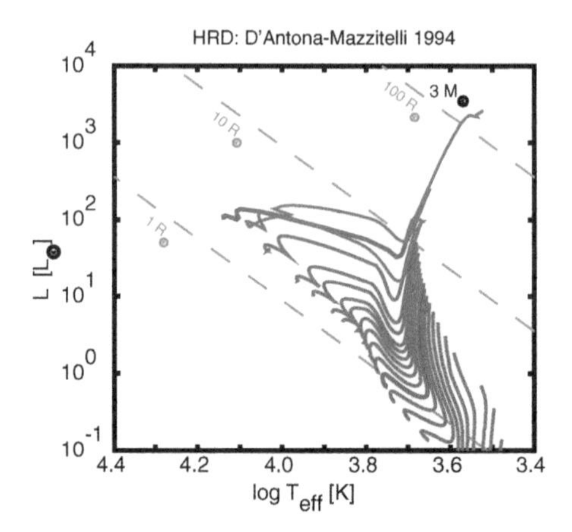

FIGURE 10.2 Evolutionary tracks in the $(\log T_{eff}, \log L/L_{\odot})$ diagram. The blue and red curves are pre-main-sequence tracks computed by D'Antona and Mazzitelli (1994). The upper and lower red curves are computed for 3 and 1 $M_{\odot}$, respectively. The orange curve for 3 $M_{\odot}$ was produced by the code STELLAR, and includes both pre-main-sequence and post-main-sequence phases.

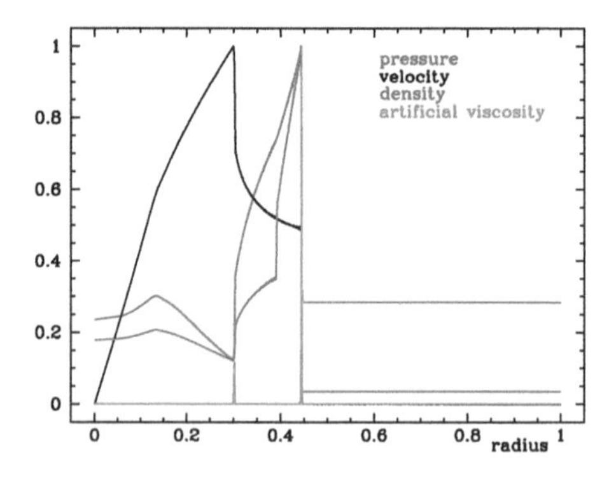

FIGURE 10.3 Illustrations of the results from the code "lh1.f" after 16,000 time steps.

6.6.3 Thin Circumstellar Disk

In Section 6.6.1, the three-dimensional problem was reduced to two dimensions by the assumption of axial symmetry. Another technique that allows a reduction in the number of effective dimensions of the simulation can be applied in the case of thin circumstellar disks. If we consider again cylindrical coordinates (R, Z, ϕ), we can integrate the equations of hydrodynamics over the Z-direction and follow the evolution of the integrated quantities in the (R, ϕ) plane only. The integration of the mass density $\rho(R, Z, \phi)$ yields the surface density (mass per unit area), $\Sigma(R, \phi)$

$$\Sigma(R, \phi) = \int_{-\infty}^{+\infty} \rho(R, Z, \phi) dZ. \tag{6.112}$$

The continuity equation becomes

$$\frac{\partial \Sigma}{\partial t} + \nabla \cdot (\Sigma \mathbf{v}) = 0, \tag{6.113}$$

where the velocity field is $\mathbf{v} = (v_r, v_\phi)$. This approximation is reasonable because in many situations the thickness of the disk $H(R)$ is only 5 to 10% of the distance to the star R. Similarly, we can integrate the pressure $P(R, Z, \phi)$ to obtain the two-dimensional pressure $p(R, \phi)$. One usually assumes that p and Σ are related by an equation of state

$$p = c_s^2 \Sigma, \tag{6.114}$$

where c_s is the local sound speed.

A circumstellar disk is generally assumed to be in near-Keplerian rotation with force balance in both the radial direction and in the vertical direction (parallel to the angular momentum vector). The gravitational field is often given just by the point-mass potential of the star, but if the disk itself has a mass that is nonnegligible compared with that of the star, the self-gravity of disk matter must also be considered (Chapter 7, Section 7.1.2). A disk can evolve through (1) gravitational interaction with an embedded object, such as a giant planet or binary companion, (2) viscous effects, (3) instabilities associated with the self-gravity of the disk, (4) turbulence induced by magnetic effects, or (5) external gravitational perturbations. The equations presented here are most relevant to the first of these situations.

It is convenient, although not absolutely necessary, to follow the evolution of the disk in a coordinate system that rotates, for example, with the constant angular velocity Ω of the circular orbit of an embedded planet at a distance R_p from the star. Thus, the coordinate system follows the orbital motion of the planet, and advection terms in the ϕ-direction in the vicinity of the planet become negligible. Let the mass of the star and the planet be, respectively, M_* and M_p. The mass of the disk is assumed to be negligible. If the coordinate system originates at the center of mass, the gravitational potential at each point $\mathbf{R}$ in the grid is:

$$\Phi = -\frac{GM_*}{|\mathbf{R} - \mathbf{R}_*|} - \frac{GM_p}{|\mathbf{R} - \mathbf{R}_p|} \tag{6.115}$$

where $\mathbf{R}_*$ and $\mathbf{R}_p$ are the radius vectors from the origin to the positions of the star and planet, respectively. The angular velocity of the coordinate system is:

$$\Omega = \sqrt{\frac{G(M_* + M_p)}{a^3}}, \tag{6.116}$$

where $a = R_* + R_p$ is the semimajor axis of the planet's orbit. The two components of the momentum equation then become

$$\frac{\partial(\Sigma v_r)}{\partial t} + \nabla \cdot (\Sigma v_r \mathbf{v}) = \Sigma R(\omega + \Omega)^2 - \frac{\partial p}{\partial R} - \Sigma \frac{\partial \Phi}{\partial R} \tag{6.117}$$

$$\frac{\partial[\Sigma R^2(\omega + \Omega)]}{\partial t} + \nabla \cdot [\Sigma R^2(\omega + \Omega))\mathbf{v}] = -\frac{\partial p}{\partial \phi} - \Sigma \frac{\partial \Phi}{\partial \phi}, \tag{6.118}$$

where the angular velocity with respect to the grid $\omega = v_\phi / R$. Note that the expression $[\Sigma R^2(\omega + \Omega)]$ is the angular momentum per unit area, entirely analogous to the quantity A that appears in Equation (6.87). In the radial equation of motion, the expression $R(\omega + \Omega)^2$ takes into account the centrifugal acceleration and the radial component of the Coriolis force. An energy equation analogous to Equation (6.92) can also be written, but it is often assumed that a polytropic relation holds: $p = K\Sigma^\gamma$. The equations are then solved according to the techniques described in Section 6.5.

6.7 EXAMPLES

6.7.1 RAYLEIGH–TAYLOR INSTABILITY

The Rayleigh–Taylor instability occurs when a heavy fluid rests on top of a light fluid in a gravitational field or, equivalently, when a heavy fluid is accelerated by a light fluid. On a dynamical time scale, fingers of heavy fluid begin to penetrate the light fluid. The growth rate of the perturbation is fastest for the smallest scales, scaling approximately as the square root of the wave number. The calculation of the development of the instability is often used as a test for numerical codes because a growth rate from a linear analysis is available for comparison and because experimental results are available (Calder et al., 2002).

An example of a calculation of the instability is shown in Figure 6.12. The initial condition is a heavy fluid, with density 2 g cm^{-3}, lying on top of a light fluid, with density 1 g cm^{-3}, in a constant gravitational field. There are no other forces. The fluids are at rest (in unstable equilibrium) except for a single-mode sinusoidal perturbation in the vertical (y)-velocity of amplitude 2.5% of the local sound speed and with wavelength $\lambda = 0.25$ cm. The boundary conditions in the x- and z- (horizontal) directions are periodic, and at the top and bottom of the box in the y-direction assume hydrostatic equilibrium. The plots show the results after 3.1 sec with increasing spatial resolution. Zones have equal values of Δx, Δy, and Δz, so the increase in the number of zones in the y-direction is proportional to that in the z-direction. It is important to note that (1) as the resolution is increased, more small-scale structure develops, so that the calculation never actually strictly converges; (2) about 25 grid points per wavelength are required to match the growth rate correctly, as obtained from linear

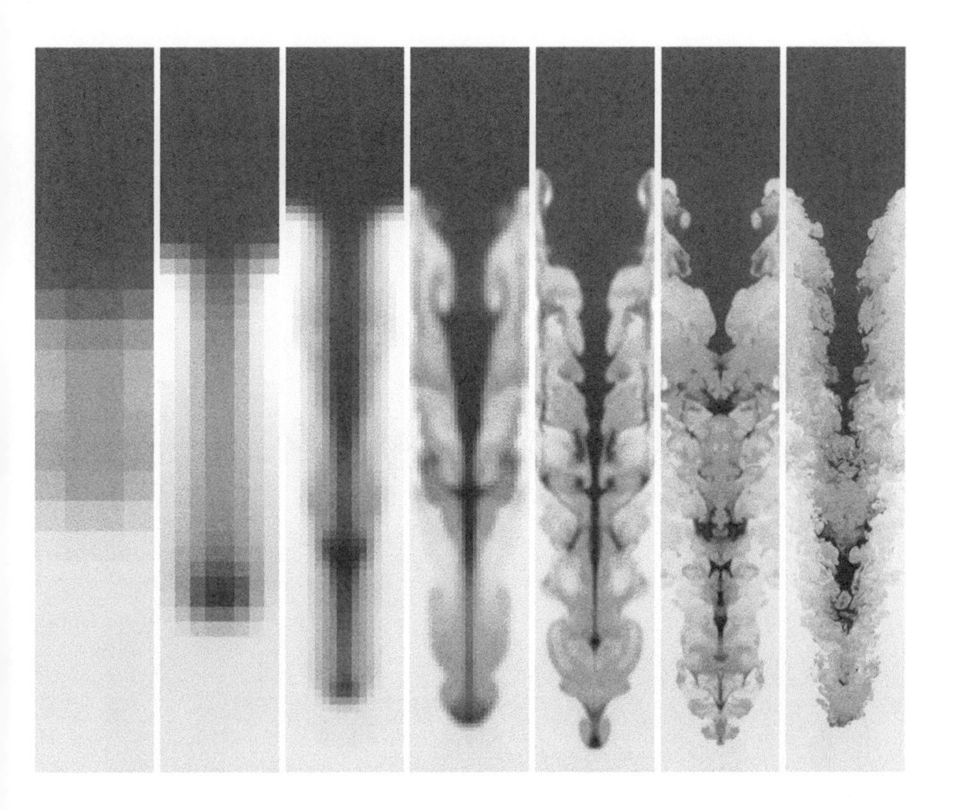

FIGURE 6.12 ***A color version of this figure follows page 212*** Three-dimensional Rayleigh–Taylor instability calculated with the piecewise-parabolic method and adaptive mesh refinement with seven different spatial resolutions. The calculations are done on a Cartesian (x, y, z) grid with gravity pointing in the downward $(-y)$ direction with a value of 1 cm s^{-2}. The greyscale indicates density, in the (y, z) plane through the center of the grid in x, with very dark corresponding to a density of 2.5 g cm^{-3} and very light corresponding to a density of 0.5 g cm^{-3}. The time is 3.1 sec after the start of the calculation in all cases. The grid has dimensions 0.25 cm × 1.5 cm × 0.25 cm, in the x-, y-, and z-directions, respectively. The number of grid points across the horizontal dimension (z) of the plane, from left to right, is 4, 8, 16, 32, 64, 128, and 256. (Images courtesy of Bruce Fryxell, based on calculations using the FLASH hydrodynamics code developed at the Center for Astrophysical Thermonuclear Flashes, University of Chicago, Chicago, IL.)

theory, and (3) the calculations only qualitatively match experimental results, although there are some difficulties in obtaining an exact comparison (Calder et al., 2002). For this classical situation, the linear stability analysis gives an exponential growth with characteristic time

$$t_{\mathrm{RT}} = \left(2\pi \frac{\lambda}{g} \cdot \frac{\rho_2 + \rho_1}{\rho_2 - \rho_1}\right)^{1/2}. \tag{6.119}$$

In the example, $t_{\mathrm{RT}} = 2.17$ sec.

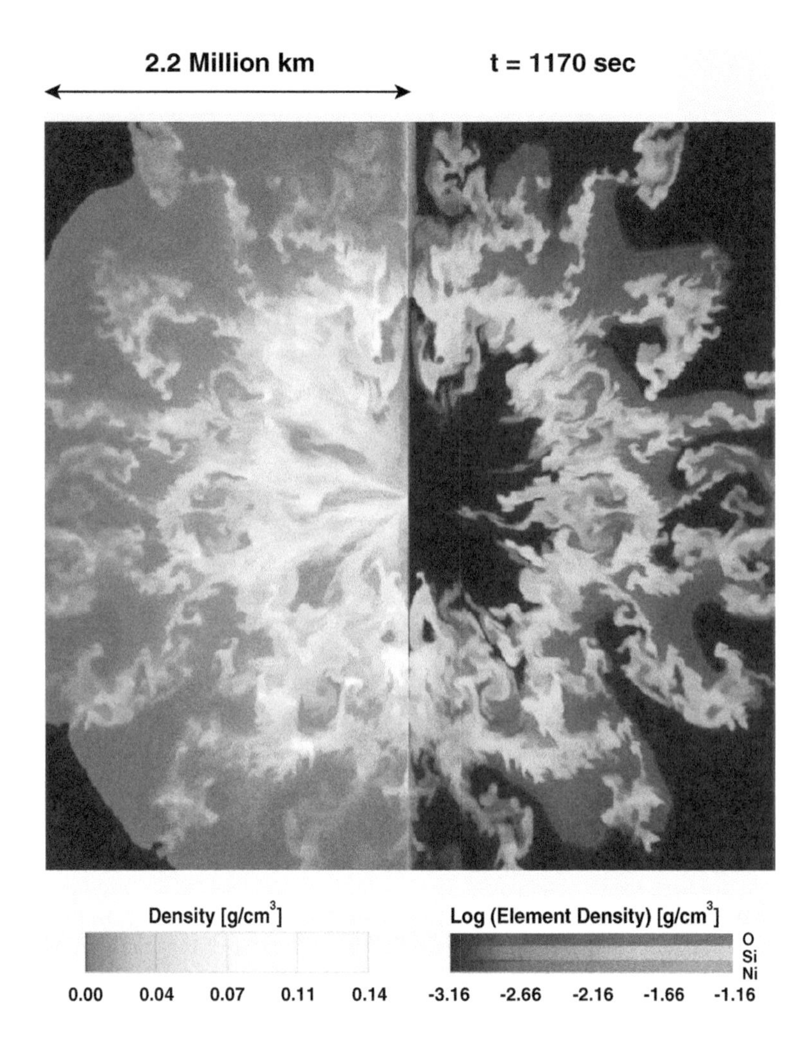

FIGURE 6.13 ***A color version of this figure follows page 212*** Hydrodynamic calculation showing Rayleigh–Taylor instability and mixing in a supernova explosion. The supernova shock has at this time expanded to a radius of 2.2×10^{11} cm, while the outer radius of the star (15 $M_\odot$) is at 3.9×10^{12} cm. The highest velocities in the outward-moving clumps are $\sim$ 4000 km s^{-1}. The left frame gives the distribution of the log of the total density on the two-dimensional axially symmetric grid. The right frame gives the partial densities of the elements oxygen, silicon, and nickel. (From Woosley, S. E. et al. (2002) *Rev. Mod. Phys.* **74**: 1015; Kifonidis, K. et al. (2003) *Astron. Astrophys.* **408**: 621. With permission.)

6.7.2 Supernova Explosion

Figure 6.13 shows a snapshot in the early development of a supernova explosion (Woosley et al., 2002). The axially symmetric calculations are performed on a grid in spherical coordinates (r, θ), and there is no assumption of symmetry across the equatorial plane. The symmetry axis is the vertical axis in the figure. The left frame shows the distribution of density when the supernova shock has reached a radius of 2.2×10^{11} cm; the right frame shows the abundances of significant elements — oxygen, silicon, nickel — at the same time. The calculations were performed by Kifonidis et al. (2003).

The computational procedure employed here consists of three stages. The first stage is a stellar evolution calculation of a spherically symmetric star of 15 $M_\odot$ and heavy element abundance 1/4 that of the Sun by Woosley et al. (1988). The evolution begins at the main sequence, proceeds through the red giant region and back to the blue supergiant region, and ends when an iron core with a radius of 3.9×10^{12} cm and a mass of 1.3 $M_\odot$ has formed. The iron core then collapses to form a neutron star, which bounces back somewhat and also emits neutrinos, which interact with the material at the outer edge of the core and drive the expansion of the outer parts of the star (Bruenn, 1993). The second stage of the calculation involves a two-dimensional axisymmetric grid with 400 radial zones (3×10^6 cm $< r < 2 \times 10^9$ cm) and 192 zones in θ (from 0 to π). Apart from an inward motion of the inner boundary, the grid structure is fixed. The Eulerian hydrodynamic equations are solved with the piecewise parabolic method for handling advection. To initiate two-dimensional effects the velocities are perturbed randomly with an amplitude of about 0.1%. The nuclear transformations in the explosion are followed with a 14-isotope network. This phase of the calculation is carried out for approximately 1 sec, after which the effects of neutrinos in driving the expansion can be neglected.

The final phase follows the shock wave and resulting Rayleigh–Taylor instabilities out to a distance beyond the initial radius of the star, with a two-dimensional code that includes adaptive mesh refinement. Four levels of refinement are used on a base grid of 48 by 12 zones, and the spatial resolution is increased by a factor of 4 in both dimensions in each refinement. The computational time is estimated to have been reduced by a factor of 5, compared to a simulation without AMR at comparable spatial resolution. The inner and outer boundaries are initially set at 10^8 cm and 4.8×10^9 cm, respectively. The outer boundary moves outward as necessary as the shock approaches it. The inner boundary is also moved gradually outward, to avoid a very short CFL time step there. The radial boundary conditions allow free outflow across both inner and outer surfaces, and the angular boundary conditions are reflecting at $\theta = 0$ and $\theta = \pi$. The simulation shows how nickel, previously synthesized in the inner regions of the star during the core collapse phase, mixes outward into the helium-rich layer of the expanding envelope, as a result of the Rayleigh–Taylor instability.

6.7.3 Protostar Collapse and Disk Formation

The collapse of a rotating interstellar cloud generally leads to a central protostar surrounded by a disk. The physical properties of such a disk are very important with regard to planet formation. Figure 6.14 shows an example, computed with a two-dimensional axisymmetric grid-based hydrodynamic code (Yorke and Bodenheimer, 1999), of the formation phase of a disk.

The calculation starts with a slowly rotating, density-peaked, two solar mass protostellar clump with a radius of 2×10^{17} cm, a temperature of 10 K, and a mean density of only 10^{-19} g cm^{-3}. Its total random kinetic energy of particle motions is smaller in absolute value than the gravitational energy, so that it is unstable to collapse. The mean density of the Sun is about 1 g cm^{-3}, thus, the protostellar phase of evolution involves an increase in mean density of nearly 17 orders of magnitude (the radius at the end of collapse is roughly five solar radii). Needless to say, to follow such a collapse numerically is quite challenging.

The calculations shown are performed in (R, Z) coordinates, thus, the hydrodynamics are treated as discussed in Section 6.6. Because of symmetry about the rotation axis and about the equatorial plane, the calculations can be done in the first quadrant of the (R, Z) plane. Two numerical procedures are used to allow the calculation to be carried out for a reasonable time, which in this case is the free fall time of the initial cloud, about 2×10^5 yr. First, a set of four nested grids is used, each grid with one-half the radial extent of the next outermost grid. Thus, the radius R_{max} of the innermost grid is 2.5×10^{16} cm. Each grid has 124×124 points, so the resolution on the innermost grid is 2×10^{14} cm or 13.3 AU. These grids are set up in the spirit of adaptive mesh refinement, except that they are fixed in space and time because it is known *a priori* that the collapse will result in concentration of material toward the center. Clearly the inner regions, down to the stellar scale of 10^{11} cm, are not resolved. In fact, at the time the calculations were made, it was not feasible to resolve the inner regions, which would result in a CFL time step limit of $\sim 10^3$ sec, and at the same time follow the entire collapse of the outer regions for more than one free fall time, $\approx 10^{13}$ sec. Thus, the second numerical procedure was to make the central grid cell into a so-called "sink" cell, which was allowed to accrete mass and angular momentum, but outflow was not allowed. This central region represents the star and inner disk. Its mass is included in the gravity calculation, and a simple model of it produces its accretion luminosity, which is used for the inner boundary condition for radiation transfer calculations. This procedure results in sufficient numerical resolution in the region where the disk is expected to form. The CFL step on the innermost grid was about 10^8 sec, and the calculation was continued for about 10^5 time steps.

The cloud begins to collapse, and initially it is optically thin to its own radiation, so it remains isothermal at 10 K until the central regions reach a density of 10^{-13} g cm^{-3} at which point they begin to heat as a result of compression (Hayashi, 1966). Beyond this point it is necessary to include radiation transfer to correctly take into account energy loss from the cloud and the increase in internal temperature. A flux-limited diffusion calculation in 2D was included with the hydrodynamics, according to the alternating-direction implicit technique described in Section 6.6.2. The assumption made in that section was that the radiation was very closely coupled to the gas; in the actual calculation the gas temperature T and the radiation temperature T_R were allowed to differ (see Chapter 9).

After about 17,000 years, a central hydrostatic core forms, which eventually builds up into a star by accretion of material from the cloud as it continues to fall toward the center. The kinetic energy of the infalling material is converted into heat just above the surface of the protostar and this energy is radiated away. This so-called accretion luminosity is responsible for heating the immediate surroundings of the protostar and even sporadically slowing down the infall. Eventually, the infalling material has too much angular momentum and it can no longer fall directly onto the protostar. A disk forms. Angular momentum transport processes allow further accretion through the

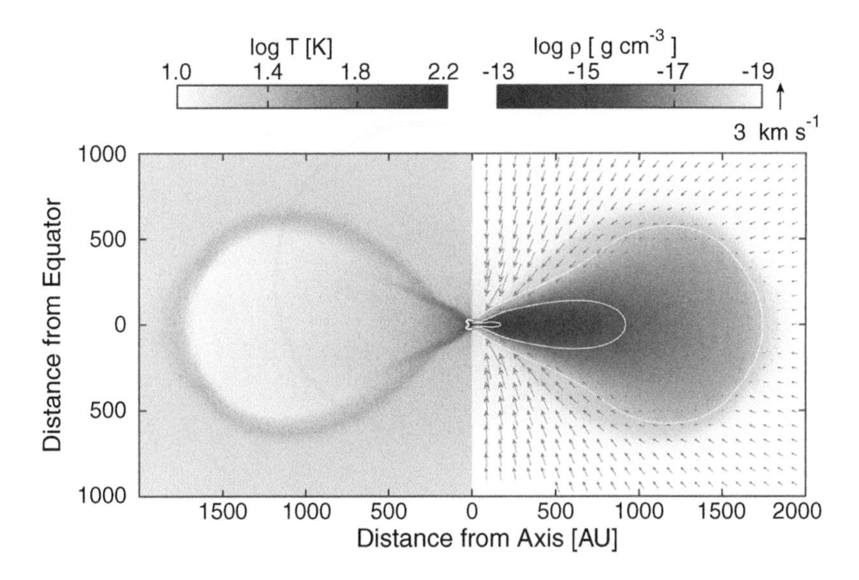

FIGURE 6.14 ***A color version of this figure follows page 212*** Two-dimensional hydrodynamic calculation with flux-limited radiation transport of the collapse of a protostar of 2 $M_\odot$. The figure shows the temperature structure (*left*) and the density structure (*right*) of the resulting disk in the (R, Z) plane, where the Z axis is the rotation axis. Arrows indicate the collapse velocity, with length proportional to speed, where the reference arrow above the plot corresponds to 3 km s^{-1}. The full computational grid extends to 13,333 AU in each direction; here only the innermost nested grid is shown.

disk. However, material in the outer disk regions gains angular momentum and the disk grows in size and mass as even more infalling gas hits the disk. Accretion shocks develop around the accretion disk, in which the infalling material, collapsing at nearly free-fall velocity, is decelerated to practically zero velocity behind the shock.

Figure 6.14 shows the situation at a time of 398,000 yr after the beginning of collapse, near the end of the calculation. The total mass in the cloud is 2.02 $M_\odot$, and the mass of the protostellar core at the time is $M_c = 1.12\ M_\odot$; it is unresolved at the origin of the plot. The radius of the protostellar core $R_c = 3.5\ R_\odot$, the accretion luminosity from the core $L_{tot} = 7.4\ L_\odot$, and the rate of accretion of disk material onto the protostar is 10^{-6} solar masses per year. The diagonal dark bands emerging from the origin in the temperature plot indicate the shock front locations. The calculation was carried to 589,000 yr, at which time 40% of the original mass is still in the disk. However, the disk evolution time scale becomes too long to follow it further with a CFL-limited hydrocode.

6.7.4 Spiral Waves in a Thin Self-Gravitating Disk

Spiral waves in disks have been extensively studied with regard to both galactic structure and planet formation. A disk with a reasonable mass compared with that of the

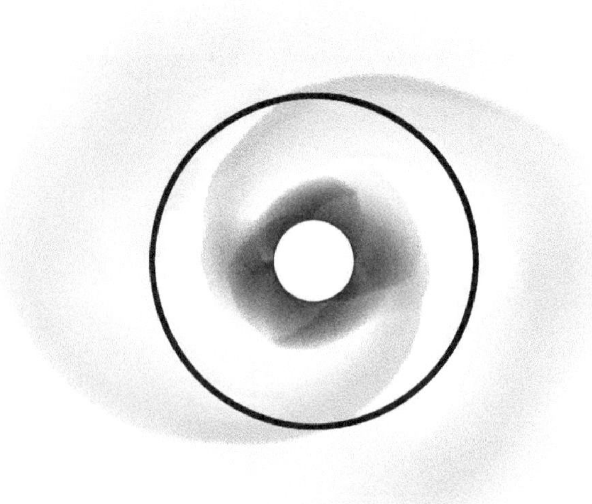

FIGURE 6.15 Generation of a two-armed spiral wave in a thin self-gravitating disk with a central star. The plot shows surface density in the (R, ϕ) plane a few rotation periods after the start of the calculation, which was performed with a two-dimensional grid-based Eulerian hydrocode. The density scale is logarithmic, with light grey corresponding to $\ln \rho = -5.5$ and dark grey to $\ln \rho = 1.5$. The circle refers to the initial outer radius of the disk. (From Laughlin, G. P. et al. (1997) *Astrophys. J.* **477**: 410. With permission.)

central object in the system can generate such waves, with two possible outcomes — saturation of the waves at a modest overdensity with respect to the surrounding matter, or continuing growth of the density in the waves, leading relatively quickly to fragmentation into self-gravitating clumps. The example shown in Figure 6.15 is a case in which the spiral mode saturates (Laughlin et al., 1997).

The disk that is shown in the figure has two-thirds the mass of the central object. It is assumed to have a Gaussian profile of surface density

$$\Sigma(R) = \Sigma_0 \exp[-(R - R_0)^2/s^2]. \tag{6.120}$$

The outer radius of the disk $R_{\max}$ is set to 1, the radius R_0 where the Gaussian peaks is 0.45, the constant s^2 is set to 0.05, and the inner edge of the disk is $R_{\min} = 0.25$. The equation of state is assumed to be a polytrope of index 1. With these assumptions, the growth of nonaxisymmetric modes in the disk can be analyzed with a linear stability analysis and, furthermore, the nonlinear interactions between the growing modes can be studied by a semianalytic method that includes second-order terms. Thus, this configuration provides an excellent check on a grid-based numerical code.

The disk is assumed to be thin in the direction parallel to the axis of rotation, and the numerical simulation is carried out in polar coordinates (R, ϕ). The grid spacing is uniform in the angular coordinate and logarithmic in the radial coordinate; there are 256 zones in each direction. The advection is handled by the second-order van Leer scheme. The self-gravitational potential of the disk is calculated with the thin-disk approximation described in Chapter 7, Section 7.1.2.

At the initial time, the disk is assumed to be azimuthally symmetric except for a random density pertubation in each zone with a maximum amplitude of only .01% of the unperturbed state, and it is in hydrostatic equilibrium. A two-armed spiral wave appears and grows, with properties in excellent agreement with the semianalytical treatment. The amplitude saturates, no fragmentation takes place, and the end result of the gravitational torques that are generated is the transfer of mass inward and angular momentum outward. The main point of this example is that a fairly straightforward, grid-based numerical solution of the equations of hydrodynamics and the Poisson equation in two space dimensions can give an accurate representation, for a particular case of an assumed disk structure, of the growth rate and structure of spiral modes, as well as of the nonlinear effects of redistribution of mass and angular momentum in the disk.

REFERENCES

Benz, W. (1991) *Late Stages of Stellar Evolution: Computational Methods in Astrophysical Hydrodynamics* (Lecture Notes in Physics, **373**) C. B. De Loore, Ed. (Berlin: Springer-Verlag), p. 259.

Berger, M. J. and Colella, P. (1989) *J. Comp. Phys.* **82**: 64.

Berger, M. J. and Oliger, J. (1984) *J. Comp. Phys.* **53**: 484.

Bowers, R. L. and Wilson, J. R. (1991) *Numerical Modeling in Applied Physics and Astrophysics* (Boston: Jones and Bartlett).

Bruenn, S. W. (1993) *Nuclear Physics in the Universe*, M. W. Guidry, and M. R. Strayer, Eds. (Bristol: Institute of Physics), p. 31.

Calder, A. C. et al. (2002) *Astrophys. J. Suppl.* **143**: 201.

Colella, P. and Woodward, P. (1984) *J. Comp. Phys.* **54**: 174.

D'Angelo, G., Henning, T., and Kley, W. (2002) *Astron. Astrophys.* **385**: 647.

Fryxell, B., Olson, K., Ricker, P., Timmes, F. X., Zingale, M., Lamb, D. Q., MacNeice, P., Rosner, R., Truran, J. W., and Tufo, H. (2000) *Astrophys. J. Suppl.* **131**: 273.

Godunov, S. K. (1959) *Mat. Sborn.* **47**: 271.

Hawley, J. F., Smarr, L. L., and Wilson, J. R. (1984) *Astrophys. J.* **277**: 296.

Hayashi, C. (1966) *Annu. Rev. Astron. Astrophys.* **4**: 171.

Kifonidis, K., Plewa, T., Janka, H.-Th., and Müller, E. (2003) *Astron. Astrophys.* **408**: 621.

Klein, R. I. and McKee, C. F. (1994) *Numerical Simulations in Astrophysics* J. Franco, S. Lizano, L. Aguilar, and E. Daltabuit, Eds. (Cambridge: Cambridge University Press), pp. 251–266.

Landau, L. D. and Lifschitz, E. M. (1959) *Fluid Mechanics* (London: Pergamon Press).

Laughlin, G. P., Korchagin, V., and Adams, F. C. (1997) *Astrophys. J.* **477**: 410.

Müller, E. (1994) *Galactic Dynamics and N-Body Simulations* (Lecture Notes in Physics, **433**) G. Contopoulos, N. K. Spyrou, and E. Vlahos Eds. (Berlin: Springer-Verlag) p. 313.

Peaceman, D. W. and Rachford, H. H. (1955) *J. Soc. Indus. Applied Math.* **3**: 28.

Potter, D. (1973) *Computational Physics* (New York: John Wiley & Sons).

Sedov, L. I. (1959) *Similarity and Dimensional Methods in Mechanics* (New York: Academic Press).

Shu, F. H. (1992) *The Physics of Astrophysics*: vol. II (Mill Valley, CA: University Science Books).

Stone, J. M. and Norman, M. L. (1992) *Astrophys. J. Suppl.* **80**: 753.

Toro, E. F. (1999) *Riemann Solvers and Numerical Methods for Fluid Dynamics: A Practical Introduction* (New York: Springer-Verlag).

Truelove, J. K., Klein, R. I., McKee, C. F., Holliman, J. H., Howell, L. H., and Greenough, J. A. (1997) *Astrophys. J.* **489**: L179.

Truelove, J. K., Klein, R. I., McKee, C. F., Holliman, J. H., Howell, L. H., Greenough, J. A., and Woods, D. T. (1998) *Astrophys. J* **495**: 821.

Tscharnuter, W. M. and Winkler, K.-H. (1979) *Comp. Phys. Comm.* **18**: 171.

van Leer, B. (1977) *J. Comp. Phys.* **23**: 276.

von Neumann, J. and Richtmyer, R. D. (1950) *J. Appl. Phys.* **21**: 232.

Woosley, S. E., Heger, A. and Weaver, T. A. (2002) *Rev. Mod. Phys.* **74**: 1015.

Woosley, S. E., Pinto, P. A., and Ensman, L. (1988) *Astrophys. J.* **324**: 466.

Yorke, H. W. and Bodenheimer, P. (1999) *Astrophys. J.* **525**: 330.

Yorke, H. W. and Kaisig, M. (1995) *Comp. Phys. Comm.* **89**: 29.

7 Poisson Equation

The Poisson equation is an example of an elliptic equation in which the solution is not explicitly time-dependent. A distribution of density ρ as a function of position in three-dimensional space is given, and it is desired to find the gravitational potential Φ at each point, from which the gravitational force can be derived. Specification of the boundary conditions is a key element of the problem. In Cartesian coordinates, the equation can be written

$$\frac{\partial^2 \Phi}{\partial x^2} + \frac{\partial^2 \Phi}{\partial y^2} + \frac{\partial^2 \Phi}{\partial z^2} = 4\pi G \rho(x, y, z), \tag{7.1}$$

where G is the gravitational constant and where boundary values Φ_B are specified.

In a cylindrical coordinate system (R, Z, ϕ), where R is the distance to the axis of the cylinder, the equation is:

$$\frac{1}{R}\frac{\partial}{\partial R}\left(R\frac{\partial \Phi}{\partial R}\right) + \frac{1}{R^2}\frac{\partial^2 \Phi}{\partial \phi^2} + \frac{\partial^2 \Phi}{\partial Z^2} = 4\pi G \rho. \tag{7.2}$$

In spherical polar coordinates (r, θ, ϕ), where r is the distance to the origin, the equation is:

$$\frac{\partial^2 \Phi}{\partial r^2} + \frac{2}{r}\frac{\partial \Phi}{\partial r} + \frac{1}{r^2}\left[\frac{1}{\sin^2\theta}\frac{\partial^2 \Phi}{\partial \phi^2} + \frac{1}{\sin\theta}\frac{\partial}{\partial \theta}\left(\sin\theta\frac{\partial \Phi}{\partial \theta}\right)\right] = 4\pi G \rho. \tag{7.3}$$

The formal solution to this equation (in Cartesian coordinates) is:

$$\Phi(x, y, z) = -\int_\tau \frac{G\rho(x', y', z')}{|\mathbf{x} - \mathbf{x}'|} dx'dy'dz' \tag{7.4}$$

where $\mathbf{x} = (x, y, z)$ is the point at which the gravity is being calculated, and τ is a finite volume containing all the mass contributing to the potential (Margenau and Murphy, 1956, Section 7.17). Implicit in this solution is the boundary condition that Φ vanish at infinity at least as fast as $1/|\mathbf{x}|$. Thus, it is usual to set the potential equal to zero at infinity. With the standard definition that the gravitational force per unit mass $\mathbf{g} = -\nabla\Phi$, it is easily shown that the potential at the surface of a sphere of radius R and mass M is

$$\Phi = -\frac{GM}{R} \tag{7.5}$$

representing the work done by the forces of gravity on a particle of unit mass as it moves, in an otherwise empty universe, from an infinite distance to the surface of the sphere.

A number of different methods are available to solve Equation (7.1) and Equation (7.4). They include:

1. Direct integration of Equation (7.4), often used in particle methods.
2. Direct integration of Equation (7.4) in combination with Fourier transforms.
3. Integration of Equation (7.4) by polynomial expansion.
4. Solution of Equation (7.1) by polynomial expansion.
5. Solution of Equation (7.1) using Fourier transforms.
6. Solution of Equation (7.1) on a grid by relaxation techniques.

In the following section (Section 7.1), we consider methods for solving the integral form of the Poisson equation by the methods (1), (2), and (3), while in Section 7.2 we outline techniques for solving the differential form of the equation by the methods (4), (5), and (6). This chapter is intended to give a sampling of the available solution methods rather than a complete catalog.

7.1 POISSON SOLUTIONS: I

7.1.1 DIRECT SUMMATION

In particle methods (see Chapter 3), the direct summation over all particles implied by Equation (7.4) is used if the number of particles is not too large. In this case, it is usually more convenient just to calculate the forces directly on each particle caused by the gravitational attraction of all other particles. The singularity when $\mathbf{x}$ approaches $\mathbf{x}'$ is generally handled by the introduction of a *softening parameter* ϵ (Binney and Tremaine, 1987, Section 2.8). Thus, the force between two particles with masses m_i and m_j is:

$$\mathbf{F}_{ij} = \frac{G m_i m_j (\mathbf{x}_j - \mathbf{x}_i)}{(\epsilon^2 + |\mathbf{x}_i - \mathbf{x}_j|^2)^{3/2}}, \tag{7.6}$$

which gives the three components of the force of the jth particle on the ith particle. For each particle (i) the forces arising from all other particles (j) must be added up. Clearly the effect of the softening approximation is to eliminate the effects of very close encounters, which result in large interparticle forces and, therefore, large accelerations and very small time steps in dynamical calculations. In the case of large stellar systems, such as entire galaxies, this method is appropriate because close encounters ("collisions") are very rare. In systems where close encounters are important, a revised method has to be used (see Chapter 3). For example, it is possible for one particle to capture another and form a close binary system, resulting in a very short time step, which drastically slows down the entire calculation. From the global point of view, it would be better simply to combine the two particles into a single particle. The quantity ϵ is arbitrary, but if one considers a volume V with N uniformly distributed particles, the softening length should be of order $[(3V)/(4\pi N)]^{1/3}$, the mean particle separation. As discussed in Chapter 3, the number of forces that has to be calculated goes as N^2, where N is the total number of particles. If N exceeds $\mathcal{O}(10^4)$, the computer time required is excessive and alternate techniques must be employed, such as the tree method (Section 3.9). This procedure, in effect, is still a direct summation method, but the number of actual calculations of the forces arising from relatively distant particles is reduced. Methods for taking into account close encounters between particles when the softening approach is not appropriate are also discussed in Chapter 3. The boundary condition is implicitly included in Equation (7.4) and no additional calculation regarding it is required.

7.1.2 Fourier Methods for Solving Equation (7.4)

Fourier methods can be used in connection with a particle calculation if a three-dimensional fixed grid is overlaid on the particle distribution (Section 3.10). At each time step, the number of particles that happen to be in each cell, combined with the masses of the particles and the cell sizes, can be used to obtain the density in each cell. Similarly, the method can be used in connection with Eulerian multidimensional hydrodynamics on a grid, in which each zone has an assigned value of density. The grid zones must be equally spaced, and the number of zones in each direction must be an integral power of 2, so that fast Fourier algorithms (such as that of Cooley and Tukey, 1965) can be used.

The basic Fourier transform is obtained as follows (Press et al., 1992; Chapter 12). Suppose a one-dimensional grid is set up with N total zones and index k running from 0 to $N-1$. If a function u_k is defined at every point on the grid, then the discrete Fourier transform of the u_k is a set of numbers $\hat{u}_n$, where n runs from 0 to $N-1$

$$\hat{u}_n = \sum_{k=0}^{N-1} u_k \exp(2\pi i k n / N) \,. \tag{7.7}$$

Note that the $\hat{u}_n$ are complex. The inverse transform, in which the u_k are recovered from the $\hat{u}_n$, is:

$$u_k = \frac{1}{N} \sum_{n=0}^{N-1} \hat{u}_n \exp(-2\pi i k n / N). \tag{7.8}$$

Other forms of the Fourier transform can be used, such as the sine transform and the cosine transform (Section 7.2.5).

These expressions are easily extended to two or three dimensions. For example, if u is a function of two indices, j and k, then the two-dimensional Fourier transform of u is $\hat{u}_{n,m}$, a complex function defined on the same two-dimensional grid as u

$$\hat{u}_{n,m} = \sum_{j=0}^{M-1} \sum_{k=0}^{N-1} u_{j,k} \exp(2\pi i k n / N) \exp(2\pi i j m / M), \tag{7.9}$$

where M and N are the number of zones in each direction. It is clear how the inverse transform is so extended (it includes the factor 1/(MN)), and also how the transforms are extended to three dimensions (see Section 3.10). Press et al. (1992, Section 12.4) provide two routines, "fourn" and "rlft3," which perform multidimensional Fourier transforms or inverse transforms on real or complex input data in an arbitrary number of dimensions.

The key to the solution of Equation (7.4) with Fourier techniques is the *Fourier convolution theorem*, as described in Chapter 3, Section 3.10. Solution of Equation (7.4) on a three-dimensional grid implies that the equation can be written symbolically as

$$\Phi(x, y, z) = -\sum G_r(x - x', y - y', z - z')\rho(x', y', z')\Delta x'\Delta y'\Delta z', \qquad (7.10)$$

where the triple sum is taken over the entire grid whose mesh spacings are $\Delta x'$, $\Delta y'$, and $\Delta z'$, respectively, in the three coordinate directions. The quantity G_r is the Green's function

$$G_r = \frac{G}{|\mathbf{x} - \mathbf{x}'|}. \qquad (7.11)$$

Equation (7.10) can also be written

$$\Phi(x, y, z) = -\sum G_r(x - x', y - y', z - z')M(x', y', z'), \qquad (7.12)$$

where M is the mass of a cell at (x', y', z'). This last equation expresses the fact that the potential is the convolution of the Green's function and the mass distribution within the object. The Fourier convolution theorem (see Binney and Tremaine, 1987, p. 93) states that the Fourier transform of a convolution is simply the product of the Fourier transforms of the two quantities convolved, namely G_r and M in this case. Thus, the following equation holds

$$\hat{\Phi} = -\hat{G}_r\hat{M}. \qquad (7.13)$$

Thus, the Fourier-transformed masses and Green's functions give the Fourier transform of the potential; the inverse Fourier transform then gives the actual potential.

Here, we illustrate the calculation of the potential for two cases, first on a two-dimensional grid used to represent a very flat, disk-like structure, and second, on a full three-dimensional grid. Disk structures are found in nature around young stars, around quasi-stellar objects and black holes, in cataclysmic variable systems, and in spiral galaxies. In such configurations, various instabilities can produce structures that are not axially symmetric but involve variations in quantities such as the density as a function of the azimuthal angle at a given radial distance. The two-dimensional approximation to the disk structure is often useful.

For this example, we use cylindrical coordinates (R, Z, ϕ), where ϕ is the angle in the plane of the disk with respect to a fixed reference direction. In many such problems, one can neglect the structure parallel to the direction of the rotation axis (Z) and consider the problem as two-dimensional in the coordinates (R, ϕ). Then the mass density ρ is replaced by the variable known as surface density

$$\Sigma(R, \phi) = \int_{-\infty}^{\infty} \rho(R, Z, \phi)dZ. \qquad (7.14)$$

In many problems, one deals with a disk with a central object, such as a star, that is very massive compared with the disk. In this case, the self-gravitational potential of the disk is unimportant, and at every point (R, ϕ) in the disk one can simply use the point-mass potential of the star

$$\Phi(R, \phi) = -GM_*/R, \qquad (7.15)$$

where R is the distance to the star and M_* its mass. However, in other problems, the central mass is unimportant or the disk mass is significant with respect to the central mass. In this case, the disk self-gravitational potential $\Phi_d(R, \phi)$ must be added to the potential of the central mass. If both disk mass and central mass are significant, once the mass distribution in the disk becomes nonaxisymmetric, the central mass will orbit about the center of mass as a result of the influence of the disk potential. This effect has to be taken into account in hydrodynamic simulations.

A relatively straightforward method for calculating $\Phi_d(R, \phi)$ given $\Sigma(R, \phi)$ is described by Binney and Tremaine (1987, pp. 96–97). First, define a logarithmic radial coordinate

$$s = \ln R. \tag{7.16}$$

Then define a *reduced surface density*

$$S(s, \phi) = e^{3s/2}\Sigma(s, \phi) \tag{7.17}$$

and a *reduced disk potential*

$$H_d(s, \phi) = e^{s/2}\Phi_d(s, \phi). \tag{7.18}$$

Binney and Tremaine (1987, Section 2.6) show that H_d and S are related by

$$H_d(s, \phi) = -\frac{G}{\sqrt{2}}\int_{-\infty}^{\infty} ds' \int_0^{2\pi} \frac{S(s', \phi')d\phi'}{\sqrt{\cosh(s - s') - \cos(\phi - \phi')}}. \tag{7.19}$$

This equation is a transformed version of Equation (7.4), applicable to a thin disk.

The next step is to set up a grid with uniform cell sizes Δs and $\Delta\phi$ in s and ϕ, respectively. The quantities H_d and S are defined in the centers of cells and are labeled (l, m), with both l and m running from 0 to $J - 1$ (however, an unequal number of zones in the two directions is possible). Then, at a given point (l, m) in the disk, the reduced disk potential becomes

$$H_d(l, m) = \sum_{l'=0}^{J-1}\sum_{m'=0}^{J-1} K(l - l', m - m')M(l', m'), \tag{7.20}$$

where M is given by

$$M(l', m') = S(s', \phi')\Delta s'\Delta\phi' \tag{7.21}$$

and

$$K(l - l', m - m') = -\frac{G}{\sqrt{2}}\frac{1}{\sqrt{\cosh(s_l - s_l') - \cos(\phi_m - \phi_m')}}. \tag{7.22}$$

The reduced potential has now been put into the form of a convolution of a Green's function K and a "reduced mass" M, so the convolution theorem can be applied.

The main difficulty with this method (and with other methods using Equation 7.4) is the singularity at $l = l'$ and $m = m'$. The usual method for dealing with this problem is to introduce a softening length ϵ as shown in Equation (7.6). In normal Cartesian coordinates, the potential is expressed as

$$\Phi(x, y, z) = -\int_\tau \frac{G\rho(x', y', z')}{\sqrt{\epsilon^2 + |\mathbf{x} - \mathbf{x}'|^2}}dx'dy'dz', \tag{7.23}$$

where τ is the volume. However, in this particular method, a different technique is used. Make the approximation, for s close to s' and ϕ close to ϕ',

$$\cosh(s - s') - \cos(\phi - \phi') \approx \frac{1}{2}(s - s')^2 + \frac{1}{2}(\phi - \phi')^2 \tag{7.24}$$

based on the first two terms of the Taylor expansions. Then it follows (Binney and Tremaine, 1987, p. 97) that

$$K(0, 0) \approx -2G \left[\frac{1}{\Delta\phi} \text{arcsinh} \left(\frac{\Delta\phi}{\Delta s} \right) + \frac{1}{\Delta s} \text{arcsinh} \left(\frac{\Delta s}{\Delta\phi} \right) \right]. \tag{7.25}$$

To solve the system, the grid is extended by a factor of 2 in each dimension, the function K is made periodic on the extended grid, and the function M is set to zero there. The Fourier transforms of K and M are carried out, and the Fourier transform of the reduced potential calculated

$$\hat{H}_d(l, m) = \hat{K}(l, m)\hat{M}(l, m). \tag{7.26}$$

The inverse transform of $\hat{H}_d(l, m)$ then gives the actual reduced potential. The actual potential Φ_d follows if one uses just that part of H_d that lies on the original grid. The method is very similar to that described for the three-dimensional example given next.

As a second example, we show how Equation (7.4) can be solved in three dimensions by a procedure developed by Hockney (1970) and Eastwood and Brownrigg (1979). The method will be illustrated for the case of a Cartesian coordinate system, with the origin in the center of the grid and with dimensions $-X \le x \le X$, $-Y \le y \le Y$ and $-Z \le z \le Z$. In each coordinate direction the number of zones is 2^n and the number of grid lines is $2^n + 1$. The density is provided for all zones.

The basic procedure is to extend the grid in each direction by a factor of 2 (Figure 7.1), so that the number of grid cells is increased by a factor of 8, and to set the density equal to zero in all of the added cells. Using periodic boundary conditions on the extended grid and an appropriate Green's function, one can obtain a correct solution to the potential on the original grid. A modified Green's function G_r is defined below. The basic sequence of operations, as before, is:

1. Calculate the Fourier transform $\hat{M}$, where the Ms are cell masses obtained from the densities and volumes.
2. Calculate the Fourier transform $\hat{G}_r$ from Green's functions.
3. Use the convolution theorem to obtain $\hat{\Phi} = -\hat{G}_r\hat{M}$.
4. Use the inverse Fourier transform to get Φ from $\hat{\Phi}$.

The method works for an unequal number of grid points $NX \ne NY \ne NZ$, but for the present example, we let the indices for the grid lines in all three directions (j, k, l) run from 0 to N, where N is a power of 2. Assume for simplicity that $\Delta x = \Delta y = \Delta z$. Eastwood and Brownrigg (1979) prove that the following procedure gives the correct potential on the grid:

- Extend the grid in the x-direction by a factor 2, so the cell at N serves as the first cell for the extended grid. Thus, there are now $2N$ lines in the

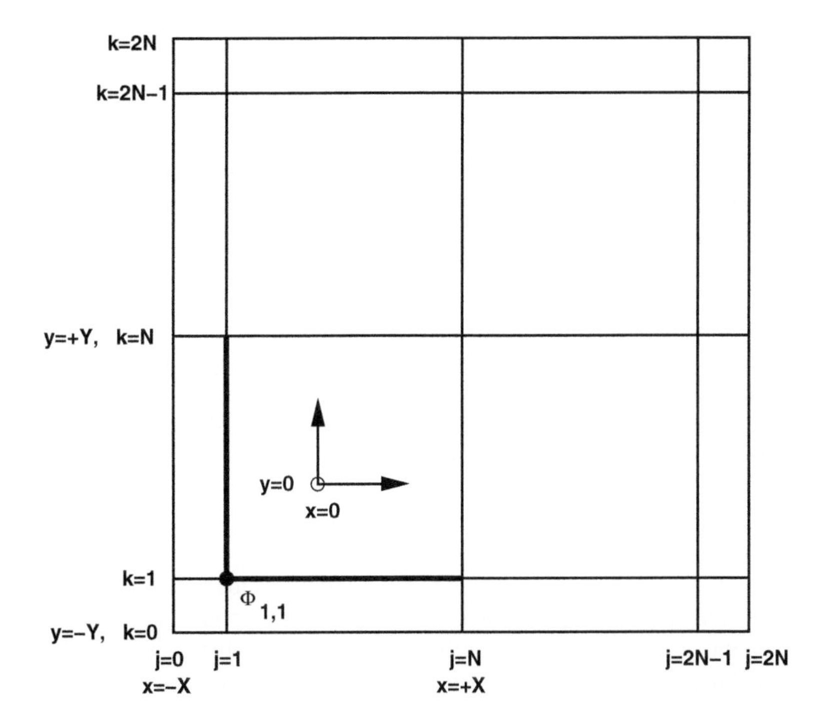

FIGURE 7.1 The (x, y) plane in a three-dimensional Cartesian coordinate system, illustrating Hockney's method of solving Equation (7.4). The origin is marked by the open circle. The actual grid extends from $-X$ to $+X$, from $-Y$ to $+Y$, and from $-Z$ to $+Z$. The extended grid, used for calculation of the potential by the Fourier convolution technique, is obtained by extending the actual grid by a factor of 2 in each dimension. The point labeled $\Phi_{1,1}$ shows that the potentials are defined at the intersections of grid lines.

x-direction, with indices running from 0 to $2N - 1$. The densities in the cells with indices $> N - 1$ are set to zero. Similarly, extend the grid in the y- and z-directions. The modified Green's functions on the entire grid are then calculated according to

$$\begin{aligned}
G_{r,j,k,l} &= \frac{G}{\Delta x}(j^2 + k^2 + l^2)^{-1/2} \\
G_{r,2N-j,k,l} &= G_{r,j,k,l} \\
G_{r,j,2N-k,l} &= G_{r,j,k,l} \\
G_{r,j,k,2N-l} &= G_{r,j,k,l} \\
G_{r,2N-j,2N-k,l} &= G_{r,j,k,l} \\
G_{r,2N-j,k,2N-l} &= G_{r,j,k,l} \\
G_{r,j,2N-k,2N-l} &= G_{r,j,k,l} \\
G_{r,2N-j,2N-k,2N-l} &= G_{r,j,k,l},
\end{aligned} \tag{7.27}$$

where the indices j, k, l run from 0 to N, with the exception that

$$G_{r,0,0,0} = \frac{G}{\Delta x}, \tag{7.28}$$

where G, as usual, is the gravitational constant. That is, the Green's function, which is defined at zone corners rather than at zone centers, corresponds to the potential arising from unit mass at the point $j = k = l = 0$. Note that the origin of the coordinate system is, in contrast, at $j = k = l = N/2$. The function is defined so that it is periodic on the extended grid. The Fourier transform of the Green's functions $\hat{G}_r$, is now calculated at all points, assuming periodic boundary conditions. The steps so far need to be calculated only once, since the grid is fixed.

- Given the density distribution for a particular model, defined on zone corners and set to zero on the extended grids, calculate $\hat{M}_{j,k,l}$ at all points from 0 to $2N - 1$ in each direction. Here $M_{j,k,l} = \rho_{j,k,l}(\Delta x)^3$.
- Use the convolution theorem to obtain $\hat{\Phi}$ at all points on the extended grid, from $\hat{M}$ and $\hat{G}_r$. The inverse Fourier transform then yields Φ at all points on the extended grid. All the values of Φ outside the original grid are meaningless and are discarded; however, the correct potentials are obtained on the original grid. It is clear that a large amount of storage is needed in 3-D to provide for the extended grids. The details of procedures to reduce the necessary storage space and computer time are provided by Hockney and Eastwood (1981, Section 6–5–4) and by Myhill and Boss (1993).

7.1.3 Self-Consistent Field

We now consider the use of polynomial expansions to solve Equation (7.4). An excellent example of this method is the so-called *self-consistent field* (SCF) method for the construction of equilibrium models for the full two-dimensional structure of rapidly rotating stars. The method was originally derived for the calculation of atomic structure by Hartree and Fock (see Hartree, 1957) and modified for astrophysical calculations of self-gravitating configurations by Ostriker and Mark (1968). The Ostriker–Mark method was derived for special cases of the equation of state; however, it also can be used to solve the full stellar structure equations for a rapidly rotating star (Jackson, 1970).

We now illustrate how the method works under the original assumptions of Ostriker and Mark. Spherical coordinates (r, θ, ϕ) as well as cylindrical coordinates (R, Z, ϕ) are used, as is the convention $\mu = \cos\theta$. The basic assumptions are (1) the models can be represented by an equation of state in which the pressure is an explicit function of density (this assumption, however, does not affect the calculation of the potential, which just requires that the density field be given); (2) the configuration is symmetric with respect to the rotation (Z) axis, i.e., the structure is independent of the azimuthal angle ϕ; and (3) the angular velocity Ω is a function only of the distance to the rotation axis $R = r(1 - \cos^2\theta)^{1/2}$, i.e., rotational velocity is constant on cylinders parallel to the rotation axis. However the angular velocity need not be constant as a function of R; the model can be differentially rotating with a prescribed rotation law specified in advance. The method has been used to calculate a wide variety of rapidly rotating configurations, for example, polytropes (Bodenheimer and Ostriker, 1973; Hachisu, 1986a), white dwarfs (Ostriker and Bodenheimer, 1968; Hachisu, 1986a), and solar mass stars with surrounding disks (Pickett et al., 1997; 2003). The method has been extended to three space dimensions, for the case of uniform rotation, which allows additional types of configurations, such as binary systems, to be calculated (Hachisu, 1986b).

The original SCF method (Ostriker and Mark, 1968) was capable of producing accurate equilibrium models with substantial rotational distortion; however, it had difficulty in converging for some assumed rotation laws when the desired equilibrium was very flattened, i.e., with a ratio of rotational energy T to gravitational potential energy W of $T/|W| > 0.25$. Hachisu (1986a) devised a modification of the method that overcame this difficulty. Discussed briefly below is his two-dimensional SCF, which can succesfully compute highly flattened, rotationally dominated configurations. A test of the quality of the equilibrium that has been obtained is the extent to which the virial theorem is satisfied. For example, Pickett et al. (1996) have defined the virial test VT to be:

$$VT = |2T + W + 3S|/|W|, \tag{7.29}$$

where $3S = 3\int PdV$ is twice the total internal energy. In principle, VT should be zero and, in typical published models, it is less than 10^{-3}.

The equation of equilibrium, representing the balance of forces arising from pressure gradients, rotation, and gravity, has both R- and Z-components and is given by

$$\frac{1}{\rho}\nabla P + \nabla\Phi = \Omega^2 R\hat{e}_R, \tag{7.30}$$

where $\hat{e}_R$ is a unit vector in the R direction. Under the third assumption above, the rotational term can also be expressed as the gradient of a potential Φ_c; for example, if the rotation is uniform

$$\Phi_c = -\frac{1}{2}R^2\Omega^2 \tag{7.31}$$

and in general

$$\Phi_c(R) = -\int_0^R \Omega^2(R')R'dR'. \tag{7.32}$$

Thus, the equilibrium equation can be written

$$\frac{1}{\rho}\nabla P = -\nabla\Phi_{\text{tot}}. \tag{7.33}$$

where $\Phi_{\text{tot}} = \Phi + \Phi_c$.

The rotation law can be specified either by $\Omega(R)$ (Hachisu, 1986a) or by $j(m_c)$ (Ostriker and Mark, 1968). In the latter representation, j is the specific angular momentum and m_c is a Lagrangian coordinate giving the fractional mass interior to the cylinder corresponding to j

$$m_c(R) = \frac{2\pi}{M_{tot}} \int_0^R R' \int_{-\infty}^{\infty} \rho(R', Z') dZ' dR', \tag{7.34}$$

where M_{tot} is the total mass, obtained by integrating Equation (7.34) out to $R_{\max}$, the outer radius of the cylinder. This form is particularly useful when one is calculating an evolutionary sequence of models, in which it is assumed that the angular momentum of each cylindrical mass element is conserved. In this case, the centrifugal potential is expressed as

$$\Phi_c(R) = -\int_0^R \frac{j^2(m_c)}{R'^3} dR'. \tag{7.35}$$

As an example, we consider a polytrope with an equation of state

$$P = K\rho^{1+1/n} \tag{7.36}$$

where $1 + 1/n = \gamma$. The equation of equilibrium (Equation 7.33) can be integrated to give

$$\begin{aligned} \int \frac{dP}{\rho} &= (1+n)\frac{P}{\rho} \\ \rho &= \left[\frac{(\Phi_{\text{surf}} - \Phi_{\text{tot}})}{K(1+n)}\right]^n, \end{aligned} \tag{7.37}$$

where the boundary condition has been incorporated that $\rho \to 0$ at the outer surface where the total potential is Φ_{surf}.

Exercise

Verify Equation (7.37).

Exercise

Write a simple numerical program to find the potential as a function of radius for a spherical (nonrotating) polytrope of index 1 (Equation 4.14 through Equation 4.16). Compare the central potential of the polytrope to that of a uniform-density sphere of the same mass and radius. The zero point for the potential is at infinity.

The essence of the SCF method is to find the equilibrium through two alternating steps:

1. Given an approximation to the density distribution, calculate the total potential at each point from Equation (7.4) and Equation (7.32).
2. Given the potential, recalculate the density according to Equation (7.37).

The two steps are repeated until the density converges, which can take from 10 up to 100 iterations depending on the strictness of convergence required and the degree of rotational distortion from the spherical case. The typical convergence criterion could be that the mean fractional change in density over the last iteration is 10^{-6}.

As far as the solution to the Poisson equation is concerned, the user may choose his/her favorite method, but the published results generally have used polynomial expansion. In spherical coordinates with symmetry with respect to the rotation axis as well as with respect to the equatorial plane, the potential is:

$$\Phi(r,\mu) = -G\int \frac{\rho(\mathbf{r}')d^3\mathbf{r}'}{|\mathbf{r}-\mathbf{r}'|} = -4\pi G\int_0^\infty dr' \int_0^1 d\mu' \times \sum_{n=0}^{\infty} f_{2n}(r',r)P_{2n}(\mu)P_{2n}(\mu')\rho(\mu',r'), \tag{7.38}$$

where the P_{2n} are even-order Legendre polynomials and

$$f_{2n}(r',r) = \frac{r'^{2n+2}}{r^{2n+1}}, \quad r' < r,$$
$$f_{2n}(r',r) = \frac{r^{2n}}{r'^{2n-1}}, \quad r' > r. \tag{7.39}$$

Here the upper expression refers to the contribution to the potential of the material inside r, while the lower expression refers to the contribution from the material outside r. This equation is explained, for example, by Ostriker and Mark (1968, Section IIIc).

The actual calculation of an equilibrium model can be done in a system of dimensionless units, if desired. The parameters of the model can be the total mass M_{tot}, the total angular momentum J, the rotation law $j(m)$ or $\Omega(R)$, and the equatorial radius R_{eq}; however, Hachisu (1986a) found that specifying instead the maximum density $\rho_{\max}$, the equatorial radius R_{eq} and the *axis ratio* R_p/R_{eq}, along with the rotation law, results in much better convergence properties. The total angular momentum is not specified; it is derived. In the polytropic case, it turns out that if n and K are specified, once $\rho_{\max}$ is given, then R_{eq} follows. Improved convergence can also be obtained for very flat configurations by undercorrecting the density obtained from Equation (7.37). That is, if ρ_i is the density array from the ith iteration, then use

$$\rho_{i,c} = (1-\delta_\rho)\rho_i + \delta_\rho \rho_{i-1} \tag{7.40}$$

where $0 \le \delta_\rho < 1$ is obtained by experiment, $\rho_{i,c}$ is the corrected value of ρ for the ith iteration, while ρ_i is the actual value from Equation (7.37).

Hachisu's method for calculating the potential is direct numerical integration of Equation (7.38). The steps are:

1. Set up a grid with equal spacing in the r-direction, with a total number of points J, an odd number.
2. Set up a grid with equal spacing in the μ-direction ($\mu = \cos\theta$) with a total number of points K, also an odd number.
3. Define the location of variables at the intersection of two grid lines: $\Phi_{j,k} = \Phi(r_j, \mu_k)$ and $\rho_{j,k} = \rho(r_j, \mu_k)$.

4. For each μ point, calculate the Legendre polynomials $P_{2n}(\mu)$ from $n = 0$ up to $n = N$, the maximum number of polynomials used, which will depend on accuracy requirements. Thus, if $N = 32$, then the highest-order Legendre polynomial is $P_{64}(\mu)$.
5. For each point r_j, calculate the functions $f_{2n}(r_l, r_j)$ for each l and for each n.
6. Calculate the quantities $D_{l,n}$, which represent the integration over μ' in Equation (7.38); the coefficients represent Simpson's rule for numerical integration

$$D_{l,n} = \sum_{k=1(2)}^{K-2} (1/6)(\mu_{k+2} - \mu_k)[P_{2n}(\mu_k)\rho_{l,k} + 4P_{2n}(\mu_{k+1})\rho_{l,k+1} + P_{2n}(\mu_{k+2})\rho_{l,k+2}] \quad (7.41)$$

where the symbol $k = 1(2)$ means that the index k starts from 1 and increases in steps of 2.
7. Calculate the quantities $E_{n,j}$, which represent the integration over r' in Equation (7.38)

$$E_{n,j} = \sum_{l=1(2)}^{J-2} (1/6)(r_{l+2} - r_l)[f_{2n}(r_l, r_j)D_{l,n} + 4f_{2n}(r_{l+1}, r_j)D_{l+1,n} + f_{2n}(r_{l+2}, r_j)D_{l+2,n}]. \quad (7.42)$$

8. Calculate the potential at each point

$$\Phi_{j,k} = -4\pi G \sum_{n=0}^{N} E_{n,j} P_{2n}(\mu_k). \quad (7.43)$$

Note that a separate explicit calculation for the boundary values is not required. The solution for the potential is continuous across the boundary, and the implicit boundary condition is that the potential goes to zero at infinite distance. The location of the surface, which is fixed at two points — the equatorial radius R_e and the polar radius R_p in Hachisu's method — is actually determined from the equilibrium calculation by the locations where $\rho \to 0$. (In practice, this can be a bit tricky.) The surface potential Φ_{surf}, needed for Equation (7.37), as well as the scaling coefficient that determines the total angular momentum, are obtained from the potentials at the two fixed points. An example of an equilibrium structure obtained by this method, consisting of a relatively slowly rotating central star and a surrounding flattened disk that is differentially rotating, is shown in Figure 7.2. The disk was assumed to have 1/8 of the total system mass and to have a surface density distribution $\Sigma(R) \propto R^{-1/2}$. The equation of state is a polytrope of index 1.5 ($\gamma = 5/3$), and the ratio of the outer radius of the disk to the polar radius of the star (R_e/R_p) was assumed to be 20.

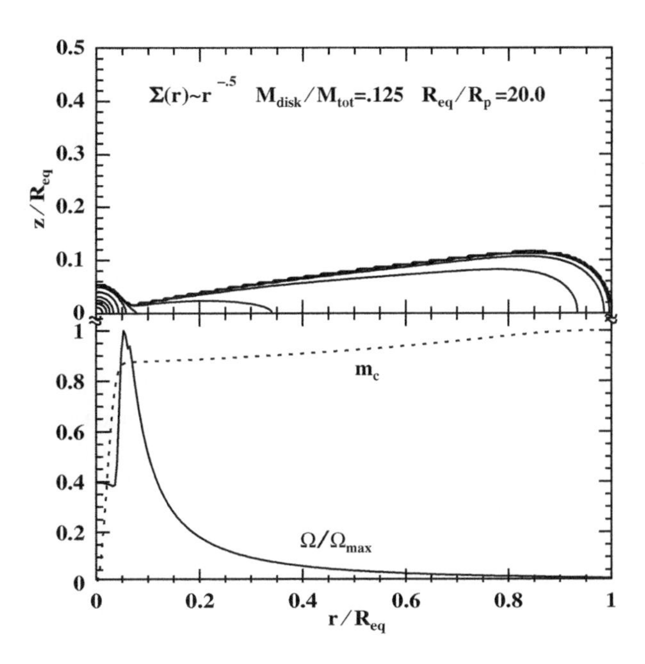

FIGURE 7.2 An equilibrium model in the (R, Z) plane (in cylindrical coordinates) with assumed symmetry with respect to the Z-axis and with respect to the equatorial plane (R-axis), calculated with the self-consistent field method. The equation of state is that of a polytrope of index 1.5. *Upper portion*: contours of equal density are plotted. *Lower portion*: the rotational velocity Ω and the cylindrical mass fraction m_c (Equation 7.34) are plotted as a function of radius, normalized to the maximum radius R_{eq}. The central region is a star with uniform rotation. The outer regions constitute a disk with an assumed surface density profile of $\Sigma \propto R^{-1/2}$. (From Pickett, B. K. et al. (2003) *Astrophys. J.* **590**: 1060. With permission.)

7.2 POISSON SOLUTIONS: II

In this section, we consider solution methods based on the differential expression of the Poisson Equation (7.1). Most of the discussion will be based on straightforward finite differencing of the equation on a grid, but the final subsection explains a technique based on polynomial expansion.

To illustrate the solution method for grid-based techniques clearly, we consider a two-dimensional grid in Cartesian coordinates (Press et al., 1992, Section 19.0)

$$\frac{\partial^2 \Phi}{\partial x^2} + \frac{\partial^2 \Phi}{\partial y^2} = 4\pi G \rho(x, y). \tag{7.44}$$

The basic difference equation becomes

$$\frac{\Phi_{j+1,k} - 2\Phi_{j,k} + \Phi_{j-1,k}}{(\Delta x)^2} + \frac{\Phi_{j,k+1} - 2\Phi_{j,k} + \Phi_{j,k-1}}{(\Delta y)^2} = 4\pi G \rho_{j,k}, \tag{7.45}$$

where j, the index in the x-direction, runs from 0 to J, k; the index in the y-direction, from 0 to K; and Δx and Δy are, respectively, the grid spacings in the x- and y-directions.

We now arrange the values of Φ into a one-dimensional sequence by letting the point (j, k) have the index m, where

$$m = j(K + 1) + k. \tag{7.46}$$

If we assume constant grid spacings $\Delta x = \Delta y$, then Equation (7.45) becomes

$$\Phi_{m+K+1} + \Phi_{m-K-1} + \Phi_{m+1} + \Phi_{m-1} - 4\Phi_m = (\Delta x)^2 4\pi G\rho_m, \tag{7.47}$$

where the ρ_m are known quantities. At the boundary $j = 0$, $\Phi_B(0, k)$ must be specified for $k = 0$ to K or equivalently $m = 0,\ 1, \ldots$ to K. At the boundary $j = J$, $\Phi_B(J, k)$ must be specified for $m = J(K+1)$ to $J(K+1)+K$. Similarly, at the boundary $k = 0$, $\Phi_B(j, 0)$ must be specified for $j = 0$ to J or $m = 0,\ K+1,\ 2(K+1), \ldots, J(K+1)$, and, finally, at $k = K$, $\Phi_B(j, K)$ must be specified for $j = 0$ to J or $m = K, 2K + 1, 3K + 2, \ldots, (J + 1)K + J$.

The two-dimensional density distribution can also be expressed as a one-dimensional array ρ_m. Then the Poisson equation can be written as a matrix equation

$$A_{lm}\Phi_m = 4\pi G\rho_m(\Delta x)^2, \tag{7.48}$$

where A_{lm} is a sparse matrix with a banded structure parallel to the diagonal (Press et al., 1992, p. 823) whose nonzero elements are either 1 or -4. The boundary values of Φ are known quantities, which can be moved over to the right-hand side and combined with the vector of densities.

One possible solution is to invert the matrix A_{lm} directly; some discussion on how this is done is given in Press et al. (1992, Chapter 19) and in Hockney and Eastwood (1981, Section 6–4). One approach to the manipulation of the matrix to put it into soluble form is *cyclic reduction* (Hockney and Eastwood, 1981, Section 6–5–1, and, in this chapter, Section 7.2.6); however, it is more usual to use the alternate methods discussed in the following subsections. They can be broadly divided into *relaxation* methods and *Fourier transform* methods.

7.2.1 Boundary Conditions

In all classes of solutions based on Equation (7.1), boundary conditions must be applied as a separate step. The usual method is to integrate over the mass, assuming that the potential can be represented by a polynomial expansion (for an alternative, see Section 7.2.7). Assume that all the mass to be considered in the evaluation of the gravitational force lies within the grid. To be more precise, the boundary must be at a radial location that is greater than that of all mass elements within the grid. If there are external mass elements, beyond the boundary radius, then that mass must be specifically taken into acccount with an extra term (Cohl and Tohline, 1999). In fact, the method is most accurate when the majority of the mass is well inside the boundary.

First, we assume an axisymmetric configuration in cylindrical coordinates (R, Z) with the distance to the origin $r = (R^2 + Z^2)^{0.5}$. A point on the boundary will be given coordinates (R, Z), while one in the interior has coordinates (R', Z') with $r' = (R'^2 + Z'^2)^{0.5}$. Then a boundary point has the potential

$$\Phi_B(R, Z) = -G \sum_{l=0}^{\infty} \frac{P_l(\cos\theta)}{r^{l+1}} Z_l \qquad (7.49)$$

where

$$Z_l = \int_{V'} \rho(R', Z')(R'^2 + Z'^2)^{l/2} P_l(\cos\alpha) dV' \qquad (7.50)$$

where $\cos\theta = Z/r$ and $\cos\alpha = Z'/r'$ and the integration is taken over the entire volume of the grid V'. If most of the mass is well within the boundary, then only a few terms in the expansion are needed. If an appreciable amount of mass is near the boundary, then enough terms in the series expansion must be taken so that the neglected terms are small, possibly up to 100 (Stone and Norman, 1992).

Exercise

Show that if the boundary values of the potential are evaluated on a spherical surface outside a mass M with $r = r_{\max}$, then the first term in the boundary value for the potential ($l = 0$) is $-GM/r_{\max}$. Show that, with symmetry with respect to the rotation axis (Z-axis) and symmetry with respect to the equatorial plane (R-axis), the ($l = 1$) term in the boundary expansion is zero. It turns out that for this assumed symmetry, all odd terms in the expansion are zero. Calculate the $l = 2$ contribution at the outer boundary points of the cylindrical grid shown in Figure 7.3, assuming the maximum radius is R_{max} and the maximum height is $Z_{max} = R_{max}$. All mass must be within a distance R_{max} of the origin.

If it is possible to introduce symmetries into the problem, then the boundary conditions may have to be modified. For example, in an axially symmetric and equatorially symmetric two-dimensional grid, the whole (R, Z) plane need not be calculated, only the first quadrant $R > 0$, $Z > 0$. Then the boundary conditions along the Z– and R–axes are, respectively,

$$\frac{\partial \Phi}{\partial R} = 0 \quad \text{and} \quad \frac{\partial \Phi}{\partial Z} = 0. \qquad (7.51)$$

These boundary conditions can be implemented by means of *ghost zones* as illustrated in Figure 7.3. One column of zones to the left of the Z-axis and one column of zones below the R-axis are added to the grid. The symmetry boundary conditions are implemented by setting $\Phi_{1,k} = \Phi_{2,k}$ and $\Phi_{j,1} = \Phi_{j,2}$. On the remaining boundaries, however, the potentials must be specified as discussed above, so that for this particular problem the boundary conditions are mixed Neumann and Dirichlet.

On a grid as shown in Figure 7.3, where ρ and Φ are defined at centers of zones, Equation (7.50) is discretized as

$$Z_l = 4\pi \sum_{j=1}^{J-1} \sum_{k=1}^{K-1} \rho_{jk} P_l(\cos\alpha_{jk})$$
$$\times R_{j+1/2}\left(R_{j+1/2}^2 + Z_{k+1/2}^2\right)^{l/2} (R_{j+1} - R_j)(Z_{k+1} - Z_k). \qquad (7.52)$$

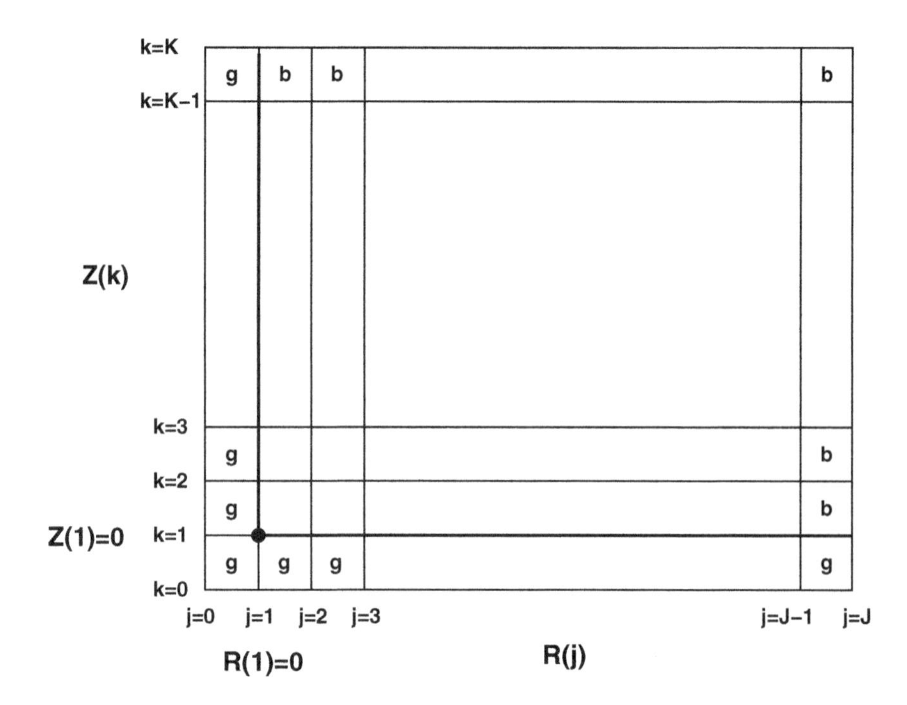

FIGURE 7.3 An (R, Z) grid in cylindrical coordinates with assumed symmetry with respect to the Z-axis and with respect to the equatorial plane (R-axis). The origin is marked by the circle. Zones to the left of and below the origin are ghost zones (marked g) and are used for symmetry boundary conditions for the gravitational potential. The uppermost row and rightmost column of zones (marked b) are used for supplying boundary conditions as calculated, for example, by expansion in Legendre polynomials.

Exercise

Discretize Equation (7.50) on an axisymmetric grid in spherical coordinates (r, θ).

The boundary calculation can be extended to three dimensions by using spherical harmonics. In spherical coordinates (r, θ, ϕ) and, again, in the case where all the mass resides within the boundary

$$\Phi_B(r, \theta, \phi) = -4\pi G \sum_{l=0}^{\infty} \sum_{m=-l}^{l} \frac{Y_{lm}(\theta, \phi)}{(2l+1)r^{l+1}} Z_{lm}, \tag{7.53}$$

where

$$Z_{lm} = \int_{V'} \rho(r', \theta', \phi')(r')^l Y_{lm}^*(\theta', \phi') dV' \tag{7.54}$$

and

$$Y_{lm} = \frac{1}{N_{lm}} \exp(im\phi) P_l^{|m|}(\cos\theta), \tag{7.55}$$

where $P_l^{|m|}$ is the associated Legendre polynomial and the normalization factor is given by

$$\frac{1}{N_{lm}} = \left(\frac{1}{2\pi}\frac{2l+1}{2}\frac{(l-|m|)!}{(l+|m|)!}\right)^{1/2}. \tag{7.56}$$

With this normalization, the functions have the property that $\int\int |Y_l^m|^2 \sin\theta d\theta d\phi = 1$, where the integral is taken over the entire surface of a sphere. The index m runs from $-l, -l+1, \cdots 0, 1, \cdots l$, where l is always a positive number. For an alternate expansion technique that is useful in cylindrical coordinates, see Cohl and Tohline (1999).

7.2.2 Alternating Direction Implicit Method

In the following subsections, we consider interior solutions to Equation (7.48). Even if the actual two-dimensional grid has only 100×100 zones, the A_{lm} matrix has a size over $10^4 \times 10^4$, so it may not be practical to invert it directly.

There are several more efficient ways to solve this system. One of them involves changing the time-independent problem into a pseudo-time-dependent problem by writing the equation

$$\frac{\partial \Phi}{\partial \tau} = \nabla^2\Phi - 4\pi G\rho, \tag{7.57}$$

where τ is the pseudo-time, unrelated to the actual timescale of the problem. This equation is solved for a sequence of values of pseudo-time steps $\Delta\tau$ until it comes to steady state. That is, ρ is considered to be a constant as a function of pseudo-time (although it may be a function of physical time), but Φ is considered to be a function of pseudo-time; it is iterated until

$$\frac{\partial \Phi}{\partial \tau} = 0 \tag{7.58}$$

at which point the Poisson equation is satisfied. Note that Equation (7.57) is formally equivalent to a diffusion equation, and a very similar method was illustrated in Chapter 6, Section 6.6.2 for the solution of the radiation diffusion equation. In general the method is simple, reliable, and accurate, but it is not as fast as some other methods. On a two-dimensional Cartesian grid, the difference equation to be solved is:

$$\frac{\Phi_{jk}^{n+1} - \Phi_{jk}^n}{\Delta\tau} = \frac{\Phi_{j+1,k} - 2\Phi_{j,k} + \Phi_{j-1,k}}{(\Delta x)^2} + \frac{\Phi_{j,k+1} - 2\Phi_{j,k} + \Phi_{j,k-1}}{(\Delta y)^2} - 4\pi G\rho_{j,k}, \tag{7.59}$$

where the time levels of the Φs on the right-hand side are specified below. These equations can be solved by the "alternating direction implicit" (ADI) method. The equation is solved implicitly to avoid stability constraints on $\Delta\tau$.

The spatial derivatives on the right-hand side are represented by two operators, Λ_x and Λ_y, corresponding to diffusion in the x- and y-directions, respectively

$$\Lambda_x(\Phi) = \frac{-2\Phi_{j,k} + \Phi_{j+1,k} + \Phi_{j-1,k}}{(\Delta x)^2} \tag{7.60}$$

$$\Lambda_y(\Phi) = \frac{-2\Phi_{j,k} + \Phi_{j,k+1} + \Phi_{j,k-1}}{(\Delta y)^2}. \tag{7.61}$$

Given a pseudo-time step $\Delta\tau$, the first step is to solve the following difference equation with time step $(\Delta\tau)/2$

$$\frac{\Phi_{jk}^{n+1/2} - \Phi_{jk}^{n}}{(\Delta\tau)/2} = \Lambda_x(\Phi^{n+1/2}) + \Lambda_y(\Phi^{n}) - 4\pi G\rho_{jk}. \tag{7.62}$$

This equation is implicit in the x-direction and explicit in the y-direction. Then, using the new Φ values generated for time $n+1/2$, one takes a second step, which is implicit in the y-direction and explicit in the x-direction

$$\frac{\Phi_{jk}^{n+1} - \Phi_{jk}^{n+1/2}}{(\Delta\tau)/2} = \Lambda_x(\Phi^{n+1/2}) + \Lambda_y(\Phi^{n+1}) - 4\pi G\rho_{jk}. \tag{7.63}$$

The first step (x-sweep) actually represents K independent solutions of a one-dimensional equation, where K is the total number of zones in the y-direction. At each fixed y, a one-dimensional equation is solved in which Φ varies only as a function of x. At each k, the equation can be written

$$A_j\Phi_{j-1,k}^{n+1/2} + B_j\Phi_{j,k}^{n+1/2} + C_j\Phi_{j+1,k}^{n+1/2} = -E_j, \tag{7.64}$$

where A_j, B_j, C_j, E_j involve constants or quantities defined at time level n. Given boundary conditions at $j = 0$ and $j = J$, this set of equations can be solved for Φ for all values of j for a given value of k by elimination and back substitution by the same method that is used to solve Equation (2.79). Note that this scheme is well suited to treat the case of mixed Dirichlet and Neumann boundary conditions: For each strip running through j, the inner boundary can be set to zero derivative and the outer boundary set to a fixed value of the potential. The background Φ^n values are left unchanged until the one-dimensional equations have been solved for all values of k. Then they are replaced by the $\Phi^{n+1/2}$ values just calculated.

Then, the second step (y-sweep) is carried out, in which J one-dimensional equations are solved, each at a fixed j; that is, equations analogous to Equation (7.64) are set up

$$A_k\Phi_{j,k-1}^{n+1} + B_k\Phi_{j,k}^{n+1} + C_k\Phi_{j,k-1}^{n+1} = -E_k, \tag{7.65}$$

where the A_k, B_k, C_k, E_k involve constants or quantities defined at time level $n+1/2$. At each j, this equation is solved for Φ at all k. Thus, the two-dimensional problem is reduced to a set of $K \times J$ one-dimensional problems.

This set of calculations is repeated with a number of pseudo-time steps, ranging from a minimum value $(\Delta\tau)_{\rm min}$ to a maximum value $(\Delta\tau)_{\rm max}$. Peaceman and Rachford (1955) showed that for a uniform grid with zone size Δx the optimum sequence of time steps is:

$$(\Delta\tau)_p = \frac{(\Delta x)^2}{4\sin^2[(2p+1)\pi/(4N)]}, \tag{7.66}$$

where N is the total number of zones and p is taken in the order $N-1, N-2, \ldots 2, 1, 0$, giving a sequence of time steps ranging from $(\Delta x)^2/4$ to $4(N\Delta x)^2/\pi^2$, that is, covering all size scales from the smallest to the largest on the grid. This expression suggests that the number of pseudo-time steps equals N, which could be a large number, but in fact in the case of large N, one can eliminate some of the larger values of p, since the function $(\Delta\tau)_p$ is slowly varying in that range.

For a nonuniform grid, Black and Bodenheimer (1975) experimented with a geometric series of pseudo-time steps, which again connect the smallest length scale (one zone) with the largest (the entire grid). Their grid was geometrical in the sense that each zone size Δx_j was a constant multiple, say, 1.05, of the previous zone size Δx_{j-1}. In analogy with Equation (7.66) the minimum and maximum time steps were

$$(\Delta\tau)_{min} = \frac{(\Delta x)^2_{min}}{4} \quad \text{and} \quad (\Delta\tau)_{max} = \frac{x^2_{max}}{4}, \tag{7.67}$$

where $(\Delta x)_{min}$ is the size of the smallest zone and x_{max} is the size of the entire grid. Then the time step sequence $(\Delta\tau)_m$, with $m = 0, 1, \ldots, M$, is given by Equation (6.110).

The appropriate value of M is determined by the requirement that the solution to the Poisson equation be correct to a given level of accuracy, as measured by the criterion

$$\chi = \frac{\nabla^2\Phi - 4\pi G\rho}{4\pi G\rho}. \tag{7.68}$$

Black and Bodenheimer (1975) found that in certain test cases, involving uniform or nonuniform spheres, where an analytical comparison is available, convergence to $\chi \approx 10^{-6}$ could be obtained with $M = 20$. For a given problem, M must be determined by experiment. The time step sequence represents approximate diffusion times across various distances; the diffusion coefficient in Equation (7.57) is unity, so the diffusion time is $(\Delta R)^2$ where ΔR is the length scale involved. Thus, $(\Delta\tau)_{max}$ is within a factor of 4 of the diffusion time across the entire grid.

7.2.3 Successive Overrelaxation

The diffusion equation form of the Poisson equation can also be solved by a technique known as *successive overrelaxation*, or SOR. It is essentially an explicit technique. Three different levels of approximation are discussed here: the *Jacobi* method, the *Gauss–Seidel* method, and SOR itself. Rewriting the basic difference Equation (7.59) and assuming $\Delta x = \Delta y$

$$\begin{aligned}\Phi^{n+1}_{j,k} - \Phi^n_{j,k} = {} & \frac{\Delta\tau}{(\Delta x)^2}\left(\Phi^n_{j+1,k} + \Phi^n_{j-1,k} + \Phi^n_{j,k+1} + \Phi^n_{j,k-1} - 4\Phi^n_{j,k}\right) \\ & - 4\pi G\rho_{j,k}\Delta\tau.\end{aligned} \tag{7.69}$$

This equation can be rewritten

$$\begin{aligned}\Phi^{n+1}_{j,k} = {} & \Phi^n_{j,k}\left(1 - \frac{4\Delta\tau}{(\Delta x)^2}\right) \\ & + \frac{\Delta\tau}{(\Delta x)^2}\left[\Phi^n_{j+1,k} + \Phi^n_{j-1,k} + \Phi^n_{j,k+1} + \Phi^n_{j,k-1} - 4\pi G\rho_{j,k}(\Delta x)^2\right].\end{aligned} \tag{7.70}$$

We now note that the explicit solution of a diffusion equation involves a limitation on the time step $\Delta\tau < 0.5(\Delta x)^2$), in one dimension, where the diffusion coefficient is 1 in this case. In two dimensions, this criterion becomes

$$\frac{\Delta\tau}{(\Delta x)^2} + \frac{\Delta\tau}{(\Delta y)^2} \le \frac{1}{2}. \tag{7.71}$$

The *Jacobi* method uses the maximum pseudo-time step allowed and, unlike the ADI method, does not use a sequence of time steps. If $\Delta\tau = 0.25(\Delta x)^2$, then Equation (7.70) becomes

$$\Phi^{n+1}_{j,k} = 0.25\left[\Phi^n_{j+1,k} + \Phi^n_{j-1,k} + \Phi^n_{j,k+1} + \Phi^n_{j,k-1} - 4\pi G\rho_{j,k}(\Delta x)^2\right]. \tag{7.72}$$

Thus, the new estimate of Φ involves an average of the four neighboring values on the grid, plus a source term. The procedure is repeated until it converges ($\partial\Phi/\partial\tau = 0$). However, an analysis by Press et al. (1992, p. 856) shows that the procedure converges too slowly to be of interest.

An improvement of about a factor of 2 in the convergence rate, which is still not sufficient, can be obtained through the *Gauss–Seidel* procedure. Instead of storing the array Φ^{n+1} at each grid point in the two-dimensional space and replacing Φ^n with Φ^{n+1} only after the entire iteration has been completed, replace $\Phi^n_{j,k}$ by $\Phi^{n+1}_{j,k}$ as soon as it has been computed. Thus, if one starts at the origin and computes from Equation (7.72), the first row of $\Phi^{n+1}_{j,k}$ values for $k = 1$ for all js ($k = 0$ are boundary values), then when the row $k = 2$ is calculated, the $\Phi_{j,k-1}$ and $\Phi_{j-1,k}$ have already been updated.

The actual SOR method, however, is to modify the $\Phi^{n+1}_{j,k}$ values, taking them to be a linear combination of $\Phi^{n+1}_{j,k}$ and $\Phi^n_{j,k}$

$$\Phi^{\text{actual}}_{j,k} = \gamma\Phi^{n+1}_{j,k} + (1-\gamma)\Phi^n_{j,k}. \tag{7.73}$$

If the parameter γ is less than 1, the procedure is known as "underrelaxation"; on the other hand, if it lies between 1 and 2, it is "overrelaxation." If it is greater than 2, the procedure is numerically unstable. Therefore, $\Phi^{n+1}_{j,k}$ is calculated according to Equation (7.72) and then $\Phi^{\text{actual}}_{j,k}$ from Equation (7.73) replaces $\Phi^{n+1}_{j,k}$. Note that in Equation (7.72) the $\Phi^n_{j,k}$ values are replaced by $\Phi^{n+1}_{j,k}$ values as soon as they are available. In principle there is an optimal value of γ, which gives the most rapid rate of convergence; however, for a given problem it usually has to be determined by trial and error. As a useful guide, Press et al. (1992) have shown that, on a uniform Cartesian grid with J grid points in the x-direction of size Δx and K grid points in the y-direction of size Δx, the optimal value is

$$\gamma = \frac{2}{1 + \left(1 - r_s^2\right)^{1/2}}, \tag{7.74}$$

where the spectral radius is given by

$$r_s = \frac{1}{2}[\cos(\pi/J) + \cos(\pi/K)]. \tag{7.75}$$

For a given problem, however, the optimal value of γ must be obtained empirically, and the advantages of SOR are realized only if it is actually close to the optimal value. Convergence is usually obtained after a number of iterations equal to the larger of J or K. A method of speeding up the convergence is to use a procedure known as *Chebyshev acceleration* in which the odd-numbered grid points and the even-numbered grid points are alternately updated and in which a variable γ as a function of an iteration number is used. A program for calculating SOR with Chebyshev acceleration is given in Press et al. (1992, p. 860).

7.2.4 MULTIGRID METHOD

The multigrid method (see Hackbusch, 1985) is an extension of the iterative solution methods (e.g., the Gauss–Seidel method is an appropriate one) with the purpose of speeding up the convergence rate. The problem with the standard methods is that while they quickly eliminate short-wavelength errors, those on the scale of the grid spacing, they are very inefficient in eliminating errors with long wavelengths, that is, on a scale of many grid cells. The reason is that one cell interacts only with its neighbors during one iteration. So, roughly, to eliminate the error at some specific wavelength, information needs to be propagated at least once across its wavelength. Thus, for a grid with hundreds of cells in each dimension, it takes a very large number of iterations to achieve convergence.

The multigrid method overcomes this problem by overlaying a series of coarser and coarser grids on top of the basic grid. For example, each grid could have twice the mesh spacing of the next finer one, down to the coarsest grid of only $3 \times 3 \times 3$ (in three dimensions). The method then involves switching between these grids according to some ordered procedure and obtaining an approximate solution on each grid. Thus, the short-wavelength errors are eliminated on the fine grids and the long-wavelength errors on the coarse grids. The simplest procedure is the so-called V-cycle: One starts with an approximate solution on the finest grid, say, one Gauss–Seidel iteration, then averages the solution onto the next coarsest grid, performs an iteration there, and then works one's way down to the very coarsest grid. One then obtains an exact solution there and works back, one grid at a time, by interpolating and performing one iteration at each level, back to the finest grid. In practice, one V-cycle can do the work of hundreds of iterations of the Gauss–Seidel method. The exact nature of the cycle to be used depends on the details of one's particular problem, and some alternatives are discussed in detail by Press et al. (1992, Section 19.6). Theoretically, the number of operations needed to solve problems with the multigrid method scales with N, the total number of grid points.

7.2.5 FOURIER TECHNIQUES

The fast Fourier transform, as discussed in Section 7.1.2, has the big advantage of speed, but its disadvantages are that (1) it requires a uniform grid, (2) only certain types of boundary conditions can be implemented, and (3) the number of grid points in each direction must be a power of 2. In Section 7.1.2, we discussed Fourier techniques for the solution of the integral form of the Poisson Equation (7.4); here we discuss the somewhat different techniques needed to solve the differential form (Equation 7.1).

The basic discrete direct and inverse Fourier transforms are given by Equation (7.7) and Equation (7.8), respectively.

As a first simple example of application of a Fourier technique, let us consider the problem of hydrodynamics discussed in Chapter 6, Section 6.6. The coordinate system is cylindrical: (R, Z, ϕ). In that subsection, axisymmetry was assumed, so that gradients in the ϕ-direction vanish. Here we consider the full 3-D problem, in which the gravitational potential Φ is desired as a function of (R, Z, ϕ). The solution domain, under the assumption of symmetry with respect to the equatorial plane, is the upper half plane, including the full range of 2π in ϕ. The boundary conditions for the potential are assumed to be provided by the method discussed in Section 7.2.1. An expansion in spherical harmonics gives the actual values of Φ on the outer edges (the sides and the top of the cylinder) of the grid (a Dirichlet boundary condition), while at the equatorial plane $Z = 0$, the Neumann condition is applied: $\partial\Phi/\partial Z = 0$. The strategy here is to do a Fourier transform in the ϕ direction and, thereby, convert the three-dimensional problem of finding the potential into a number of two-dimensional problems. The fact that the boundary condition in the ϕ-direction is periodic $[\Phi(R, Z, 0) = \Phi(R, Z, 2\pi)]$ enables the Fourier transform to be done simply.

The Poisson equation in this coordinate system is given by Equation (7.2). If the grid points in R, Z, and ϕ run from $j = 0 \ldots\ldots J$, $k = 0 \ldots\ldots K$, and $l = 0 \ldots\ldots L$, respectively, and if both ρ and Φ are defined at the centers of cells, then a standard differencing procedure yields

$$\begin{aligned} &A_j\Phi_{j+1,k,l} + B_j\Phi_{j-1,k,l} + A_k\Phi_{j,k+1,l} + B_k\Phi_{j,k-1,l} + C_j\Phi_{j,k,l+1} \\ &\quad + C_j\Phi_{j,k,l-1} - (A_j + B_j + 2C_j + A_k + B_k)\Phi_{j,k,l} = 4\pi G\rho_{j,k,l}, \end{aligned} \tag{7.76}$$

where

$$\begin{aligned} &A_j = \frac{R_{j+1}}{R_{j+1/2}\Delta R_{j+1/2}\Delta R_{j+1}}, \qquad B_j = \frac{R_j}{R_{j+1/2}\Delta R_{j+1/2}\Delta R_j} \\ &A_k = (\Delta Z_{k+1/2}\Delta Z_{k+1})^{-1}, \qquad B_k = (\Delta Z_{k+1/2}\Delta Z_k)^{-1} \\ &C_j = (R_{j+1/2}\Delta\phi)^{-2}, \end{aligned} \tag{7.77}$$

where uniform grid spacing $\Delta\phi$ is assumed in the ϕ direction, and

$$\begin{aligned} &\Delta R_{j+1/2} = R_{j+1} - R_j \qquad \Delta R_{j+1} = R_{j+3/2} - R_{j+1/2} \\ &\Delta Z_{k+1/2} = Z_{k+1} - Z_k \qquad \Delta Z_{k+1} = Z_{k+3/2} - Z_{k+1/2}. \end{aligned} \tag{7.78}$$

In this example, the grid in the R- and Z-directions does not have to be uniformly spaced. The grid setup in the (R, Z) plane at a given angle ϕ is shown in Figure 7.4.

Then Φ and ρ are expanded in Fourier series in the ϕ direction, assuming periodic boundary conditions $\Phi_{j,k,0} = \Phi_{j,k,L}$

$$\Phi_{j,k,l} = \sum_{m=0}^{L/2} \left[\Phi^{(1)}_{j,k,m}\cos(m\phi_l) + \Phi^{(2)}_{j,k,m}\sin(m\phi_l)\right], \tag{7.79}$$

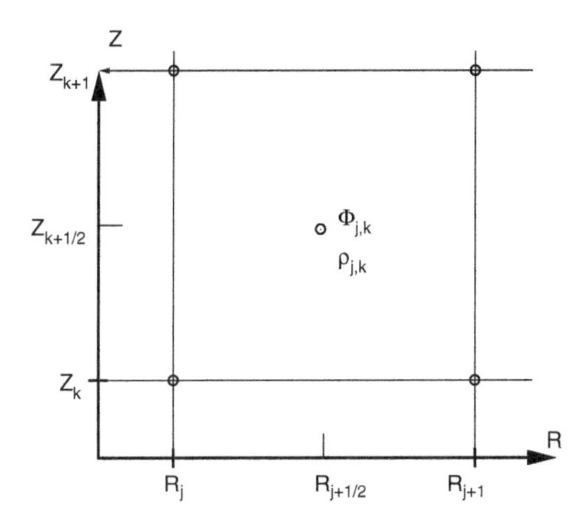

FIGURE 7.4 A typical cell in an (R, Z) grid in cylindrical coordinates showing the locations of the discretized coordinates and the gravitational potential Φ and the density ρ.

where

$$\phi_l = \frac{2\pi l}{L} \tag{7.80}$$

and the transformed potentials $\Phi^{(1)}$ and $\Phi^{(2)}$ are to be solved for.

When this series is inserted into Equation (7.76), the terms involving $(l+1)$ and $(l-1)$ can be combined with those involving l, at a given value of (j, k) through the assumption that the ϕ direction is uniformly zoned $(\Delta\phi = 2\pi/L)$ and the trigonometric relations

$$\begin{aligned} \cos(m\phi_{l+1}) + \cos(m\phi_{l-1}) &= 2\cos(m\Delta\phi)\cos(m\phi_l) \\ \sin(m\phi_{l+1}) + \sin(m\phi_{l-1}) &= 2\cos(m\Delta\phi)\sin(m\phi_l). \end{aligned} \tag{7.81}$$

Combining terms with the same value of $m\phi_l$ in the Fourier expansion, one obtains from Equation (7.76) a set of L individual equations, each being two-dimensional with indices (j, k), in the transformed potentials $\Phi^{(1)}$ and $\Phi^{(2)}$. Away from the Z-axis, that is, for $j > 1$ and for $q = 1, 2$

$$\begin{aligned} A_j\Phi^{(q)}_{j+1,k,m} &+ B_j\Phi^{(q)}_{j-1,k,m} + A_k\Phi^{(q)}_{j,k+1,m} + B_k\Phi^{(q)}_{j,k-1,m} \\ &- (A_j + B_j + A_k + B_k - 2[\cos(m\Delta\phi) - 1]C_j)\Phi^{(q)}_{j,k,m} \\ &= 4\pi G\rho^{(q)}_{j,k,m}, \end{aligned} \tag{7.82}$$

where $\rho^{(q)}$ is the transformed density, obtained in the same way as the transformed potential, and where the index m runs from 0 to $L/2$ for $q = 1$ and from 1 to $L/2 - 1$ for $q = 2$. For the zones along the Z-axis, $j = 1$, the coefficient of the quantity $\Phi^{(q)}_{j-1,k,m}$ is zero, and the coefficient of the quantity $\Phi^{(q)}_{j,k,m}$ becomes

$$-[A_j + (1 - (-1)^m)B_j + A_k + B_k - 2[\cos(m\Delta\phi) - 1]C_j].$$

The reason for this exception is that the zone at $(j-1,k,l)$ is on the other side of the Z-axis from the zone at (j,k,l) so, consistent with the Neumann boundary condition, $\Phi(j-1,k,l) = \Phi(j,k,l+L/2)$ corresponding to an angle $\phi_l + \pi$. Thus, in the Fourier expansion, $\Phi^{(q)}(j-1,k,m)$ can be replaced by $(-1)^m\Phi^{(q)}(j,k,m)$.

The three-dimensional problem has now been reduced to L two-dimensional problems, $L/2+1$ equations for the $\Phi^{(1)}$ and $L/2-1$ equations for the $\Phi^{(2)}$, which can be solved by a variety of methods (further Fourier transforms are not practical because of the boundary conditions). For example, Tohline (1980) and Pickett et al. (1996) use cyclic reduction (see below) based on routines of Swarztrauber and Sweet (1975). Once the transformed potentials $\Phi^{(q)}$ have been obtained, then the actual potentials $\Phi_{j,k,l}$ are obtained from Equation (7.79). Fast Fourier transforms, therefore, are involved in this final step as well as in the initial transformation of the density to $\rho^{(1)}$ and $\rho^{(2)}$.

We now return to the two-dimensional problem in Cartesian coordinates, with the difference scheme expressed by Equation (7.45). The procedure is discussed by Press et al. (1992, Section 19.4). The method works best when the boundary conditions are periodic

$$\Phi_{j,k} = \Phi_{j+J,k} = \Phi_{j,k+K}.$$

In the typical astrophysical problem, these conditions are not met. Press et al. also describe what to do if the problem has Dirichlet boundary conditions, for example, $\Phi_{j,k} = 0$ along all boundaries of the grid, or if it has Neumann boundary conditions $\nabla\Phi = 0$ along all boundaries. In a typical astrophysical problem, such as the grid setup shown in Figure 7.3, the boundary conditions are mixed, with Dirichlet conditions at the boundaries $j = J$ and $k = K$ and Neumann conditions at the inner boundaries $j = 0$ and $k = 0$. Such boundary conditions are more easily handled by other techniques. But the basic Fourier procedure is the same in all cases, and it is illustrated here for the case of (uniform) $\Delta x = \Delta y$ and periodic boundary conditions:

1. Compute the two-dimensional Fourier transform of the density

$$\hat{\rho}_{mn} = \sum_{j=0}^{J-1}\sum_{k=0}^{K-1} \rho_{jk}\exp(2\pi imj/J)\exp(2\pi ink/K). \tag{7.83}$$

2. Calculate the Fourier transform of the potential from the relation

$$\hat{\Phi}_{mn} = \frac{2\pi G\hat{\rho}_{mn}(\Delta x)^2}{\cos(2\pi m/J)+\cos(2\pi n/K)-2}. \tag{7.84}$$

3. Calculate the actual potential from the inverse Fourier transform

$$\Phi_{jk} = \frac{1}{JK}\sum_{m=0}^{J-1}\sum_{n=0}^{K-1} \hat{\Phi}_{mn}\exp(-2\pi imj/J)\exp(-2\pi ink/K). \tag{7.85}$$

Exercise

Prove Equation (7.84) by substituting Equation (7.85) and the corresponding inverse Fourier transform for the density into Equation (7.45).

For Dirichlet boundary conditions, where $\Phi = 0$ on all boundaries, one uses a sine series

$$\Phi_{jk} = \frac{4}{JK} \sum_{m=1}^{J-1} \sum_{n=1}^{K-1} \hat{\Phi}_{mn} \sin(\pi mj/J) \sin(\pi nk/K). \tag{7.86}$$

Note that the sine transform is its own inverse, except for the factor $\frac{4}{JK}$ in this case. If the boundary conditions are not zero but are provided, for example, by a series expansion (Equation 7.49), then the known boundary values that appear on the left side of Equation (7.45) can be moved to the right-hand side and treated as known constants. The resulting system is equivalent to one with zero boundary conditions and it can be solved using Equation (7.86).

For Neumann boundary conditions, where $\nabla\Phi = 0$ on all boundaries, one uses a cosine series

$$\Phi_{jk} = \frac{4}{JK} \sum_{m=0}^{J} \sum_{n=0}^{K} \hat{\Phi}_{mn} \cos(\pi mj/J) \cos(\pi nk/K), \tag{7.87}$$

with the added provision that the terms for $m = 0$, $n = 0$, $m = J$, and $n = K$ are multiplied by 0.5. Note again that the cosine series is its own inverse except for the factor $\frac{4}{JK}$ in the inverse. If the gradients are not zero but finite values, the corresponding boundary terms again can be brought to the right-hand side of Equation (7.45) and the equivalent problem with zero gradients solved. Further details on the use of both Dirichlet and Neumann boundary conditions are given in Press et al. (1992, Section 19.4).

7.2.6 Cyclic Reduction

A Poisson equation of the form of Equation (7.2) or Equation (7.3), which has a first-derivative term with a nonconstant coefficient, may be better solved by the technique of *cyclic reduction*. The details of this technique are described in Press et al. (1992, Section 19.4), so it will be summarized only briefly here.

The standard difference representation of the Poisson equation in two space dimensions in Cartesian coordinates and constant grid spacing $\Delta x = \Delta y$ is:

$$\Phi_{j+1,k} + \Phi_{j-1,k} - 4\Phi_{j,k} + \Phi_{j,k-1} + \Phi_{j,k+1} = 4\pi G\rho(\Delta x)^2. \tag{7.88}$$

Again, k runs from 0 to K and j runs from 0 to J. For a given j, a total of K such equations can be written. The procedure starts at the center of the j-grid, that is, with $j = J/2$ (assuming J is even) in the above equation. Using matrix methods, these equations can be transformed into another set of K equations, of the same form as the original equations, in each of which $\Phi_{j+2,k}$, $\Phi_{j-2,k}$, $\Phi_{j,k}$, $\Phi_{j,k+1}$, and $\Phi_{j,k-1}$ appear, resulting in the elimination of terms with $\Phi_{j+1,k}$ and $\Phi_{j-1,k}$. The total number of equations in the j-direction has been reduced by a factor of 2. This process can be repeated a number of times; for example, the next step would be to obtain equations (at each k) relating $\Phi_{j+4,k}$, $\Phi_{j-4,k}$, $\Phi_{j,k}$, $\Phi_{j,k+1}$, and $\Phi_{j,k-1}$. The process works its way outward until the boundaries are reached, with one final equation (for each k) involving boundary values $\Phi_{J,k}$, $\Phi_{0,k}$, as well as $\Phi_{J/2,k}$, $\Phi_{J/2,k+1}$, and $\Phi_{J/2,k-1}$.

Since the boundary values are known, one is left with a set of equations for a single $j = J/2$, each with terms involving $k, k+1, k-1$. These equations are in standard tridiagonal form and can be solved by the standard technique (Equation 2.79) to get $\Phi_{J/2}$ for all k.

The solution for other values of j proceeds in the reverse order, back down the hierarchy of systems that has been set up. Suppose, for example, that $J = 32$, so we have obtained the solution for $j = 16$. There are two sets of equations at the next level down, one involving $j = 0$, $j = 8$, and $j = 16$; the other involving $j = 16$, $j = 24$, and $j = 32$. In the first set, the values at $j = 0$ and $j = 16$ are known, so there remains a tridiagonal system to be solved at $j = 8$. Similarly, the set at $j = 24$ can be solved. Going to the next level down, there are four sets of equations that can be solved for $j = 4$, $j = 12$, $j = 20$, and $j = 28$. Proceeding down the chain, the solutions are obtained for all values of j, through $J - 1$ solutions of tridiagonal systems. The specifics of a numerically stable version of this procedure, as devised by Buneman (1969), are given in Hockney and Eastwood (1981, Section 6–5–1).

A variation of this method known as the *FACR method*, also described in Press et al. (1992), is a combination of Fourier analysis and cyclic reduction. It offers economies in computational time over either the Fourier method or the cyclic reduction method used individually.

7.2.7 Polynomial Expansions in Three Dimensions

We now consider a different procedure, using polynomial expansions, to solve the Poisson Equation (7.1). Assume we are working in three space dimensions in a spherical coordinate system (r, θ, ϕ). We solve the differential equation (7.3) directly through an expansion in spherical harmonics (Boss, 1980)

$$\Phi(r, \theta, \phi) = \sum_{l=0}^{L} \sum_{m=-l}^{l} \Phi_{lm}(r) Y_{lm}(\theta, \phi), \tag{7.89}$$

$$\rho(r, \theta, \phi) = \sum_{l=0}^{L} \sum_{m=-l}^{l} \rho_{lm}(r) Y_{lm}(\theta, \phi), \tag{7.90}$$

where $Y_{lm}(\theta, \phi)$ is given by Equation (7.55). The density coefficients that are needed in Equation (7.90) are obtained by numerical integration from

$$\rho_{lm}(r) = \int_0^{2\pi} \int_0^{\pi} \rho(r, \theta, \phi) Y_{lm}^*(\theta, \phi) \sin\theta d\theta d\phi. \tag{7.91}$$

When the expansions of Equation (7.89) and Equation (7.90) are substituted into Equation (7.3), the variables separate and the orthonormality condition of the spherical harmonics results in an equation as a function of radius only

$$\frac{d^2\Phi_{lm}(r)}{dr^2} + \frac{2}{r}\frac{d\Phi_{lm}(r)}{dr} - \frac{l(l+1)}{r^2}\Phi_{lm}(r) = 4\pi G \rho_{lm}(r). \tag{7.92}$$

This one-dimensional equation may be transformed into difference equations of the form

$$A_j \Phi_{j-1} + B_j \Phi_j + C_j \Phi_{j+1} = -E_j, \tag{7.93}$$

one for each combination of (l, m), where the coefficients are known quantities. These equations can be solved by the method of elimination and back substitution that is described in connection with Equation (2.79).

Exercise

Calculate the coefficients A_j, B_j, C_j, and E_j for Equation (7.93).

Thus, the coefficients Φ_{lm} are obtained for values of (l, m) up to some cutoff value, say $l \leq 16$, in which case the appropriate maximum m values are $-16 \leq m \leq 16$. If the configuration is symmetric with respect to the equatorial plane, then only even values of $l + m$ have to be included. A boundary condition must be applied at the outer edge, which is assumed to be a spherical surface of radius R. If one assumes that the density vanishes outside R, then Equation (7.92) can be solved to give

$$\Phi_{lm}(r) = \text{constant} \cdot r^{-(l+1)} \tag{7.94}$$

leading to the boundary condition at R

$$\frac{d\Phi_{lm}(r)}{dr} + \frac{l+1}{r}\Phi_{lm}(r) = 0. \tag{7.95}$$

Then the potential at all points in the three-dimensional mesh is obtained from Equation (7.89).

7.3 TEST OF THE POTENTIAL

Any numerical program must be tested against known analytical solutions to assess its accuracy. In the case of a potential solver, it is simple to derive the potential as a function of radius inside a sphere of uniform or nonuniform density and compare it with numerical results. The real test of a program in two or three dimensions is to compare with the potential of a nonspherical object. The most suitable available test is the uniform-density ellipsoid in 3-D or the corresponding oblate or prolate spheroid in 2-D. Such tests, of course, are invaluable for making decisions as to which potential solver is most suitable for a given problem.

For an internal point (x, y, z) in an ellipsoid of uniform density ρ, whose semiaxes are a, b, and c, the potential is (Ramsey, 1981, Section 7.35)

$$\Phi = \pi G\rho abc(Ax^2 + By^2 + Cz^2 - D), \tag{7.96}$$

where

$$A = \int_0^\infty [(a^2+u)^3(b^2+u)(c^2+u)]^{-1/2}du$$

$$B = \int_0^\infty [(a^2+u)(b^2+u)^3(c^2+u)]^{-1/2}du$$

$$C = \int_0^\infty [(a^2+u)(b^2+u)(c^2+u)^3]^{-1/2}du$$

$$D = \int_0^\infty [(a^2+u)(b^2+u)(c^2+u)]^{-1/2}du. \tag{7.97}$$

For an oblate spheroid at an interior point (x, y, z) with semiaxes $a = b > c$, these expressions reduce to a much simpler form

$$\Phi = -\frac{2\pi G\rho a^2 c}{\sqrt{a^2 - c^2}}\left(1 - \frac{x^2 + y^2 - 2z^2}{2(a^2 - c^2)}\right)\sin^{-1}\sqrt{1 - c^2/a^2} - \frac{\pi G\rho}{a^2 - c^2}(c^2x^2 + c^2y^2 - 2a^2z^2). \tag{7.98}$$

For a prolate spheroid at an interior point (x, y, z) with semiaxes $a = b < c$

$$\Phi = -\frac{2\pi G\rho a^2 c}{\sqrt{c^2 - a^2}}\left(1 + \frac{x^2 + y^2 - 2z^2}{2(c^2 - a^2)}\right)\sinh^{-1}\sqrt{c^2/a^2 - 1} + \frac{\pi G\rho}{c^2 - a^2}(c^2x^2 + c^2y^2 - 2a^2z^2). \tag{7.99}$$

These uniform-density structures actually provide a rather strict test for a potential solver, since much of the mass is near the outer boundary and there is a discontinuity in density at the edge; thus, the overall accuracy is controlled by the accuracy of the procedure used to get the boundary conditions. If the boundary potentials are obtained by expansion in Legendre polynomials, the added precision obtained by including more terms in that expansion must be weighed against the additional computational time required. If the configuration to be studied is centrally condensed, with much of

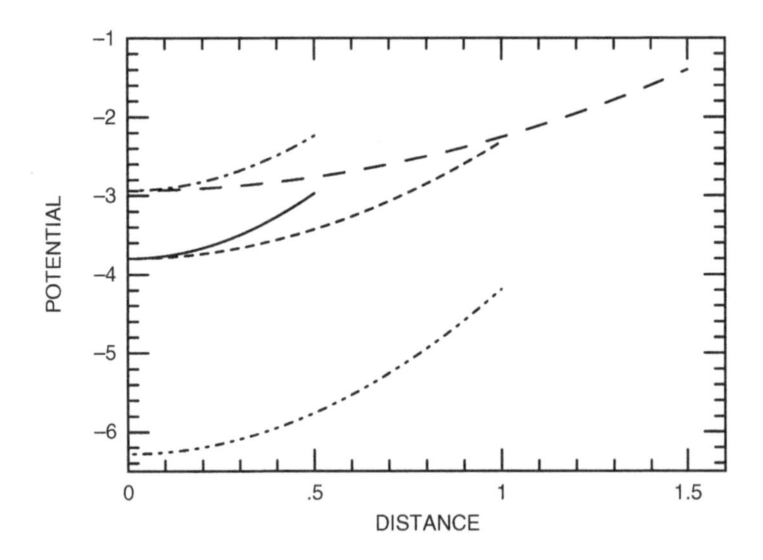

FIGURE 7.5 The gravitational potential of a uniform-density oblate spheroid as a function of distance from the center along the short axis with length $c = 0.5$ (*solid line*) and along the long axis with length $a = 1$ (*short dashed line*). The potential of a uniform-density prolate spheroid is shown along the long axis with length $c = 1.5$ (*long dashed line*) and along the short axis with length $a = 0.5$ (*dot dashed line*). The potential of a uniform-density sphere of radius 1 is shown by the *dot-dot-dashed line* (lowest curve). The density ρ and the gravitational constant G are taken to be 1.

the mass well away from the boundary, then the precision of the boundary condition is of less importance and the accuracy of the potential will be better than that deduced for the uniform-density configuration. In any case, the precision of the potential must be quite good, .01 to .001%, because the derivative of the potential is generally used to obtain the forces. Note also that the spheroidal analytical solutions can be used to test 3-D potential solvers if one changes the orientation of the Z-axis of the spheroid with respect to the grid coordinate system. Figure 7.5 plots the potential as a function of radius for a uniform sphere, along the major and minor axes for uniform-density prolate and oblate spheroids. Further information on the potential external to these structures may be found in Chandrasekhar (1969, Chapter 3) and Cohl and Tohline(1999). In Chandrasekhar's book, there is also a discussion of the potential of ellipsoids with nonuniform density.

Exercise

Show analytically that the central potential of a uniform-density sphere ($\rho = 1, R = 1, G = 1$) is -2π.

Exercise

As an example of a configuration with a nonuniform density, calculate analytically the potential as a function of radius r inside a sphere with density distribution

$$\rho(r) = \rho_c \left(1 - \frac{r^2}{R^2}\right), \tag{7.100}$$

where R is the total radius and ρ_c the central density. Set $G = R = \rho_c = 1$.

Exercise

Show that the internal potential as a function of distance to the center of a uniform-density ellipsoid (Equation 7.96) reduces to that of a uniform-density sphere in the limit when $a = b = c$.

REFERENCES

Binney, J. and Tremaine, S. (1987) *Galactic Dynamics* (Princeton, NJ: Princeton University Press).

Black, D. C. and Bodenheimer, P. (1975) *Astrophys. J.* **199**: 619

Bodenheimer, P. and Ostriker, J. P. (1973) *Astrophys. J.* **180**: 159.

Boss, A. P. (1980) *Astrophys. J.* **236**: 619.

Buneman, O. (1969) *SUIPR Report No. 294*, (Stanford, CA: Stanford University).

Chandrasekhar, S. (1969) *Ellipsoidal Figures of Equilibrium* (New Haven: Yale University Press).

Cohl, H. S. and Tohline, J. E. (1999) *Astrophys. J.* **527**: 86.

Cooley, J. W. and Tukey, J. (1965) *J. Math. Comp.* **19**: 297.

Eastwood, J. W. and Brownrigg, D. R. K. (1979) *J. Comp. Phys.* **32**: 24.

Hachisu, I. (1986a) *Astrophys. J. Suppl.* **61**: 479.
Hachisu, I. (1986b) *Astrophys. J. Suppl.* **62**: 461.
Hackbusch, W. (1985) *Multi-Grid Methods and Applications* (New York: Springer-Verlag).
Hartree, D. R. (1957) *The Calculation of Atomic Structure* (New York: John Wiley & Sons).
Hockney, R. W. (1970) *Meth. Comp. Phys.* **9**: 135.
Hockney, R. W. and Eastwood, J. W. (1981) *Computer Simulation Using Particles* (New York: McGraw–Hill).
Jackson, S. (1970) *Astrophys. J.* **161**: 579.
Margenau, H. and Murphy, G. (1956) *The Mathematics of Physics and Chemistry*, 2nd ed. (Princeton, NJ: Van Nostrand).
Myhill, E. A. and Boss, A. P. (1993) *Astrophys. J. Suppl.* **89**: 345.
Ostriker, J. P. and Bodenheimer, P. (1968) *Astrophys. J.* **151**: 1089.
Ostriker, J. P. and Mark, J. W.-K. (1968) *Astrophys. J.* **151**: 1075.
Peaceman, D. W. and Rachford, H. H. (1955) *J. Soc. Indust. Appl. Math.* **3**: 28.
Pickett, B. K., Durisen, R. H. and Davis, G. A. (1996) *Astrophys. J.* **458**: 714.
Pickett, B. K., Durisen, R. H., and Link, R. (1997) *Icarus* **126**: 243.
Pickett, B. K., Mejía, A. C., Durisen, R. H., Cassen, P. M., Berry, D. K., and Link, R. P. (2003) *Astrophys. J.* **590**: 1060.
Press, W. H., Teukolsky, S. A., Vetterling, W. T. and Flannery, B. P. (1992) *Numerical Recipes in Fortran: The Art of Scientific Computing,* 2nd ed. (Cambridge: Cambridge University Press).
Ramsey, A. S. (1981) *Newtonian Attraction* (Cambridge: Cambridge University Press).
Stone, J. M. and Norman, M. L. (1992) *Astrophys. J. Suppl.* **80**: 753.
Swarztrauber, P. and Sweet, R. (1975) *NCAR–TN/IA–109.*
Tohline, J. E. (1980) *Astrophys. J.* **235**: 866.

8 Magnetohydrodynamics

Simulations of many astrophysical processes must include effects associated with the presence of magnetic fields. Indeed, it is hard to imagine an interstellar cloud, an accretion disk, a pulsar, or a stellar atmosphere without magnetic fields, which significantly influence or even dominate their behavior. The solar atmosphere is a prime example: magnetic fields are significant in generating solar activity, heating the corona, and driving the solar wind. As another example, the generating mechanism for astrophysical jets is thought to be associated with the interaction of rotation and magnetic fields. Most of the universe is plasma and wherever there is a plasma, there is a magnetic field. As the plasma moves, it drags field lines along, bending and compressing them. The field, in turn, influences the motions of the plasma through the Lorentz force. As a result, a feedback develops, generating a multitude of interesting phenomena. This extremely rich area of astrophysical research is so complex that we will not be able to go beyond the very essentials. Our intention is to provide the reader with just the basic insight into the problems of numerical magnetohydrodynamics (MHD). We approach these problems with the help of the ZEUS code, whose hydrodynamical part was discussed in Chapter 6, Section 6.5. The presentation of the MHD techniques is followed by a few simple, but, in our opinion, very illustrative examples of MHD effects — from the converging motion of a fluid permeated by an azimuthal field to the famous magnetohydrodynamical instability responsible for the angular momentum transfer in accretion disks (Balbus and Hawley, 1991), mentioned already in Chapter 1, Section 1.5.

8.1 BASIC ASSUMPTIONS AND DEFINITIONS

Throughout this chapter, we shall assume that the fluid is nonviscous and has a negligible resistivity; thus, we consider "ideal" MHD. In other words, the *magnetic Reynolds number*, $R_{eM} = Lv/\eta_e$, is very large (L and v are the characteristic length and characteristic velocity, respectively, and η_e is the electric resistivity, Equation (1.145)). Extensions to nonideal situations are discussed by Stone (1999). With these assumptions, the MHD equations (1.148) reduce to

$$\frac{\partial \rho}{\partial t} + \nabla \cdot (\rho \mathbf{v}) = 0 \tag{8.1}$$

$$\frac{\partial \rho \mathbf{v}}{\partial t} + \nabla \cdot (\rho \mathbf{v}\mathbf{v}) = -\nabla P - \frac{1}{8\pi}\nabla(B^2) + \frac{1}{4\pi}(\mathbf{B}\cdot\nabla)\mathbf{B} \tag{8.2}$$

$$\frac{\partial e}{\partial t} + \nabla \cdot (e\mathbf{v}) = -P\nabla\cdot\mathbf{v} \tag{8.3}$$

$$\frac{\partial \mathbf{B}}{\partial t} = \nabla \times (\mathbf{v}\times\mathbf{B}), \tag{8.4}$$

where, as in Chapter 6, Section 6.5, the internal energy is used instead of the total energy. We shall assume that the MHD flow is axially symmetric; however, all three components of velocity and magnetic field will be accounted for (the "2 1/2-D" approximation). As in Chapter 6, Section 6.6.1, we shall use cylindrical coordinates (R, Z, Φ) in which the set (Equation 8.1 to Equation 8.4) takes the form

$$\frac{\partial \rho}{\partial t} + \nabla \cdot (\rho \mathbf{v}) = 0 \tag{8.5}$$

$$\begin{aligned}\frac{\partial(\rho u)}{\partial t} + \boldsymbol{\nabla} \cdot (\rho u \mathbf{v}) = &-\frac{\partial P}{\partial R} + \frac{A^2}{\rho R^3} + \frac{1}{4\pi} B_Z \frac{\partial B_R}{\partial Z} \\ &- \frac{1}{4\pi}\left(B_Z \frac{\partial B_Z}{\partial R} + \frac{B_\Phi}{R}\frac{\partial(R B_\Phi)}{\partial R}\right)\end{aligned} \tag{8.6}$$

$$\begin{aligned}\frac{\partial(\rho w)}{\partial t} + \boldsymbol{\nabla} \cdot (\rho w \mathbf{v}) = &-\frac{\partial P}{\partial Z} + \frac{1}{4\pi} B_R \frac{\partial B_Z}{\partial R} \\ &- \frac{1}{4\pi}\left(B_R \frac{\partial B_R}{\partial Z} + B_\Phi \frac{\partial B_\Phi}{\partial Z}\right)\end{aligned} \tag{8.7}$$

$$\frac{\partial A}{\partial t} + \nabla \cdot (A\mathbf{v}) = \frac{1}{4\pi}\left(B_R \frac{\partial(R B_\Phi)}{\partial R} + R B_Z \frac{\partial B_\Phi}{\partial Z}\right) \tag{8.8}$$

$$\frac{\partial e}{\partial t} + \boldsymbol{\nabla} \cdot (e\mathbf{v}) = -P\nabla \cdot \mathbf{v} \tag{8.9}$$

$$\frac{\partial B_R}{\partial t} = \frac{\partial \mathcal{E}_\Phi}{\partial Z} \tag{8.10}$$

$$\frac{\partial B_Z}{\partial t} = -\frac{1}{R}\frac{\partial(R\mathcal{E}_\Phi)}{\partial R} \tag{8.11}$$

$$\frac{\partial B_\Phi}{\partial t} = \frac{\partial \mathcal{E}_Z}{\partial R} - \frac{\partial \mathcal{E}_R}{\partial Z}, \tag{8.12}$$

where the *electromotive force*

$$\boldsymbol{\mathcal{E}} \equiv \mathbf{v} \times \mathbf{B} \tag{8.13}$$

has been introduced as an auxiliary vector field. As before, A is the angular momentum per unit volume. Note that some terms cancel out when the magnetic pressure gradient $\boldsymbol{\nabla}(B^2)$ and the magnetic tension $(\mathbf{B} \cdot \nabla)\mathbf{B}$ (see Chapter 1, Section 1.7) are explicitly calculated. As a result, the expressions for the R and Z components of the magnetic pressure gradient (terms in brackets on the right-hand side of Equation 8.6 and Equation 8.7) contain only two derivatives instead of three. Note also that because of the assumed symmetry the Φ component of the magnetic pressure gradient is equal to 0.

The linearized equations of hydrodynamics (Equation 2.61) allowed us to calculate the characteristic velocity with which pressure perturbations propagate across the fluid. Similarly, from the linearized MHD equations, one could derive the characteristic velocity with which magnetic field perturbations propagate across a conducting medium. As the procedure is too complex to be repeated here, readers interested in a detailed derivation are referred to the excellent textbook by Priest (1987). For our purposes, it will be sufficient to know that in the limit where the magnetic pressure greatly exceeds the gas pressure, a perturbation of the magnetic field excites *shear* and/or *compressional Alfvén waves*. Shear Alfvén waves are driven by magnetic tension (henceforth, following the widespread tradition, we shall refer to them simply as Alfvén waves). They are transverse waves and sometimes it is useful to think of them as vibrations of magnetic field lines. Their important property is that in a uniform medium they do not cause any density or pressure perturbations. A particular case of a shear wave is the *torsional* wave, for which only the azimuthal component of the perturbed field is different from zero. The compressional Alfvén waves are longitudinal waves driven by magnetic pressure and, as such, they are accompanied by density and pressure perturbations of the medium. Both shear and compressional waves propagate at a speed

$$v_A = \frac{B}{\sqrt{4\pi\rho}}, \tag{8.14}$$

called the *Alfvén speed*. In low-density media, even if the magnetic field is not very strong, v_A may be much larger than the velocity of sound (c_s). In such cases, the time step must be much shorter than indicated by the CFL (Courant-Friedrichs-Lewy) condition (Equation 6.81). For a coordinate x, the proper adjustment of the time step is guaranteed by the modified condition

$$\Delta t \leq \frac{\Delta x}{|u| + \sqrt{c_s^2 + v_A^2}}, \tag{8.15}$$

in which the Alfvén velocity is explicitly included.

To fully account for magnetic effects in a practical way, the basic algorithm of ZEUS is supplemented with the *method of characteristics* algorithm and the *constrained transport* algorithm, which, when combined, are referred to as MOC-CT (Hawley and Stone, 1995). These procedures are discussed in the next two sections. For the sake of completeness, it should be noted that when the magnetic pressure is of the same order as the gas pressure, another family of waves becomes important, namely *magnetoacoustic* or *magnetosonic* waves. However, since Stone and Norman (1992) argue that they are automatically taken into account by the standard hydrodynamic ZEUS procedures, there is no need to discuss them here, and the interested reader is again referred to the textbook by Priest (1987).

In ZEUS, all components of **B** are centered the same way as their corresponding components of velocity. However, the components of $\mathcal{E}$ are centered differently: $\mathcal{E}_R$ and $\mathcal{E}_Z$ are centered at $(j + 1/2, k)$ and $(j, k + 1/2)$, respectively, and $\mathcal{E}_\Phi$ is centered at (j, k), i.e., at the corner of the cell rather than in its middle (Figure 8.1; compare to Figure 6.10). Equation (8.5) to Equation (8.9) closely resemble Equation (6.88) to Equation (6.92) without gravity. The only other difference is additional source terms in the momentum equations, representing the components of the Lorentz force. Thus, the advection terms on the left-hand sides of Equation (8.5) to Equation (8.9) can be treated in exactly the same way as described in Section 6.5.2.

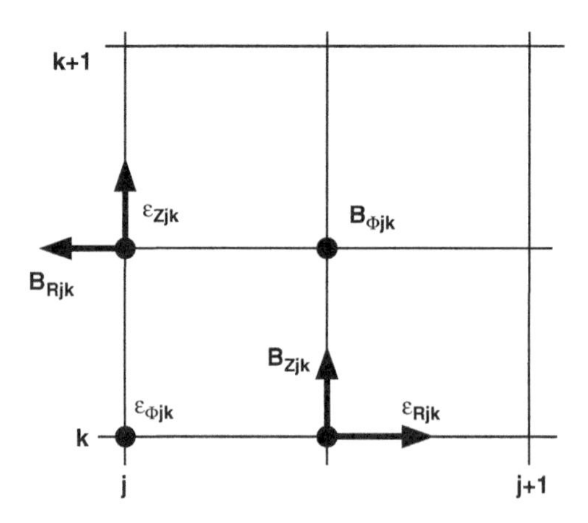

FIGURE 8.1 Schematic diagram of a two-dimensional (R, Z) grid of the ZEUS code indicating the centering of MHD variables. $\mathcal{E}_\Phi$ and B_Φ are perpendicular to the plane of the figure.

8.2 MHD SOURCE TERMS

The Lorentz terms in Equation (8.6) to Equation (8.8) are accounted for by a non-conservative procedure, conceptually similar to the one described in Section 6.5.1. However, as one might expect, the corresponding algorithms are much more complex than in a purely hydrodynamical case.

Let us begin with Equation (8.6) to Equation (8.7). The Lorentz terms

$$\frac{1}{4\pi}\left(B_Z\frac{\partial B_Z}{\partial R} + \frac{B_\Phi}{R}\frac{\partial(RB_\Phi)}{\partial R}\right) \qquad \text{and}$$

$$\frac{1}{4\pi}\left(B_R\frac{\partial B_R}{\partial Z} + B_\Phi\frac{\partial B_\Phi}{\partial Z}\right), \tag{8.16}$$

are related, respectively, to the $R-$ and $Z-$ components of the magnetic pressure gradient, but note that the first of the two equations also includes the tension term B_Φ^2/R. They are treated in the same way as the components of the gas pressure gradient in Equation (6.68) to Equation (6.69); however, before they are evaluated, B_R and B_Z must be properly centered via appropriate averaging.

Exercise

Write the difference equations for the partial updates of the momenta ρu and ρw arising from the Lorentz terms (Equation 8.16). This update is explicit, using only magnetic field quantities available at time step n.

The remaining Lorentz terms

$$\frac{1}{4\pi}B_Z\frac{\partial B_R}{\partial Z} \qquad \text{and}$$

$$\frac{1}{4\pi}B_R\frac{\partial B_Z}{\partial R} \tag{8.17}$$

are related, respectively, to the $R-$ and $Z-$ components of the magnetic tension, and as such they do not have analogs in the equations of hydrodynamics. Their difference representations in ZEUS are discussed in detail by Stone and Norman (1992). Here we shall limit ourselves to a brief outline of the method, taking the Z-component as an example.* Since its difference representation must be centered at the same location as the vertical momentum, i.e., at $(j+1/2, k)$, we have

$$B_R \frac{\partial B_Z}{\partial R} \approx \frac{\langle B_R \rangle_{j+1/2,k}}{\Delta R} \left(B^{n+1/2}_{Z,j+1,k} - B^{n+1/2}_{Z,j,k} \right), \tag{8.18}$$

where $\langle B_R \rangle$ is the appropriately averaged value of B_R, and $B^{n+1/2}_{Z,j,k}$ stands for the corner-centered value of B_Z, for stability reasons advanced to the middle of the time step. This particular form of the difference equation was obtained after lengthy experimentation, and it was chosen because it acceptably describes the propagation of Alfvén waves.

The advanced value of B_Z is a solution of the *Alfvén wave characteristic equation*

$$\frac{D}{Dt} w \pm \frac{1}{\sqrt{4\pi\rho}} \frac{D}{Dt} B_Z = 0, \tag{8.19}$$

where

$$\frac{D}{Dt} = \frac{\partial}{\partial t} \mp \frac{B_R}{\sqrt{4\pi\rho}} \frac{\partial}{\partial R}. \tag{8.20}$$

To illustrate the practical meaning of Equation (8.19), consider a grid point (j, k) to which (transverse) Alfvén waves arrive along the R-direction, influencing both $w_{j,k}$ and $B_{Z,j,k}$ (Figure 8.2). The waves that arrive from the left propagate at

$$v_A^+ = B_{R,j-1,k} / \sqrt{4\pi\rho_{j-1,k}},$$

while those from the right at

$$v_A^- = -B_{R,j,k} / \sqrt{4\pi\rho_{j,k}}.$$

At each $t^n < t < t^{n+1}$, the locations

$$R^+ = R_{j,k} - v_A^+(t - t^n) \quad \text{and} \quad R^- = R_{j,k} - v_A^-(t - t^n)$$

define the *domain of dependence* — the region that within t–t^n has been able to communicate via Alfvén waves with the point (j, k). The edges of the domain of dependence and the point t on the time-axis define, in turn, two straight lines called *characteristics* (in Figure 8.2, they are marked with C^+ and C^-). They are lines in the (R, t) plane with slopes v_A^+ and v_A^-, respectively. Assuming that B_Z and w vary linearly within grid cells, one can calculate their values at R^+ and R^- at time t^n (B_Z^+, w^+, B_Z^- and w^- in Figure 8.2). For that purpose, the monotonized interpolation procedure described in Chapter 6, Section 6.3.2 is used in ZEUS, with B_Z or w substituted for ρ and v_A^+ or v_A^- substituted for v.

---

* Note that Stone and Norman place the coordinates in the order (Z, R, Φ), while we use the standard (R, Z, Φ).

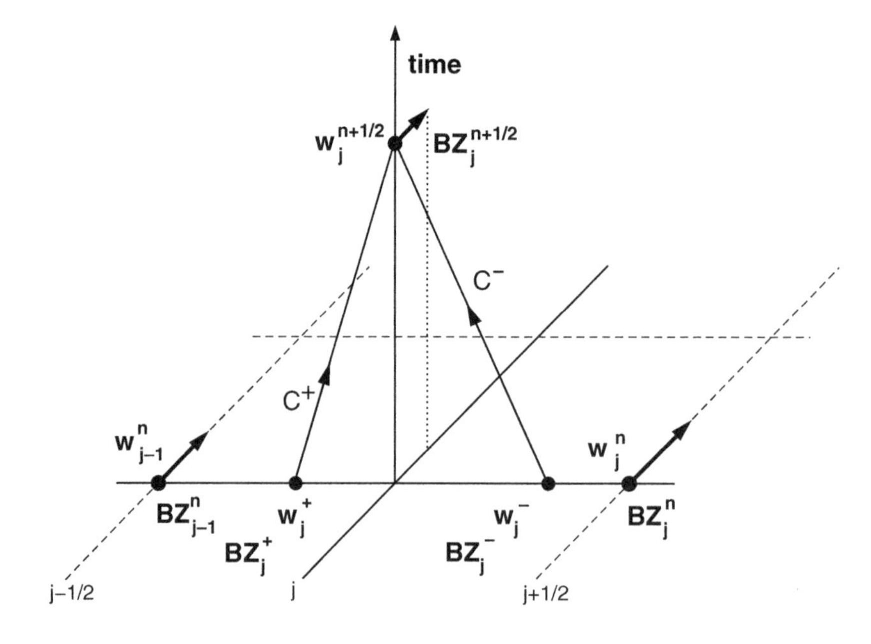

FIGURE 8.2 Method of characteristics: characteristics C^+ and C^- are lines in space–time along which Alfvén waves propagate. The waves cause the magnetic field to evolve to the same value of ${B_Z}^{n+1/2}$ both from ${B_Z}^+$ along C^+, and from ${B_Z}^-$ along C^-. Similarly, the velocity evolves to $w^{n+1/2}$ both from w^+ along C^+ and w^- along C^-. Subscripts k, referring to the Z-coordinate, are omitted for clarity.

During half of the time step $\Delta t/2 = t - t^n$, B_Z and w evolve simultaneously to the same values $B^{n+1/2}_{Z,j,k}$ and $w^{n+1/2}_{j,k}$ from B^+_Z and w^+ along C^+, and from B^-_Z and w^- along C^-. Differencing Equation (8.19) along each characteristic, we get a set of two equations with two unknowns

$$\begin{aligned} w^{n+1/2}_{j,k} - w^+ - \frac{B^{n+1/2}_{Z,j,k} - B^+_Z}{\sqrt{4\pi\rho_{j-1,k}}} &= 0 \quad \text{along } C^+, \\ w^{n+1/2}_{j,k} - w^- + \frac{B^{n+1/2}_{Z,j,k} - B^-_Z}{\sqrt{4\pi\rho_{j,k}}} &= 0 \quad \text{along } C^-, \end{aligned} \tag{8.21}$$

from which we calculate $B^{n+1/2}_{Z,j,k}$ and $w^{n+1/2}_{j,k}$. Note that values of the density at time level n are used. The momentum ρw is then updated in time through Equation (8.18). The procedure just outlined is referred to as the *method of characteristics* or MOC.

Exercise

Write the difference equation for the partial update of the momentum in the R-direction, ρu, arising from the first term in Equation (8.17). What are the two simultaneous equations that are solved to get B_R at time level $n + 1/2$?

The description of the source terms is completed with a discussion of Equation (8.8). Again, the correct treatment of Alfvén waves requires that the MOC be used. Stone and Norman (1992) derive the characteristic equation

$$\frac{D\Omega}{Dt} \mp \frac{1}{R^2\sqrt{4\pi\rho}} \frac{D(RB_\Phi)}{Dt} = 0, \tag{8.22}$$

where Ω is the angular velocity v_Φ/R. The minus and plus signs refer, respectively, to the characteristics C^+ and C^-. This equation then leads to the analogs of Equations (8.21)

$$\Omega_{j,k}^{n+1/2} - \Omega_{j,k}^{+} - \frac{[(RB_\Phi)_{j,k}^{n+1/2} - (RB_\Phi)_{j,k}^{+}]}{(R^2\sqrt{4\pi\rho})^{+}} = 0.$$

$$\Omega_{j,k}^{n+1/2} + \Omega_{j,k}^{-} + \frac{[(RB_\Phi)_{j,k}^{n+1/2} - (RB_\Phi)_{j,k}^{-}]}{(R^2\sqrt{4\pi\rho})^{-}} = 0. \tag{8.23}$$

The equations are solved simultaneously at all grid points for the partially time-advanced values $\Omega_{j,k}^{n+1/2}$ and $(RB_\Phi)_{j,k}^{n+1/2}$. The angular momentum change from the first term on the right-hand side of Equation (8.8) is then calculated

$$\frac{A_{j,k}^{\mathbf{n+1}} - A_{j,k}^{\mathbf{n}}}{\Delta t} = \frac{1}{4\pi} \frac{(B_{R,j+1,k} + B_{R,j,k})}{2(R_{j+1} - R_j)} \left[(RB_\Phi)_{j+1,k}^{n+1/2} - (RB_\Phi)_{j,k}^{n+1/2}\right] \tag{8.24}$$

This update is derived from the propagation of Alfvén waves in the R–direction. The second term on the right-hand side of Equation (8.8), which can be written as $B_Z \frac{\partial}{\partial z}(RB_\Phi)$, is obtained for a very similar characteristic equation for Alfvén waves propagating in the Z–direction. The resulting second update for the angular momentum is:

$$\frac{A_{j,k}^{\mathrm{n+1}} - A_{j,k}^{\mathbf{n+1}}}{\Delta t} = \frac{1}{4\pi} \frac{(B_{Z,j,k+1} + B_{Z,j,k})}{2(Z_{k+1} - Z_k)} \left[(RB_\Phi)_{j,k+1}^{n+1/2} - (RB_\Phi)_{j,k}^{n+1/2}\right]. \tag{8.25}$$

In a parallel procedure, the update $\Omega_{j,k}^{n+1/2}$ is used to update B_Φ, as described in the following section.

8.3 SOLVING THE INDUCTION EQUATION

Unlike the equations of continuity, momentum, and energy, the induction equation (8.4) cannot be written as a transport equation for a conserved quantity and, as such, it cannot be split into one-dimensional sweeps using a procedure analogous to that described in Section 6.3.4. Thus, the magnetic field must be updated with an entirely different scheme. On the other hand, stringent requirements for numerical schemes used to solve the induction equation are set by Maxwell's equations, which demand that $\mathbf{B}$ be a divergence-free field at all times. An arbitrary scheme does not necessarily guarantee that the condition $\nabla \cdot \mathbf{B} = 0$ is satisfied. A nonzero divergence of $\mathbf{B}$ means that magnetic monopoles have been created, resulting in unphysical accelerations along field lines. Spurious accelerations can be avoided by transforming the right-hand sides of the momentum equations in such a way as to eliminate all terms that depend on the divergence of **B**; in fact, we did just that when we explicitly calculated the components of magnetic tension and magnetic pressure gradient in Equation (8.6) to Equation (8.8). However, the $\nabla \cdot \mathbf{B} = 0$ condition also implies (through the divergence theorem) that the magnetic flux through a closed surface is always zero, and a good numerical scheme should satisfy this constraint as accurately as possible.

ZEUS solves the induction equation with the help of a multidimensional scheme of *constrained transport* (CT), which by construction enforces the divergence-free constraint and ensures the conservation of the total magnetic flux through the surface of the computational domain. To explain its workings, let us write the general induction equation in integral form using Stokes's theorem

$$\frac{\partial \Phi_{m,S}}{\partial t} = \oint_C (\mathbf{v} \times \mathbf{B}) \cdot d\hat{\mathbf{l}} = \oint_C \boldsymbol{\mathcal{E}} \cdot d\hat{\mathbf{l}}, \tag{8.26}$$

where $\Phi_{m,S}$ is the flux through surface S bounded by contour C. Each two-dimensional grid cell (j, k) may be thought of as the base of an elementary volume, whose third dimension goes in the Φ direction, which has components of **B** and $\boldsymbol{\mathcal{E}}$ located, respectively, on its faces and its edges (Figure 8.3). Now, for each face the last integral in Equation (8.26) reduces to a sum

$$\sum_{i=1}^{4} \mathcal{E}_i \Delta_i,$$

where $\mathcal{E}_i$ is the electromotive force at the ith edge, and Δ_i is the length of the ith edge taken with a positive or negative sign depending on whether the edge is oriented parallel or antiparallel to the direction of integration. For example, the finite-difference equation for the evolution of the flux through the front face of the volume shown in Figure 8.3 takes the form

$$\Phi_m^{n+1} = \Phi_m^n + \Delta t(\mathcal{E}_{\Phi\, j,k} R_j \Delta\Phi - \mathcal{E}_{\Phi\, j+1,k} R_{j+1} \Delta\Phi), \tag{8.27}$$

where contributions from $\mathcal{E}_R$ cancel out because of axial symmetry, and $\Delta\Phi$ is an arbitrary azimuthal extent of the elementary volume.

In Figure 8.3, the total flux through the surface of the elementary volume is the sum of fluxes through four vertical faces (again, fluxes through the lower and upper faces cancel out due to axial symmetry). In such a sum, each contribution from the vertical edge of the elementary volume appears twice with a different sign. As a result, the total flux does not change after the magnetic field has been updated; in particular, if it is equal to zero initially, it will stay zero within the machine accuracy throughout the simulation. The total flux through the surface of the computational domain will also be conserved. In that sense, CT is analogous to the conservative advection scheme discussed in Section 6.3.

Exercise

Calculate explicitly the change in the flux through each of the six faces of the solid shown in Figure 8.3. Show that they sum to zero.

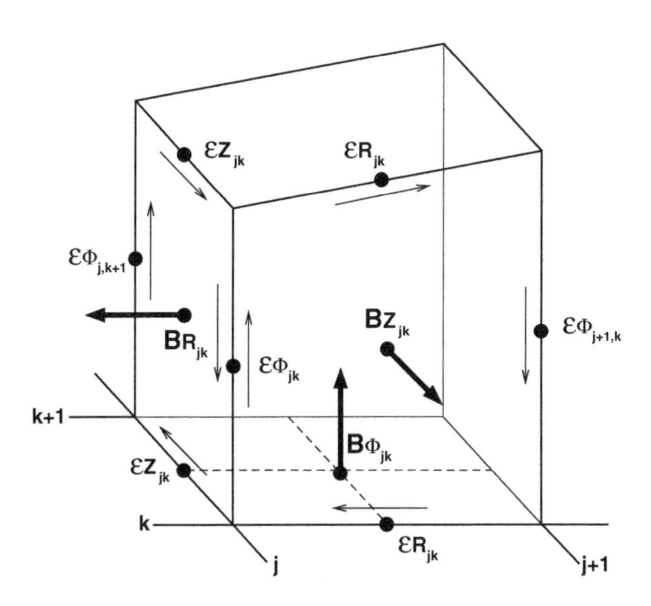

FIGURE 8.3 Constrained transport method for evolving **B**. Thin arrows indicate the direction of the integral of the electromotive force along the edges of each face of the elementary volume. When the total flux through the surface of the elementary volume is calculated, the contribution from each edge is counted twice with a different signs. Note that because of assumed axial symmetry, the azimuthal extent of the elementary volume can be arbitrary.

From the construction of the algorithm, it is evident that an originally divergence-free field evolved with CT will remain divergence-free *no matter how the electromotive force is calculated.* In practice, for accuracy and stability reasons, the scheme used to solve Equation (8.13) should yield the components of $\boldsymbol{\mathcal{E}}$ centered at $t^{n+1/2}$. In ZEUS, this is achieved by using the solutions $\mathbf{v}^{n+1/2}$ and $\mathbf{B}^{n+1/2}$ of the characteristic Equation (8.19) in which the co-moving derivative (Equation 8.20) is replaced by

$$\frac{D^*}{D^*t} \equiv \frac{\partial}{\partial t} + \left(u \mp \frac{B_R}{\sqrt{4\pi\rho}}\right)\frac{\partial}{\partial R} \tag{8.28}$$

for an Alfvén wave propagating in the $R-$ direction. The starred derivative in Equation (8.28) includes the velocity of the fluid u and, thus, it accounts for the field advected into the grid cell. In the case of Lorentz source terms, including u was not necessary because the momentum equation is solved with an operator-split scheme, and the forces are effectively applied in a Lagrangian frame of reference in which the only agent modifying the field is Alfvén waves.

Thus, the slopes of the characteristics in this case are $u \pm v_A$. The same Equations (8.21) are used to solve for the time-centered values $w_{j,k}^{n+1/2}$ and $B_{Z,j,k}^{n+1/2}$ for waves propagating in the $R-$direction, and for $u_{j,k}^{n+1/2}$ and $B_{R,j,k}^{n+1/2}$ for waves propagating in the $Z-$direction. The difference in this case is that the velocities used to determine the positions R^+ and R^-, which define the domain of dependence, are $u + B_R/\sqrt{4\pi\rho}$ and $u - B_R/\sqrt{4\pi\rho}$. Once this procedure has been done in both directions, the electromotive force in the $\Phi-$direction can be calculated:

$$\mathcal{E}^{n+1/2}_{\Phi,j,k} = u^{n+1/2}_{j,k} B^{n+1/2}_{Z,j,k} - w^{n+1/2}_{j,k} B^{n+1/2}_{R,j,k} . \tag{8.29}$$

The essence of the CT method is to evolve the flux in time through Equation (8.27) and its analog in the $R-$ direction. In practice in this coordinate system, this is equivalent to straightforward differencing of Equation (8.10) and Equation (8.11) to obtain the field components B_R^{n+1} and B_Z^{n+1}. The flux through the left face in Figure 8.3 is $B_R R dZ d\Phi$, and the time evolution of B_R is obtained from

$$B^{n+1}_{R,j,k} = B^n_{R,j,k} + \Delta t \frac{\left(\mathcal{E}^{n+1/2}_{\Phi,j,k+1} - \mathcal{E}^{n+1/2}_{\Phi,j,k}\right)}{Z_{k+1} - Z_k}. \tag{8.30}$$

The flux through the front face in Figure 8.3 is $B_Z R_{j+1/2} dR d\Phi$, and the time evolution of B_Z is obtained from

$$B^{n+1}_{Z,j,k} = B^n_{Z,j,k} - \Delta t \frac{\left(R_{j+1}\mathcal{E}^{n+1/2}_{\Phi,j+1,k} - R_j \mathcal{E}^{n+1/2}_{\Phi,j,k}\right)}{R_{j+1/2}(R_{j+1} - R_j)}. \tag{8.31}$$

In cylindrical coordinates, the magnetic field vector naturally splits into the *poloidal components* (B_R, B_Z), also called *poloidal field*, and the *toroidal component* B_Φ, also called *toroidal field*. While the update of the poloidal field clearly requires the MOC-CT scheme, the toroidal field (which in axially symmetric cases is divergence-free at all times) may be evolved with a different method, based on similarities between Equation (8.8) and Equation (8.12). Since $\nabla \cdot \mathbf{B} = 0$, the latter may be transformed into

$$\frac{\partial B_\Phi}{\partial t} + \frac{\partial (u B_\Phi)}{\partial R} + \frac{\partial (w B_\phi)}{\partial Z} = R B_R \frac{\partial}{\partial R}\left(\frac{v_\Phi}{R}\right) + R B_Z \frac{\partial}{\partial Z}\left(\frac{v_\Phi}{R}\right), \tag{8.32}$$

where the left-hand side describes passive advection of the toroidal field, and the right-hand side is the source term accounting for the effects of torsional waves. In order to optimize the treatment of those waves, the right-hand sides of Equation (8.32) and Equation (8.8) are simultaneously evaluated using MOC.

Exercise

Derive Equation (8.32).

The characteristic equation is Equation (8.22). For the update of B_Φ and A, a directionally split algorithm is used, as indicated in Equation (8.24) and Equation (8.25). First, corrections due to torsional waves propagating in R are calculated and applied in the same way as explained in Section 8.2; then the same procedure is repeated for Z. In the first case the difference operator D/Dt in Equation (8.22) is given by Equation (8.28) and, in the second one, by an analogous expression in which R is replaced by Z. These procedures give, at all grid points, the values of $\Omega^{n+1/2}_{j,k}$ and

$(RB)^{n+1/2}_{\Phi,j,k}$. The two equations for updating B_Φ, analogous to Equation (8.24) and Equation (8.25) are:

$$\frac{B^{n+1}_{\Phi,j,k} - B^{n}_{\Phi,j,k}}{\Delta t} = \frac{R_{j+1/2}(B_{R,j+1,k} + B_{R,j,k})}{2(R_{j+1} - R_j)}\left(\Omega^{n+1/2}_{j+1,k} - \Omega^{n+1/2}_{j,k}\right) \tag{8.33}$$

and

$$\frac{B^{n+1}_{\Phi,j,k} - B^{n+1}_{\Phi,j,k}}{\Delta t} = \frac{R_{j+1/2}(B_{Z,j,k+1} + B_{Z,j,k})}{2(Z_{k+1} - Z_k)}\left(\Omega^{n+1/2}_{j,k+1} - \Omega^{n+1/2}_{j,k}\right). \tag{8.34}$$

The updated B_Φ is advected. However, although it is cell-centered, its advection must not be performed in the same way as the advection of mass, internal energy, or angular momentum. This is because the left-hand side of Equation (8.32) is not an axially symmetric continuity equation in three dimensions, but rather a two-dimensional continuity equation on the (R, Z) plane. Thus, instead of areas (Equation 6.96), the fluxes of B_Φ may only contain the lengths of cell edges, ΔR and ΔZ. Maintaining the consistent advection approach, ZEUS interpolates and monotonizes the quantity B_Φ/ρ rather than B_Φ itself. The flux of B_Φ through the right edge of the cell (j, k) is given by

$$\mathcal{F}^{B_\Phi}_{R(j+1,k)} = u_{j+1,k}\overline{\rho}\overline{\left(\frac{B_\Phi}{\rho}\right)}\Delta Z, \tag{8.35}$$

where barred quantities are interpolated and monotonized. Analogous expressions define the remaining fluxes. The differences in the fluxes across the cell in the $R-$ and $Z-$ directions then give the updates to B_Φ as a result of advection. Thus, the operator-split contribution to the advection in the $R-$direction is:

$$\left(B^{n+1}_{\Phi,j,k} - B^{n}_{\Phi,j,k}\right)\frac{\Delta R \Delta Z}{\Delta t} = -\left(\mathcal{F}^{B_\Phi}_{R(j+1,k)} - \mathcal{F}^{B_\Phi}_{R(j,k)}\right). \tag{8.36}$$

Exercise

Write the difference equations for the partial update of the toroidal field arising from the advection in the $Z-$direction.

Exercise

If you were solving the MHD equations on a full 3-D grid in (R, Z, Φ), how would you treat the advection of the $\Phi-$ component of the magnetic field?

Elegant and efficient as it is, MOC-CT suffers from spurious *reconnection*. Physical reconnection is a process that occurs in a medium with nonzero resistivity, leading to topological rearrangement of magnetic field lines and rapid release of the energy stored in the field. Basically, oppositely directed field lines are brought close together. As a result of the reconnection, sharp changes in the direction of the field are removed and the energy lost from the field is transferred to the plasma in the form of Joule heat. The combined effects of thermal pressure and tension of rearranged field lines can accelerate the plasma to very high velocities. Numerous examples of such phenomena are observed in the upper solar atmosphere; in fact, reconnection is thought to be the main agent that drives coronal mass ejections and is responsible for the ultra-high temperatures of the coronal plasma.

The spurious numerical reconnection occurs whenever oppositely directed fluxes are carried into the same cell. Such fluxes annihilate in a completely uncontrolled way and the associated magnetic energy is lost from the grid (no Joule heat is generated). The numerical reconnection does not affect fields in stationary media and, as such, it cannot be formally associated with the presence of resistivity. However, a *numerical resistivity*

$$\eta_{num} = \Delta v_A \Delta l \tag{8.37}$$

can be assigned to it, where Δl is the distance between grid points, and Δv_A is the characteristic variation of the Alfvén velocity between grid points. The magnitude of η_{num} sets a lower limit for any physical resistivity whose results could be reliably modeled with MOC-CT.

8.4 INITIAL AND BOUNDARY CONDITIONS

The initial magnetic field $\mathbf{B}_0$ may not contain monopoles, i.e., it must obey the condition $\nabla \cdot \mathbf{B}_0 = 0$. A straightforward way to obtain such a field is by using a *vector potential* $\mathbf{A}_m$ such that

$$\mathbf{B}_0 = \nabla \times \mathbf{A}_m. \tag{8.38}$$

In axially symmetric configurations

$$B_R = \frac{\partial A_\Phi}{\partial Z}$$

and

$$B_Z = -\frac{1}{R}\frac{\partial (R A_\Phi)}{\partial R}, \tag{8.39}$$

while B_Φ may be arbitrary. As an example, consider an axially symmetric gaseous torus rotating around a central body and embedded in a medium of constant density ρ_0 such that the constant density surface $\rho = \rho_0$ defines the surface of the torus. It is easy to verify that the vector potential

$$A_\Phi(R, Z) = \omega_0 \rho(R, Z); \quad \rho > \rho_0$$

and

$$A_\Phi(R, Z) = 0; \quad \rho \le \rho_0, \tag{8.40}$$

where ω_0 is a constant, generates a poloidal field whose vectors are parallel to constant density surfaces within the torus.

Unwanted magnetic monopoles can also be introduced into the computational domain through inconsistent boundary conditions for the poloidal field. We shall avoid this possibility if we apply the boundary conditions to the electromotive force rather than to the field itself, and evolve the field in ghost zones the same way as in the active part of the grid. The poloidal field is generated by the azimuthal component of the electromotive force $\mathcal{E}_\Phi$, and by analogy to Chapter 6, Section 6.5.3, we can specify:

- *Reflecting Boundary Condition*: $\mathcal{E}_\Phi$ in the ghost cells is equal to the negative of $\mathcal{E}_\phi$ in the corresponding active cells.
- *Inflow Boundary Condition*: $\mathcal{E}_\Phi$ in the ghost cells is set to a prescribed value, which may vary in time.
- *Outflow Boundary Condition*: $\mathcal{E}_\Phi$ is copied from the first active cell into the corresponding ghost cells.
- *Periodic Boundary Condition*: $\mathcal{E}_\Phi$ in a ghost cell is equal to $\mathcal{E}_\Phi$ in the corresponding active cell at the opposite side of the grid.

The toroidal field fulfills the same boundary conditions as the scalars ρ and e. A special procedure is needed, however, when the outflow B.C. is applied, and the outflow velocity is smaller than V_A. In this case, the outgoing Alfvén waves have to be prevented from backscattering into the grid. This is achieved by setting gradients of all quantities equal to zero along the incoming (i.e., $\mathbf{C}^-$) characteristic.

8.5 EXAMPLES AND EXERCISES

8.5.1 Contraction of a Magnetized Ring

This simple example illustrates the fact that magnetohydrodynamics is often counterintuitive and far more complicated than hydrodynamics. Originally, the uniform medium is at rest in a cylindrical reference frame. Then, a small rectangular area in the (R, Z) plane is filled with a homogeneous toroidal field B_Φ, and the internal energy density e of the medium in that area is multiplied by a factor

$$\xi = \frac{P_g}{P_g + \frac{B_\Phi^2}{8\pi}}, \tag{8.41}$$

where P_g is the pressure of the medium. Thus, a magnetized ring is generated, whose internal pressure is equal to the ambient pressure. In hydrodynamics, the uniformity of pressure across the grid would suppress motions. However, we see that the medium starts moving toward the symmetry axis and eventually collides with it (Figure 8.4). Note that this motion reflects the contraction of the magnetized ring.

Exercise

Explain the behavior illustrated in Figure 8.4. Hint: consider the right-hand side of Equation (8.6). How does the energy of the magnetic field change in time?

8.5.2 Propagation of a Jet with a Helical Field

Astrophysical jets originating from accretion disks are likely to contain helical magnetic fields. If the disk field is mostly vertical, it will be twisted into a helix by the

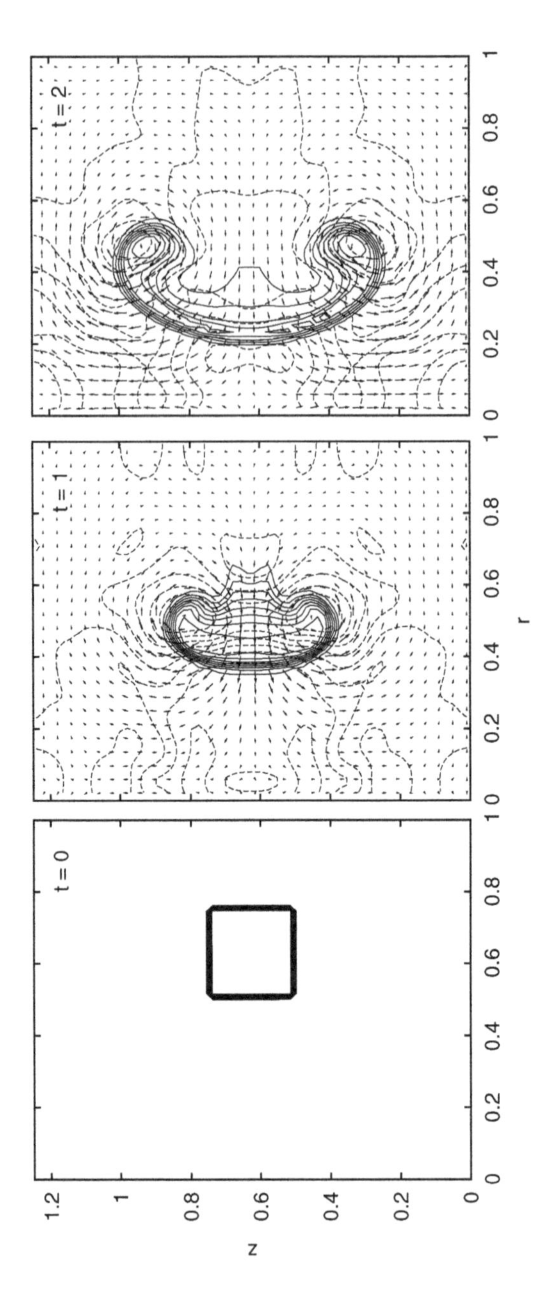

FIGURE 8.4 A magnetized ring filled with a uniform toroidal field, originally at rest and in pressure equilibrium with the ambient medium, contracts toward the symmetry (Z) axis. A time-sequence of meridional cuts through the ring is shown, starting from the initial configuration in the left panel. *Solid lines*: contours of constant toroidal field; *broken lines*: contours of constant total pressure; *arrows*: velocity field. Units of time and distance are arbitrary.

rotation of the disk and, if it is mostly toroidal, it will be stretched into a helix by the flow of the gas in the jet. Another counterintuitive conclusion from the MHD equations is that an originally non-rotating jet with a frozen-in helical field begins to rotate upon propagating through the ambient medium. As illustrated in Figure 8.5, this effect results from the propagation of torsional Alfvén waves.

Exercise

At $t = 0$, the only unbalanced force acting on the jet in Figure 8.5 is the tension of the toroidal component. Based on Equation (8.6) through Equation (8.13), explain how ZEUS generates rotational motions within the jet.

8.5.3 Magnetic Buoyancy Instability

Consider an isothermal, gravitationally stratified medium with a constant gravitational acceleration g directed along the z-axis. The medium, originally in hydrostatic equilibrium in a Cartesian reference frame, is permeated by a magnetic field $\mathbf{B}$ whose lines are parallel to the y-axis (i.e., $\mathbf{B} = (0, B_y, 0)$). The support against gravity is partly provided by the magnetic pressure, and the equilibrium condition reads

$$\frac{d}{dz}\left(P_g + \frac{B_y^2}{8\pi}\right) = -\rho g, \qquad (8.42)$$

where ρ and P_g are the density and pressure of the medium, respectively.

Imagine now a part of a field line displaced upwards, e.g., as a result of a small velocity perturbation. The line is frozen into mass elements, which, upon being displaced, expand in order to achieve pressure equilibrium with their new surroundings. When their new density is lower than the ambient density, they tend to continue moving upwards. While doing so, they stretch the field line whose tension opposes the upward motion. At the same time, the displaced medium tends to slide along the field line downwards, and the density contrast between the rising elements and their surroundings tends to increase. These three effects are the basis of the linear *magnetic buoyancy instability* (often referred to as *Parker instability*; Parker, 1979), which causes the field to at least partly escape from the stratified region.

Perturbations in the (y, z) plane include all effects just mentioned and excite *undulating* modes of the instability. When P_g and P_m are comparable, and

$$\frac{dB_y}{dz} < 0, \qquad (8.43)$$

such modes grow for wavelengths larger than

$$\lambda_m \approx 4\pi h,$$

where h is the density scale height in the absence of magnetic field (Priest, 1987). Note that the criterion (Equation 8.43) is slightly modified when the stabilizing influence of normal buoyancy is accounted for. Wavelengths shorter than λ_m are stabilized by the magnetic tension. The evolution of an undulating mode is illustrated in Figure 8.6.

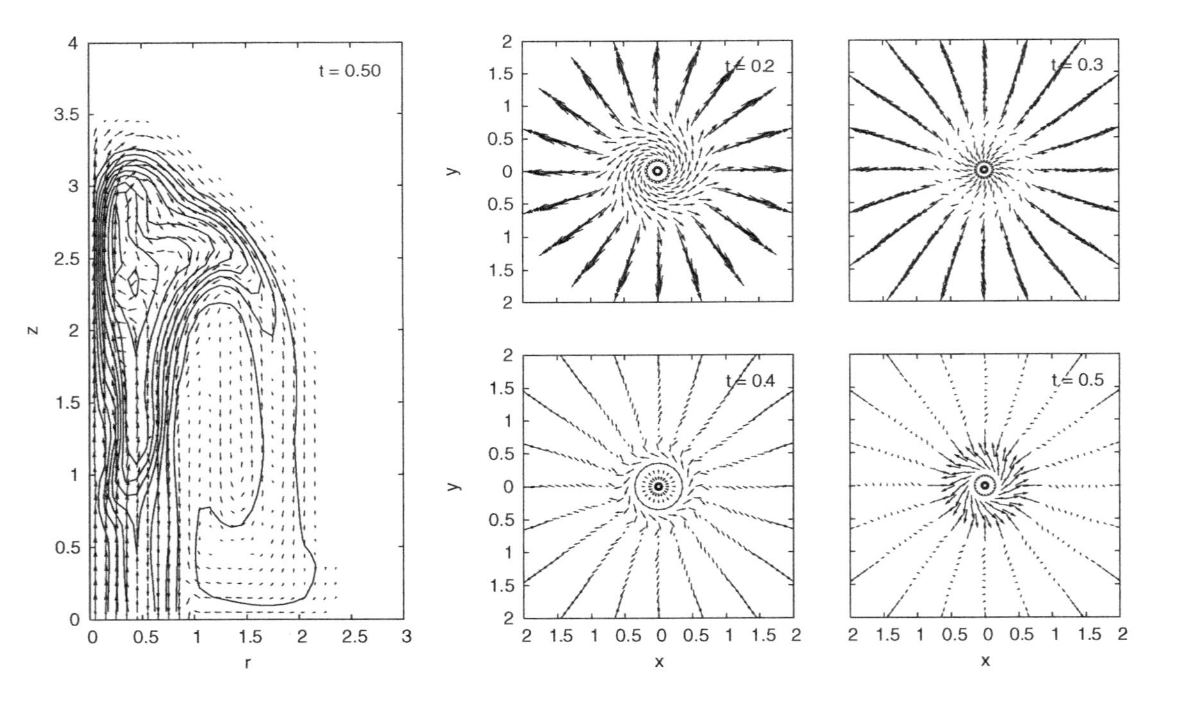

FIGURE 8.5 Cylindrical jet with a frozen-in helical field (simulation based on the JETINIT setup provided with the ZEUS code). At $t = 0$ the magnetized jet material begins to flow into the grid through its lower boundary with only v_Z different from zero. *Left*: vectors of poloidal field (arrows) and contours of constant toroidal field at $t = 0.5$. Contour levels are 0.5, 1.0, 1.5 . . . 5.5. *Right*: velocity field in a cross section of the jet at $z = 1.0$ shown at indicated evolutionary times. Units of time and distance are arbitrary.

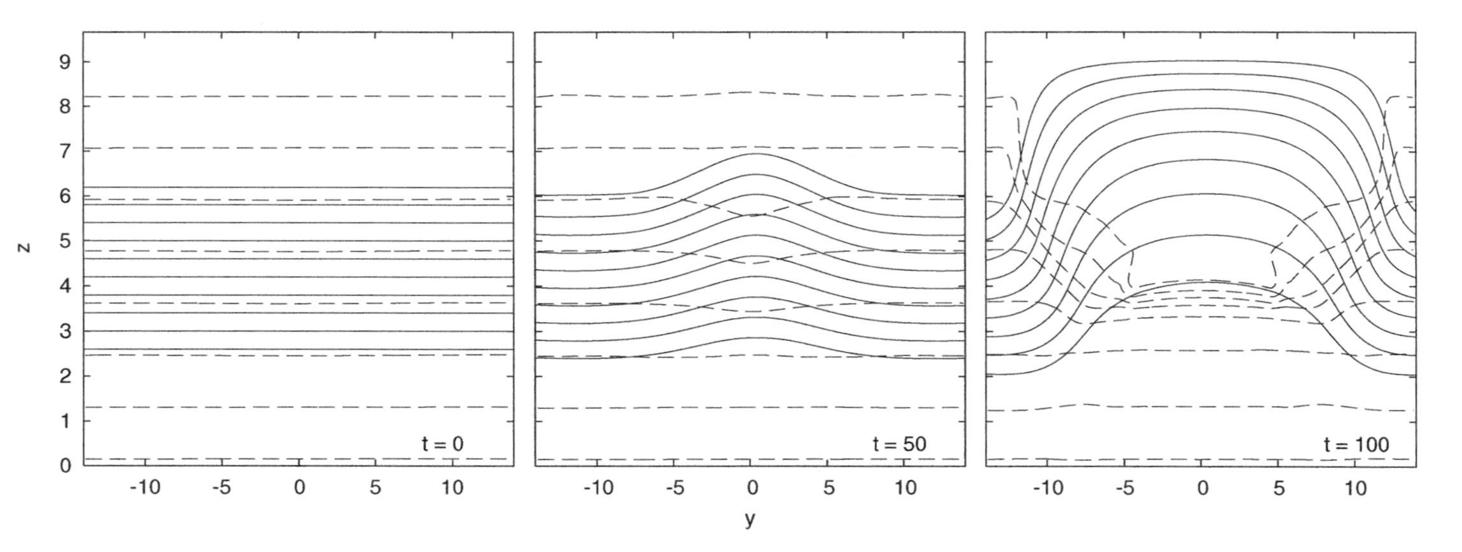

FIGURE 8.6 A stratified gaseous atmosphere, originally ($t = 0$) in hydrostatic equilibrium in a constant gravitational field pointed downward, contains a frozen-in magnetic field whose vectors are parallel to the (y, z) plane of the figure. The magnetized region is cooler than the rest of the atmosphere, so as to keep the total pressure smooth in z. A very small sinusoidal density perturbation (1%) is sufficient to excite an undulating mode of the magnetic buoyancy instability. *Solid curves*: field lines; *dashed curves*: contours of constant density. Units of time and distance are arbitrary.

The pattern is repeated for other values of x in the direction perpendicular to that shown.

Perturbations in the (x, z) plane, which displace field lines but do not bend them, excite *interchange* modes of the instability. In other words, an entire field line, which extends in the y-direction, is displaced upwards at $x + \Delta x$ by a small amount, as compared with the corresponding field line at x. For such modes, all wavelengths are unstable provided that

$$\frac{d}{dz}\left(\frac{B_y}{\rho}\right) < 0 \tag{8.44}$$

and the shortest wavelengths grow most rapidly (note that Equation 8.44 is a more stringent requirement than Equation 8.43). The evolution of interchange modes is illustrated in Figure 8.7.

The magnetic buoyancy instability is not entirely analogous to the hydrodynamical Rayleigh–Taylor instability (see Chapter 6, Section 6.7.1), in which a heavier fluid lies on top of a lighter fluid in a gravitational field. The magnetic field is considered to be the "light" fluid in this case, but in ideal MHD the matter is not free to move across field lines, only along them. However, the general effect is for the field to move upward and the matter downward. The magnetic pressure gradient causes the density to decrease more slowly with z than in the absence of the field. As a result, the gravitational potential energy is raised and the stratified region becomes unstable.

8.5.4 Magnetorotational Instability

A magnetized disk is subject to a linear magnetorotational instability (MRI) that amplifies the magnetic field and, in the nonlinear regime and in three space dimensions, develops into magnetoturbulence. The instability was first mentioned by Velikhov (1959) and Chandrasekhar (1960). The application to accretion disks was examined by a linear analysis (Balbus and Hawley, 1991), and by a two-dimensional numerical simulation extending into the nonlinear regime (Hawley and Balbus, 1991). In the linear regime, the numerical simulations give the same growth rates as the analytic linear analysis, an important test of the method. Hawley, Gammie, and Balbus (1995) extended the numerical simulations to three space dimensions. MRI operates under very general conditions, and the simplest system in which it develops is an axisymmetric differentially rotating disk permeated by a uniform vertical field B_Z. Consider two fluid elements, m_1 and m_2, originally at the same distance R from the axis. The elements are on the same vertical field line, with m_1 placed above m_2. If they are displaced in R such that m_1 is now at $R - \delta R$ and m_2 at $R + \delta R$, the line is stretched and develops tension. The tension decelerates m_1 and accelerates m_2, thus reducing the angular momentum of m_1 and increasing that of m_2. As a result, for both elements δR increases. The field line is stretched even more and the process continues provided that

$$(kv_A)^2 < -R\frac{d\Omega^2}{dR}, \tag{8.45}$$

i.e., when the magnetic tension generated by the displacement is smaller than the tidal force associated with that displacement. In principle, for a given B_Z (and corresponding Alfvén speed v_A) a wavelength $\lambda = 2\pi/k$ large enough to fulfill Equation (8.45)

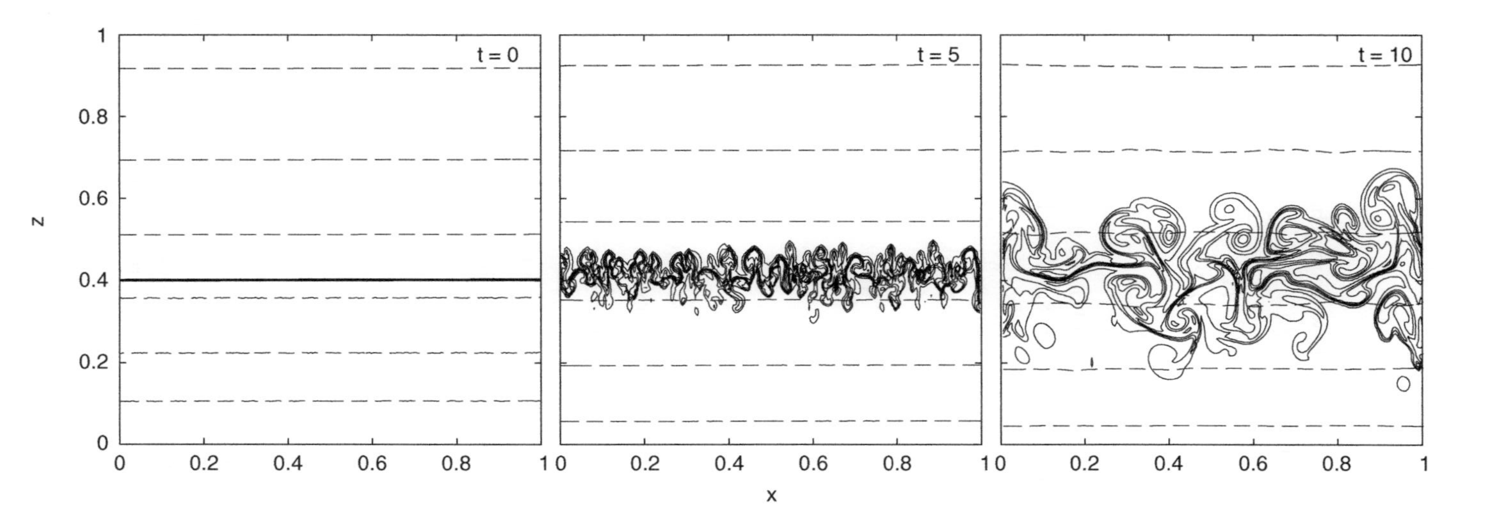

FIGURE 8.7 The stratified atmosphere of Figure 8.6 is shown here in the (x, z) plane, with the frozen-in field perpendicular to the plane of the figure. The magnetized region extends below $z = 0.4$. In contrast to Figure 8.6, here the temperature is constant across the computational domain, and the density of the magnetized region is adjusted so as to keep the total pressure smooth in z. Perturbations due to machine roundoff error are sufficient to excite chaotic interchange modes of the magnetic buoyancy instability, which evolve into magnetoturbulence. *Solid curves*: contours of constant field; *dashed curves*: contours of constant total pressure. Units of time and distance are arbitrary.

can be found, indicating that all disks with

$$\frac{d\Omega}{dR} < 0$$

are unstable. However, the shortest unstable wave may be longer than the total thickness h of the disk, which happens when

$$v_A^2 > -R\frac{h^2}{4\pi^2}\frac{d\Omega^2}{dR}. \tag{8.46}$$

In a thin and isothermal Keplerian disk, where $h \sim \sqrt{2}c_s/\Omega$ (see, e.g., Balbus and Hawley, 1991), Equation (8.46) yields $v_A >\sim c_s$ and the MRI is quenched when the magnetic pressure exceeds approximately half of the thermal pressure. The development of the instability is illustrated in Figure 8.8. Note how the angular momentum is removed from the inner disk and added to the outer disk, thus resulting in inward mass flow in the inner disk. It is important to know that the MRI grows on a dynamical timescale, i.e., extremely rapidly. In a Keplerian disk its most unstable wavelength

$$\lambda^\star = \frac{8\pi}{\sqrt{15}}\frac{v_A}{\Omega}$$

grows on a timescale as short as

$$\tau_{min} = \frac{8\pi}{3\Omega},$$

that is, faster than one orbital period. Such rapid growth ensures that the field will be regenerated faster than it is lost by the buoyancy instability.

Exercise

Repeat the simulation illustrated in Figure 8.8 with a loop of the poloidal field immersed into the disk instead of the purely vertical field permeating the disk. Generate the loop with the help of the vector potential (Equation 8.40), and make sure that $\nabla \cdot \mathrm{B}$ is negligible before you start the calculations. Caution: to generate a loop contained within the grid an additional field varying in R and Z must be used, which will play the role of the density in Equation (8.40).

In summary, the requirements for the instability in a disk include (1) the magnetic field must be coupled to the matter, (2) the angular velocity must decrease outwards, (3) there must be an initial weak poloidal field present, and (4) the energy density in the field must be less than the thermal energy density. The consequences of the instability are (1) amplification of the field and development of the toroidal component, (2) magnetic turbulence, and (3) transport of angular momentum outward, allowing an accretion flow of mass inward. The instability will eventually saturate when the magnetic energy density approaches that of the thermal energy; however, in a numerical simulation, depending on resolution, numerical reconnection in the turbulent flow may limit the growth of the field.

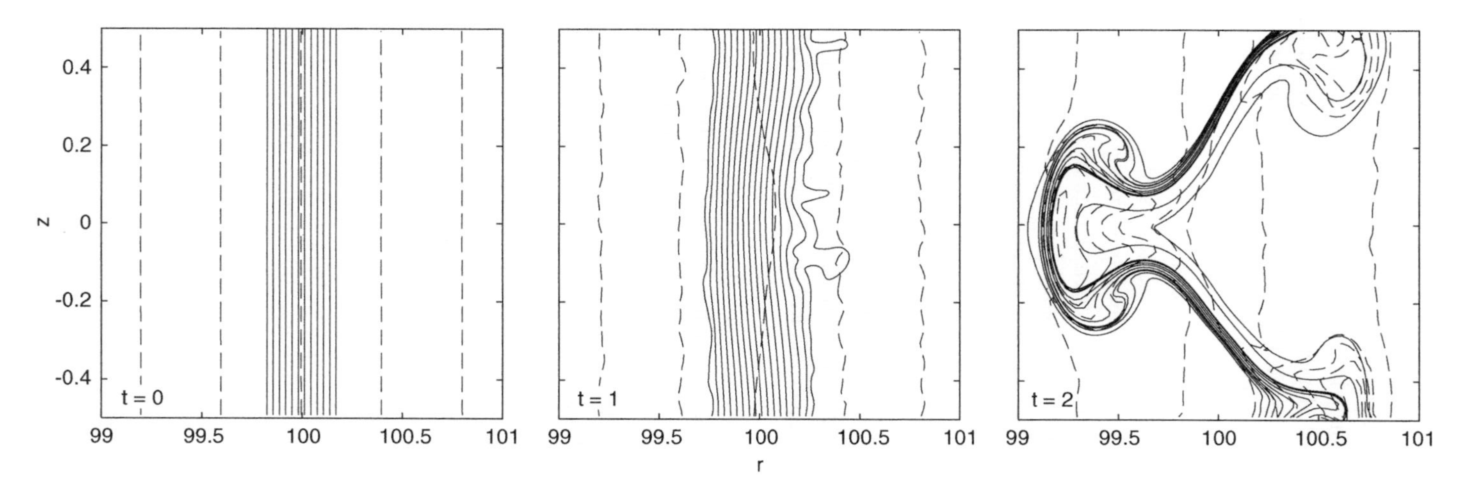

FIGURE 8.8 The magnetorotational instability: a simple case studied by Hawley and Balbus (1991). The computational domain of the axisymmetric simulation is a small section of the (R, Z) plane of a Keplerian accretion disk. The vertical component of gravity is neglected, and initial pressure and density are constant across the domain. At $t = 0$, part of the domain is filled with a uniform vertical field. Simultaneously, the temperature is adjusted such that the total pressure across the domain remains constant. *Solid curves*: field lines; *dashed curves:* contours of constant angular momentum, with angular momentum increasing to the right at $t = 0$. At time $t = 2$, material near $Z = 0$ has lost angular momentum and is moving inward (to the left), while material near $Z = \pm 0.4$ has gained angular momentum and is moving outward (to the right). Units of time and distance are arbitrary.

8.6 CONCLUDING REMARKS

The solution of magnetohydrodynamic problems in astrophysics can be extremely complicated, and numerical methods are required. It may not be possible to have a single code that can accommodate the wide range of problems that are encountered; the numerical method must often be adapted to the particular problem at hand. The method described here is applicable to ideal MHD. It can be extended to situations in which effects of resistivity are important or where the degree of ionization is very low, resulting in only weak coupling of the magnetic field to the matter.

The main difficulties in obtaining numerical solutions to the MHD equations are (1) the existence of a considerable variety of waves, such as Alfvén waves or fast and slow magnetosonic waves, whose propagation must be correctly represented; (2) the CFL condition on the time step can be very restrictive, e.g., in a gas of low density where the Alfvén speed is very high; (3) the condition div $\mathbf{B} = 0$ must be strictly enforced; and (4) the unavoidable effects of numerical reconnection must somehow be dealt with.

The ZEUS procedure outlined here is just one example of a numerical treatment. In some problems of star formation in which a Lagrangian rather than a Eulerian treatment could have advantages, an SPH formulation including magnetic fields is promising and is currently under development (Price and Monaghan, 2004). A three-dimensional MHD spectral method has been used to successfully model the generation of the Earth's magnetic field (Glatzmaier and Roberts, 1996; Olson and Glatzmaier 1995).

There may be advantages (and disadvantages) to using a vector potential A_m (Equation 8.38) in the difference equations rather than the field **B**. Thus, for example, the magnetic induction equation (8.4) is replaced by the equivalent equation for $\partial \mathbf{A}_m/\partial t$ (Bowers and Wilson, 1991, Chapter 8). The principal advantage is that the constraint $\nabla \cdot \mathbf{B} = 0$ is automatically satisfied. The main disadvantage is that numerical second derivatives are required in the evaluation of the Lorentz forces in the momentum equation.

The ZEUS code has provided a reliable scheme for solving astrophysical problems with magnetic fields, it has been widely used and it is relatively easy to understand by the user. However, in some situations, it is not particularly accurate (Falle, 2002), so for some problems more complicated and more accurate methods are being developed. Examples of advanced techniques, whose details will not be discussed here, include a Riemann solver (briefly introduced in Chapter 6, Section 6.1.3) modified for MHD and including a divergence-free method for magnetic field updating, as well as adaptive mesh refinement (Ryu and Jones, 1995; Balsara, 2004; Collins and Norman, 2004).

REFERENCES

Balbus, S. A. and Hawley, J. F. (1991) *Astrophys. J.* **376**: 214.
Balsara, D. (2004) *Astrophys. J. Suppl.* **151**: 149.
Bowers, R. L. and Wilson, J. R. (1991) *Numerical Modeling in Applied Physics and Astrophysics* (Boston: Jones and Bartlett).

Chandrasekhar, S. (1960) *Proceedings of the National Academy of Sciences* **46**: 253.
Collins, D. C. and Norman, M. L. (2004) American Astronomical Society Meeting #205, abstract #153.14.
Falle, S. A. E. G. (2002) *Astrophys. J.* **577**: L123.
Glatzmaier, G. and Roberts, P. H. (1996) *Science* **274**: 1887.
Hawley, J. F. and Balbus, S. A. (1991) *Astrophys. J.* **376**: 223.
Hawley, J. F., Gammie, C. F. and Balbus, S. A. (1995) *Astrophys. J.* **440**: 742.
Hawley, J. F. and Stone, J. M. (1995) *Comp. Phys. Comm.* **89**: 127.
Olson, P. and Glatzmaier, G. (1995) *Phys. Earth Planet. Inter.* **92**: 109.
Parker, E. N. (1979) *Cosmical Magnetic Fields* (Oxford, U.K.: Oxford University Press).
Price, D. J. and Monaghan, J. J. (2004) *Mon. Not. R. Astr. Soc.* **348**: 123.
Priest, E. R. (1987) *Solar Magnetohydrodynamics* (Dordrecht: D. Reidel).
Ryu, D. and Jones, T. W. (1995) *Astrophys. J.* **442**: 228.
Stone, J. M. (1999) *J. Comp. Appl. Math.* **109**: 261.
Stone, J. M. and Norman, M. L. (1992) *Astrophys. J. Suppl.* **80**: 791.
Velikhov, E. P. (1959) *Sov. Phys. JETP* **36**: 995.

9 Radiation Transport

The basic concepts relating to radiation are introduced in Chapter 1, Section 1.6. Clearly almost all information about astronomical sources is delivered to the observer in the form of radiation, and a wide variety of applications of these basic concepts have been derived to deduce the physical properties of sources ranging from planets to stars to galaxies to quasars to the cosmological background. In this introductory chapter, we explain how the radiation transport equations are solved numerically in a few special cases. In Section 9.1 is a simple example of how one solves the equation of radiation transport for continuum radiation along one particular direction. The next section applies the method of Section 9.1 to the self-consistent calculation of the temperature and emergent spectral energy distribution of a spherically symmetric structure whose density distribution is prescribed. Then the classic problem of the calculation of the structure of a stellar atmosphere in hydrostatic equilibrium, with full frequency dependence, is discussed in Section 9.3. Such solutions, which still involve just one space dimension, have many modern astrophysical applications.

The full frequency-dependent radiative transfer problem in more than one space dimension has to be solved as a function of up to six variables plus the time. It is not even close to being feasible to solve the complete problem in full generality numerically. Nevertheless, there are numerous astrophysical problems that require solving the equations of hydrodynamics coupled with the equation of radiation transfer. The equations of radiation hydrodynamics are given in Section 9.4. A commonly used technique is to average over frequency and to solve the equations in the flux-limited diffusion approximation (Section 1.6.4). In Section 9.4, this method is discussed for the particular case of hydrodynamics in two space dimensions. The same method is easily extended to three dimensions. Even though the time-dependent solution of these equations does not give information as a function of frequency, it is possible to get approximate emergent spectra at particular times in the evolution, by means of the technique of Section 9.1. This procedure is discussed in Section 9.5. The final section gives an example of a full three-dimensional calculation of hydrodynamics plus radiation transport, and how the results can be compared with observations.

9.1 SOLVING THE RAY EQUATION FOR THE CONTINUUM

We consider here the solution of the simple, time-independent radiation transfer equation (1.91), which describes the change in intensity of radiation along a particular direction $\hat{s}$

$$\frac{\partial I_\nu}{\partial s} = -\epsilon_\nu \rho (I_\nu - S_\nu), \tag{9.1}$$

where ϵ_ν includes attenuation of the ray from both scattering and true absorption and $S_\nu = j_\nu/\epsilon_\nu$ (where j_ν is the emission coefficient) is the so-called *source function*. Note that under conditions of local thermodynamic equilibrium, where scattering is unimportant, S_ν reduces to B_ν, the black-body intensity or Planck function. An equation of the form of Equation (9.1) is known as a *ray equation* to distinguish it from one of the form of Equation (1.104), which is known as a *moment equation*. We assume in this example that the function S_ν is known, so the numerical integration of Equation (9.1) is straightforward. If S_ν depends in some way on the radiation intensity itself, then an iterative method must be used to derive it (see Section 9.2 and Section 9.5).

Equation (9.1) has the formal solution, integrating from point a to point b along a single ray with distance element ds

$$I_\nu^b = I_\nu^a \exp\left(-\int_a^b \epsilon_\nu \rho ds\right) + \int_a^b S_\nu \exp\left(-\int_s^b \epsilon_\nu \rho ds'\right) \epsilon_\nu \rho ds. \quad (9.2)$$

We define the element of optical depth as $d\tau_\nu = \epsilon_\nu \rho ds$, so the total optical depth between point a and some point s along the direction $\hat{s}$ is:

$$\tau_\nu(s) = \int_a^s \epsilon_\nu \rho ds'. \quad (9.3)$$

Exercise

Derive Equation (9.2) from Equation (9.1) using the integrating factor $\exp(\tau_\nu)$.

Now assume, for the moment, that S_ν varies linearly as a function of τ_ν between points a and b

$$S_\nu(\tau_\nu) = S_\nu(a) + [S_\nu(b) - S_\nu(a)]\frac{\tau_\nu}{\tau_\nu(b)}. \quad (9.4)$$

Then the formal solution becomes

$$\begin{aligned} I_\nu(b) = {} & I_\nu(a)\exp(-\tau_b) + S_\nu(a)[(1 - \exp(-\tau_b))/\tau_b - \exp(-\tau_b)] \\ & + S_\nu(b)[1 - (1 - \exp(-\tau_b))/\tau_b], \end{aligned} \quad (9.5)$$

where $\tau_b = \tau_\nu(b)$.

Exercise

Derive Equation (9.5) from Equation (9.2) and Equation (9.4). Note that the quantities $S_\nu(a)$, $S_\nu(b)$, and τ_b may be considered to be known constants.

In the limits of $\tau_\nu(b) << 1$ and $\tau_\nu(b) >> 1$, it is necessary to use alternate forms of Equation (9.5) in order to avoid calculating $\exp[-\tau_\nu(b)]$ or to avoid subtracting two nearly equal numbers

$$\begin{aligned} I_\nu(b) = {} & I_\nu(a)\left[1 - \tau_b + \tau_b^2/2\right] + S_\nu(a)\left[\tau_b/2 - \tau_b^2/3 + \tau_b^3/8\right] \\ & + S_\nu(b)\left[\tau_b/2 - \tau_b^2/6 + \tau_b^3/24\right] \end{aligned} \quad (9.6)$$

for $\tau_\nu(b) \ll 1$ and

$$I_\nu(b) = S_\nu(a)/\tau_b + S_\nu(b)[1 - 1/\tau_b] \tag{9.7}$$

for $\tau_\nu(b) \gg 1$. The procedure now is to set up a grid of points along the direction $\hat{s}$, each one of which corresponds to a distance interval (a, b). One starts the integration at $\tau_\nu = 0$, where $S_\nu(0)$ is given and proceeds, using Equation (9.5), Equation (9.6) or Equation (9.7) from point to point, with the result being the intensity at the end point of the ray $\hat{s}$. The grid points are set up so there is a relatively small change in τ_ν and S_ν in all cases between points a_j and $b_j = a_{j+1}$; then the approximation of the linear variation of the source function should be quite adequate. The procedure is very simple if one knows $S_\nu(\tau_\nu)$ along $\hat{s}$, for example, if one knows the temperature and density at all points and if the source function S_ν is the Planck function, which is just a function of temperature at a given frequency. One would think one could simply integrate Equation (1.91) by setting up a grid of J points with index j running from 0 to J

$$I_{\nu,j+1} = I_{\nu,j} - \epsilon_{\nu,j}\rho_j(I_{\nu,j} - B_{\nu,j})(s_{j+1} - s_j), \tag{9.8}$$

integrating from point to point in an explicit fashion. But it turns out this is much less accurate and less stable than Equation (9.5).

9.2 SOLUTION FOR FREQUENCY-DEPENDENT RADIATION TRANSFER IN SPHERICAL SYMMETRY

The discussion in the previous section describes how to integrate from grid point to grid point along a given line of sight provided the extinction coefficient ϵ_ν and the source function S_ν are known at each grid point. Here, we consider the application of such a ray-tracing solution to a more general problem, where both the extinction coefficient and the source function depend in a complicated fashion on the radiation intensity. Thus, one faces the dilemma of needing to know the solution to the radiation transfer equation in all directions before the ray equation can be solved along a single line of sight. In general, an iterative procedure must be developed for the self-consistent determination of radiation intensity, source function, and extinction coefficient. We now consider a method for accomplishing this for the special case of time-independent continuum transfer in spherical symmetry. The method can be used, for example, to determine the observable intensity of radiation from an astrophysical object as a function of position and frequency. As a specific example, we take a spherical source with a given luminosity L and given temperature T_s, surrounded by a spherical envelope whose density as a function of distance is provided. The problem is to determine the temperature distribution in the envelope and the observable radiation properties.

The coordinates we consider are the radial direction with unit vector $\hat{e}_r$ and the angle θ between an arbitrary direction $\hat{s}$ and $\hat{e}_r$. Let $\mu = \cos\theta$. Define the new variable

$$p = r(1 - \mu^2)^{1/2} \tag{9.9}$$

Then Equation (1.91) becomes (Hummer and Rybicki, 1971)

$$\mu\frac{\partial I_\nu^\pm}{\partial r} = \mp\epsilon_\nu\rho\left(I_\nu^\pm - S_\nu\right), \tag{9.10}$$

where the intensity has to be split into two components I_ν^+ and I_ν^- because the variable p does not distinguish between $\mu > 0$ and $\mu < 0$. Thus, $I_\nu^\pm$ refers to directions where μ is positive or negative. The source function will be taken to include both true absorption and isotropic scattering

$$S_\nu = \frac{\kappa_\nu B_\nu + \sigma_\nu J_\nu}{\kappa_\nu + \sigma_\nu}. \tag{9.11}$$

The time-independent moment equations (1.110) and (1.111) become, when integrated over frequency

$$\frac{\partial(fJ)}{\partial r} + (3f - 1)\frac{J}{r} + \kappa_H \rho H = 0 \tag{9.12}$$

$$\frac{\partial H}{\partial r} + \frac{2H}{r} + \kappa_J \rho J - \kappa_P \rho B = 0, \tag{9.13}$$

where

$$J = \int_0^\infty J_\nu d\nu, \qquad H = \int_0^\infty H_\nu d\nu, \qquad f = \frac{1}{J}\int_0^\infty K_\nu d\nu, \tag{9.14}$$

and where

$$\kappa_H = \frac{1}{H}\int_0^\infty \epsilon_\nu H_\nu d\nu, \quad \kappa_J = \frac{1}{J}\int_0^\infty \kappa_\nu J_\nu d\nu, \tag{9.15}$$

$$\kappa_P = \frac{1}{B}\int_0^\infty \kappa_\nu B_\nu(T) d\nu, \quad B = \int_0^\infty B_\nu(T) d\nu. \tag{9.16}$$

Equation (9.12) and Equation (9.13) are the exact equations for spherically symmetric, nonrelativistic, time-independent, but frequency-dependent radiation transfer. The problem is to devise a scheme for supplying the correct values for f, κ_H, and κ_J, where f is the Eddington factor. The procedure will be:

1. Start with an assumed temperature distribution and assumed source function, for example, B_ν. Given (T, ρ) at each point, calculate ϵ_ν for various frequencies.
2. Go back to the frequency-dependent ray equation (9.10) and solve it for various frequencies for I_ν, with the current estimates of the source function and absorption and scattering coefficients, along a set of lines of sight, each with constant p.
3. Use the intensity thus obtained to calculate κ_H, κ_J, and f.
4. Solve the moment equations simultaneously with the appropriate energy equations to obtain improved estimates for the temperature $T(r)$ and $J(r)$. Use these to recalculate the source function S_ν and the absorption and scattering coefficients κ_ν and σ_ν.
5. Solve the ray equations again with these new estimates, and continue by iteration until the solution converges.

The advantage of this procedure is that if one is combining the radiative transfer with the time-dependent solution of the hydrodynamic equations, then once a converged solution has been obtained, the mean absorption coefficients and f usually change only slowly with time, so for subsequent hydrodynamic time steps the convergence of the radiation transfer solution is very rapid.

The first phase of the iteration (steps (2) and (3) above) involves the combination of (1) the two ray equations (9.10) for I_ν^+ and I_ν^-, (2) the definitions of frequency-integrated quantities given by Equation (9.14) to Equation (9.16), and (3) the definitions of the fundamental radiation quantities J_ν, H_ν, and K_ν from Equation (1.96), Equation (1.97), and Equation (1.99). The second phase (step 4) includes the moment Equation (9.12) and Equation (9.13) along with the appropriate energy equation. We now describe the numerical solution for the first phase.

The (r, p) space is discretized into a radial (r) grid with index j and a p grid with index k. Note from Figure 9.1 that the lines of constant p are straight, parallel lines intersecting the lines of constant r and pointing, perhaps, to a distant observer. Thus, $p_1 = 0$ and $p_K = r_{\max} = r_J$. Note that $r_1 \neq 0$; it represents the outer boundary of a central source with a fixed temperature. Given an estimate for the source function and given the density at each grid point, the ray equations (9.10) are solved along

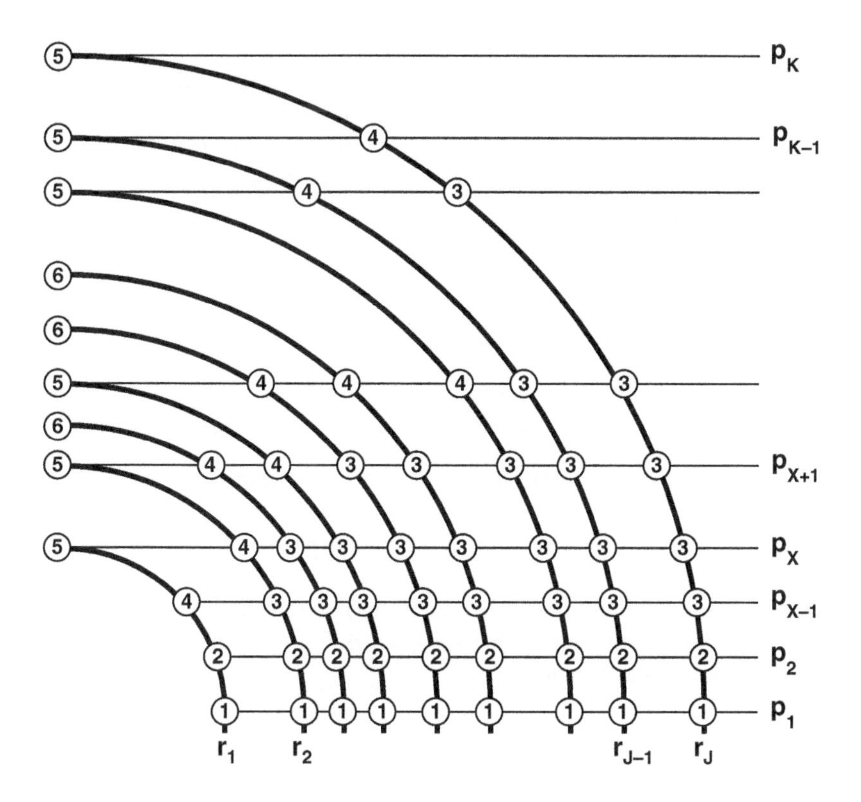

FIGURE 9.1 The setup of grid points in the (r, p) plane for the solution of the spherically symmetric radiative transfer problem. The numbers on the points correspond to the different cases that determine the coefficients $a_{k,j}$, $b_{k,j}$, $c_{k,j}$ that appear in Equation (9.19) to Equation (9.21).

each value of p from $p = 0$ to $p = r_J$, integrating, for a given p, from $\mu_{j,k} \cdot r_j$ to $\mu_{j+1,k} \cdot r_{j+1}$ in sequence using Equation (9.5), Equation (9.6) or Equation (9.7). Here,

$$\mu_{j,k} = \left(1 - \frac{p_k^2}{r_j^2}\right)^{1/2}. \tag{9.17}$$

The integration is done for each frequency independently.

Boundary conditions must be provided: For I_ν^- this condition is provided at $r_{\max}$ for each p and could be the Planck function B_ν at the outer temperature of the structure. For I_ν^+, one can use $I_\nu^-(r = p, p) = I_\nu^+(r = p, p)$. In other words, for a given p, one starts the integration at the outer boundary and works inward to get I_ν^-, then works back outwards to get I_ν^+. This boundary condition corresponds to symmetry at the inner boundary (left-hand edge of Figure 9.1). Of course, if the configuration is very optically thick, the intensity will be very nearly isotropic, so I_ν^+ will be very close to I_ν^- in a larger region. In addition, for all points on the inner radius r_1 (i.e., $p_k < r_1$) the following boundary condition is applied

$$I_\nu^+ - I_\nu^- = B_\nu(T_s), \tag{9.18}$$

where T_s is the temperature of the central source.

The result is two three-dimensional arrays, $I_{j,k,l}^+$ and $I_{j,k,l}^-$, giving the intensities at each point along each ray in the outward and inward directions, respectively, at each frequency (l). The next step is to calculate the moments that are integrations over angle which reduce to sums over p, at each radial point and each frequency

$$J_{jl} = \frac{1}{2}\sum_k \left(I_{j,k,l}^+ + I_{j,k,l}^-\right) a_{k,j} \tag{9.19}$$

$$H_{jl} = \frac{1}{2}\sum_k \left(I_{j,k,l}^+ - I_{j,k,l}^-\right) b_{k,j} \tag{9.20}$$

$$K_{jl} = \frac{1}{2}\sum_k \left(I_{j,k,l}^+ + I_{j,k,l}^-\right) c_{k,j}. \tag{9.21}$$

Thus, for example, J_{jl} is the mean intensity for the jth radial point at the lth frequency. The integration weights $a_{k,j}$, $b_{k,j}$, and $c_{k,j}$ depend on the distribution of the p grid points for each radial grid point, and do not depend on frequency, that is, they are functions of (r_j, p_k). They are calculated by making a linear (or higher-order) assumption for the variation of $I^+(p)$ and $I^-(p)$ and integrating from p_k to p_{k+1} using the same assumption for all three weights, a, b, and c. The following relations must be satisfied for each value of j

$$\sum_k a_{k,j} = 1 \tag{9.22}$$

$$\sum_k b_{k,j} = \frac{1}{2} \tag{9.23}$$

$$\sum_k c_{k,j} = \frac{1}{3}. \tag{9.24}$$

The weights corresponding to the points in Figure 9.1 labeled (1) through (6) were constructed by Yorke (1980, Table 3) to take advantage of the symmetry properties of $I^{\pm}(r, p)$ at $p = 0$ and $p = r$. Since they are complicated, we give just two examples. Points of type (2) ($k = 2$) (Figure 9.1) were constructed assuming a linear relation for $I_\nu^{\pm}(p)$ between $\mu_1 = (1 - p_2^2/r_j^2)^{1/2}$ and $\mu_2 = (1 - p_3^2/r_j^2)^{1/2}$ (that is, $I^+ = A^+ + b^+\mu$), and a quadratic relation with μ between $\mu = 1$ and μ_1

$$\begin{aligned}
a_{k,j} &= \frac{p_2^2}{3r_j^2}\frac{(2+\mu_1)}{(\mu_1+1)^2} + \frac{p_3^2 - p_2^2}{2r_j^2}\frac{1}{(\mu_1+\mu_2)} \\
b_{k,j} &= \frac{p_2^2}{4r_j^2} + \frac{(p_3^2 - p_2^2)}{6r_j^2}\left(1 + \frac{\mu_1}{\mu_1+\mu_2}\right) \\
c_{k,j} &= \frac{p_2^2}{15r_j^2}\left(\frac{2+\mu_1}{(1+\mu_1)^2} + 3\mu_1\right) + \frac{p_3^2 - p_2^2}{12r_j^2}\left(\frac{\mu_1^2+\mu_2^2}{\mu_1+\mu_2} + 2\mu_1\right).
\end{aligned} \tag{9.25}$$

Points of type (5) occur at the inner boundary where $r_j = p_k$. Here, $a_{kj} = 2\mu_0/3$, $b_{kj} = 0$, $c_{kj} = 2\mu_0^3/15$, where $\mu_0 = (1 - p_{k-1}^2/r_j^2)^{1/2}$. Type (6) points do not occur when, for each radial grid point j, there is exactly one "impact parameter" p; if they do occur, the weights are the same as for type (5).

The next step is to integrate over all frequencies to obtain J, H, f, κ_J, and κ_H at each radial point. The corresponding sums are:

$$J_j = \sum_l g_l J_{j,l} \tag{9.26}$$

$$H_j = \sum_l g_l H_{j,l} \tag{9.27}$$

$$f_j = \sum_l g_l K_{j,l}/J_j \tag{9.28}$$

$$\kappa_{J,j} = \sum_l g_l \cdot \kappa_{j,l} J_{j,l}/J_j \tag{9.29}$$

$$\kappa_{H,j} = \sum_l g_l \cdot \epsilon_{j,l} H_{j,l}/H_j. \tag{9.30}$$

The frequency weights g_l depend on the distribution of frequency grid points ν_l. For example, 40 frequency points could be used over the range 3×10^{15} to 3×10^{13} Hz, and Simpson's rule can be used to determine the frequency weights.

In general the second major step, the solution of the moment equations, must be carried out in connection with the equations of radiation hydrodynamics (Section 9.4) in which the radiation and matter are treated as separate fluids. Here, we describe a very simple formulation for this step in which the spherical structure is time-independent and there is no energy production in the region of interest, so the luminosity $L = (4\pi r)^2 H$ is constant. Then Equation (9.13) becomes

$$\kappa_J J = \kappa_P B, \tag{9.31}$$

which must be solved in connection with Equation (9.12). The object of this part of the calculation is to solve for the updated values of temperature T and mean intensity J as a function of spherical radius.

Denote by $b^{-1}(x)$ the inverse of the function $b(T) = \kappa_P B$, so the temperature T at each grid point j is given by

$$T_j = b^{-1}(\kappa_{J,j} J_j). \tag{9.32}$$

Then we note that with the definition of an integrating factor

$$g(r) = \exp \int_0^r [3 - 1/f(r')]/r' dr' \tag{9.33}$$

Equation (9.12) becomes

$$\frac{\partial}{\partial r}(gfJ) + \frac{g\kappa_H L}{(4\pi r)^2} = 0. \tag{9.34}$$

For the numerical solution, we integrate this equation from grid point to grid point in the radial direction; between two grid points r_{j-1} and r_j, we obtain

$$J_j = g_j^{(1)} J_{j-1} + g_j^{(2)} \kappa_{H,j-1/2} L, \tag{9.35}$$

where

$$g_j^{(1)} = \frac{f_{j-1}}{f_j} \exp\left[-\int_{r_{j-1}}^{r_j} [3 - 1/f(r')]/r' dr'\right] \tag{9.36}$$

$$g_j^{(2)} = -\frac{1}{(4\pi)^2 f_j} \int_{r_{j-1}}^{r_j} \left[\exp\left[\int_{r_j}^{r''} (3 - 1/f(r'))/r' dr'\right]\right] \Big/ r''^2 dr'', \tag{9.37}$$

where $\kappa_{H,j-1/2}$ is an appropriate mean value of κ_H between r_j and r_{j-1}. The integrations for $g_j^{(1)}$ and $g_j^{(2)}$ are performed assuming f varies linearly between r_{j-1} and r_j. The Js are obtained from Equation (9.35) one by one starting at the center, where the central boundary condition is obtained from Equation (9.26). Then the temperatures follow from Equation (9.32). With the new temperature distribution, the frequency-dependent absorption coefficients are recalculated, and the solution procedure returns to the integration of the ray equations for all p. The procedure converges more rapidly if the above-described numerical integration for recalculation of J is performed. One could, in principle, merely use the Js that come out of Equation (9.26).

An example of the results of a frequency-dependent calculation of the emerging radiation from a dusty envelope surrounding a main-sequence star is shown in Figure 9.2. The opacity is calculated from the properties of the dust. Note that, as the optical depth of the envelope increases, the longer wavelength radiation is enhanced.

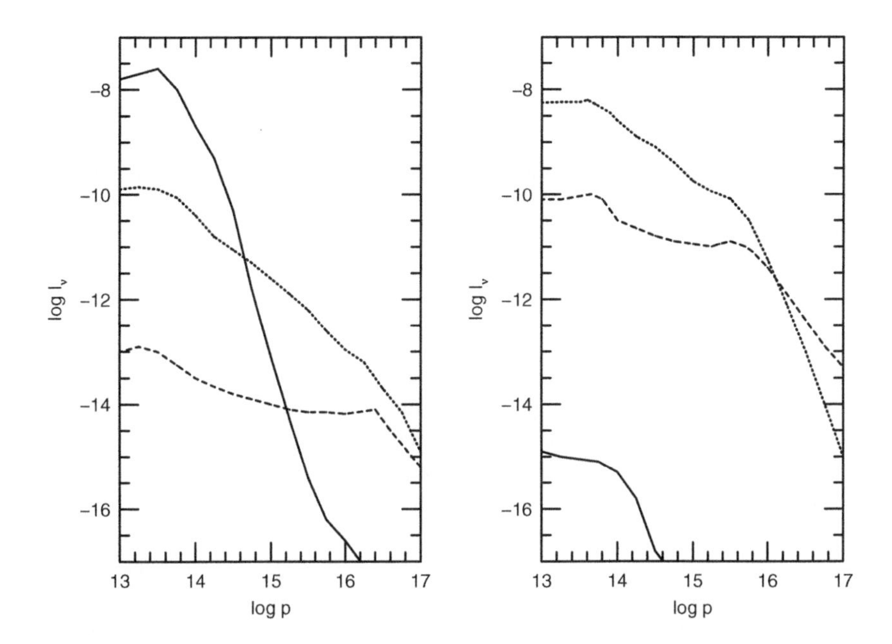

FIGURE 9.2 An example of the results of calculations of frequency-dependent radiative transfer in the spherical case. A central source with temperature 16,600 K and luminosity 10^3 $L_\odot$ is surrounded by an envelope with density distribution $\rho \propto r^{-3/2}$. The emerging intensity I_ν^+ is plotted across the observable disk of the object as a function of p. *Solid curves*: 2.5 microns; *dotted curves*: 29 microns; *dashed curves*: 125 microns. The optical depth in the envelope at 0.1 microns is 1 in the left panel and 1000 in the right panel. (Adapted from Yorke, H. and Shustov, B. (1981) *Astron. Astrophys.* **98**: 125.)

9.3 FREQUENCY-DEPENDENT STELLAR ATMOSPHERES

The purpose of stellar atmosphere calculations is to develop physically realistic models of the emerging spectrum from a star, so that, by comparison with observations, one can examine the basic parameters of stars and their changes with evolution. The input quantities for a particular model are log g, the acceleration of gravity at the surface of the star, log $T_{\rm eff}$, the photospheric ($\tau = 2/3$) temperature, and the chemical composition, in particular, the abundances of elements heavier than helium. The output of a model, which may be calculated in combination with a full stellar model or independently, is the emerging radiation flux as a function of frequency, the colors (for example, based on broadband filters U, B, and V), and the detailed profiles of spectral lines. In this section, we outline one particular procedure for calculation of the atmospheric structure and radiation properties with full frequency dependence.

The usual assumptions employed in the calculation of an atmosphere are (1) the atmospheric vertical thickness is very small compared with the radius of a star, so that the atmosphere is treated as plane parallel layers; (2) the atmosphere is in hydrostatic equilibrium; (3) the radiated flux $\sigma_B T_{\rm eff}^4$ is constant through the atmosphere, that is, there is no energy production; (4) the atmosphere is in local thermodynamic equilibrium; and (5) physical quantities such as density vary as a function of depth only, not as a function of the spherical coordinates (θ, ϕ). There are some calculations, however, that relax one of the last two assumptions.

Two levels of approximation are commonly used for the calculation of radiative transfer in stellar atmospheres. For the calculation of stellar evolution, the atmosphere forms the surface boundary condition, and for many purposes it is sufficient to use the frequency-independent diffusion equation (1.117) relating the radiative flux and the temperature gradient

$$\mathbf{F} = -\frac{c}{3\kappa_R \rho}\nabla(aT^4) \tag{9.38}$$

where κ_R is the Rosseland mean opacity defined by Equation (1.115). In the outer layers, this equation takes the form given by Equation (5.33). One starts near optical depth $\tau = 0$ and numerically integrates the radiation transfer equation inward, along with the equation of hydrostatic equilibrium and the equation of state, obtaining T, ρ, P as a function of τ. The integration is terminated when a predetermined optical depth or total inwardly integrated mass has been reached. A procedure for integration of a radiative grey (frequency-independent) atmosphere is given in Chapter 5, Section 5.4.

Once such a simplified model has been calculated, it is possible to use the given temperature-density structure plus frequency-dependent absorption and scattering coefficients to obtain the flux as a function of frequency emerging from the atmosphere (see Gray, 1992, Chapter 7)

$$F_\nu(\tau_\nu = 0) = 2\pi \int_0^\infty S_\nu(\tau_\nu) E_2(t_\nu) dt_\nu, \tag{9.39}$$

where $E_2(x)$ is the second exponential integral

$$E_2(x) = \int_1^\infty \frac{\exp(-xt)}{t^2} dt, \tag{9.40}$$

S_ν is the source function, and τ_ν is the optical depth defined by Equation (1.93). One can also compute the profiles of spectral lines based upon the calculated temperature-density structure (Mihalas, 1978, Chapters 8, 9, 10).

However, if one wishes to make detailed comparisons with observations of particular objects, the model itself must be calculated with full frequency dependence. The elaborate theory of constructing such model atmospheres is covered in the book by Mihalas (1978) and in a report by Kurucz (1970). The model now becomes two-dimensional, with independent variables, depth (τ_{ref}) and frequency, where τ_{ref} is the optical depth at some reference frequency. The overall procedure for calculating the model is as follows:

- Provide a starting guess for temperature and density as a function of depth, and derive all required physical quantities, including pressure, degree of ionization, and opacity as a function of frequency. The first guess can be obtained from a grey model or from a nongrey (frequency-dependent) model with similar input parameters.
- Calculate the total radiation flux, integrated over all frequencies, at all depths in this model. The constraint that must be satisfied by the final

model is that this flux, at all depths, must be constant and equal to the prescribed value $\sigma_B T_{\text{eff}}^4$.
- Iteratively, adjust the temperature at each depth, while maintaining hydrostatic equilibrium until the flux constraint is met.

The key element of this procedure is to efficiently produce a temperature correction. A number of such procedures have been used, as described by Mihalas (1978, Section 7.2); here, we give just one example, a procedure described by Kurucz (1970), based on the Avrett and Krook method (1963). The method uses the first two moments of the radiation field

$$H_\nu = \frac{1}{2}\int_{-1}^{1} \mu I_\nu d\mu \quad \text{and} \quad K_\nu = \frac{1}{2}\int_{-1}^{1} \mu^2 I_\nu d\mu, \tag{9.41}$$

where $\mu = \cos\theta$. The plane-parallel transfer equation (1.92), when multiplied by μ and integrated over all solid angles, becomes

$$\frac{dK_\nu}{d\tau_\nu} = H_\nu, \tag{9.42}$$

where $d\tau_\nu = (\kappa_\nu + \sigma_\nu)\rho dz$, and z is the vertical (downward) direction in the atmosphere.

If we define $H = \int H_\nu d\nu$ as the current value of this moment at a particular depth z in an approximate model of the atmosphere, and

$$H_{\text{target}} = \frac{\sigma_B T_{\text{eff}}^4}{4\pi} \tag{9.43}$$

as effectively one of the input parameters for a particular atmosphere, then the object of the procedure is to force $\Delta H = H_{\text{target}} - H$ to zero at all depths.

A new variable is introduced to replace τ

$$dM = \rho dz \tag{9.44}$$

and Equation (9.42) is written

$$\frac{dK_\nu}{dM} = (\kappa_\nu + \sigma_\nu)H_\nu. \tag{9.45}$$

The procedure now is to add first-order perturbations to a number of the basic variables and eventually to solve for

$$\Delta T = -\frac{dT}{dM}\Delta M, \tag{9.46}$$

which is the correction to the present temperature at z that will hopefully bring better agreement with the constant flux condition $H = H_{\text{target}}$. The method is not mathematically rigorous, but it is computationally practicable and it works. (The Avrett–Krook procedure itself is mathematically rigorous, but it involves more computer time.) Denote by a subscript “0” the current value of a variable, and let $\epsilon_\nu = \kappa_\nu + \sigma_\nu$; then the perturbed variables become

$$K_\nu = K_{\nu,0} + \Delta K_\nu, \qquad H_\nu = H_{\nu,0} + \Delta H_\nu,$$
$$\epsilon_\nu = \epsilon_{\nu,0} + \Delta\epsilon_\nu, \qquad M = M_0 + \Delta M. \tag{9.47}$$

Note that the radiation quantities are perturbed and also the mass scale.

Now substitute the perturbed expressions into Equation (9.45), drop second-order terms, and divide by ϵ_ν

$$\frac{1}{\epsilon_\nu}\frac{d\Delta K_\nu}{dM} = \Delta H_\nu + \frac{\Delta\epsilon_\nu}{\epsilon_\nu} H_\nu + H_\nu \frac{d\Delta M}{dM}. \tag{9.48}$$

Integrate over all frequencies and make the approximation

$$\int \frac{1}{\epsilon_\nu}\frac{d\Delta K_\nu}{dM} d\nu = \int \frac{\Delta\epsilon_\nu}{\epsilon_\nu} H_\nu d\nu. \tag{9.49}$$

The result is

$$H - H_{\text{target}} = H \frac{d\Delta M}{dM}, \tag{9.50}$$

which can be integrated inward from the surface, where $\Delta M = 0$, to give

$$\Delta M = \int_0^M \frac{H(M') - H_{\text{target}}}{H(M')} dM' \tag{9.51}$$

after which $\Delta T(M)$ follows from Equation (9.46).

Once the revised temperature distribution and the revised mass distribution have been obtained, the total pressure P is obtained from the equation of hydrostatic equilibrium

$$\frac{dP}{dM} = g \tag{9.52}$$

and since, in the very outer layers of the star, g is effectively constant, $P = gM$. This calculation is made very simple because of the way M is defined. The total pressure is:

$$P = P_{\text{gas}} + P_{\text{rad}}, \tag{9.53}$$

the sum of the gas pressure and the radiation pressure. Given the gas pressure and T, the equation of state is used to calculate the number densities of various atoms, ions, molecules, and electrons and the total density ρ. This calculation can be fairly complicated if ionization or dissociation is occurring. The number densities and T then allow calculation of the absorption coefficient κ_ν and the scattering coefficient σ_ν at each frequency and at each depth. The information is now available to calculate new values of $H(M)$ for the next iteration.

The formal solution to the plane parallel transfer equation (Mihalas, 1978, Section 2.2) gives the radiation field $J_\nu(\tau_\nu)$, $H_\nu(\tau_\nu)$, and $K_\nu(\tau_\nu)$, where τ_ν is the optical depth measured inward along the normal to the surface at frequency ν.

$$
\begin{aligned}
J_\nu(\tau_\nu) &= \frac{1}{2}\int_0^{\tau_\nu} S_\nu(t_\nu)E_1(\tau_\nu - t_\nu)dt_\nu + \frac{1}{2}\int_{\tau_\nu}^{\infty} S_\nu(t_\nu)E_1(t_\nu - \tau_\nu)dt_\nu \\
H_\nu(\tau_\nu) &= -\frac{1}{2}\int_0^{\tau_\nu} S_\nu(t_\nu)E_2(\tau_\nu - t_\nu)dt_\nu + \frac{1}{2}\int_{\tau_\nu}^{\infty} S_\nu(t_\nu)E_2(t_\nu - \tau_\nu)dt_\nu \\
K_\nu(\tau_\nu) &= \frac{1}{2}\int_0^{\tau_\nu} S_\nu(t_\nu)E_3(\tau_\nu - t_\nu)dt_\nu + \frac{1}{2}\int_{\tau_\nu}^{\infty} S_\nu(t_\nu)E_3(t_\nu - \tau_\nu)dt_\nu .
\end{aligned}
\tag{9.54}
$$

Each radiation quantity is written as the sum of two contributions, the first arising from an integration over all layers above τ_ν and the other from integration over all layers below τ_ν. The exponential integrals are defined

$$
E_n(x) = \int_1^{\infty} \frac{\exp(-xy)}{y^n} dy = \int_0^1 \exp(-x/\mu)\mu^{n-2} d\mu \tag{9.55}
$$

and the source function is normally given by Equation (9.115).

The depth dependence of physical quantities in the atmosphere is defined by a set of N discrete points, each characterized by a mass (actually surface density) M_j. The optical depth at each point at each frequency must now be obtained

$$
\tau_{\nu,j} = \int_0^{M_j} \epsilon_\nu dM . \tag{9.56}
$$

If the source function is simply $S_\nu = B_\nu$, the calculation of the quantities given in relations (9.54) is a straightforward numerical integration. However, if the source function is a function of J_ν (Equation 9.115), the first of these relations becomes an integral equation for J_ν, and the solution is more complicated. Numerous approaches to this problem have been devised (Mihalas, 1978, Chapter 6) based either on an integral–equation approach or a differential–equation approach. Here, we review briefly the integral–equation approach used in the standard Kurucz model atmosphere program (Kurucz, 1969).

The technique is simply to take the first of Equations (9.54) and to represent the run of the source function by an approximate interpolation formula. The integral is replaced by a sum, and one is able to reduce the problem to a set of linear algebraic equations for J_ν at each of the discrete depth points in the atmosphere.

The first step of the process is to use a quadratic polynomial to represent the source function between two particular depth points τ_j and τ_{j+1}

$$
S_\nu(t) = \sum_{k=1}^{3} t^{k-1} \sum_{i=1}^{N} C_{jki} S_i , \tag{9.57}
$$

where the C_{jki} are just functions of the discrete optical depths τ_{j-1}, τ_j, τ_{j+1}, and τ_{j+2}. Then the contribution to the lth value of J_ν arising from the interval between τ_j and τ_{j+1} may be written

$$
J_{\nu,l,j} = \sum_{k=1}^{3} \left[\frac{1}{2} \int_{\tau_j}^{\tau_{j+1}} t^{k-1} E_1 |t - \tau_l| dt \right] \sum_{i=1}^{N} C_{jki} S_i . \tag{9.58}
$$

After analytic evaluation of the integral and some further manipulation (see Kurucz, 1969), it can be shown that the vector $\mathbf{J}$ of the mean intensity at all depths can be expressed as

$$\mathbf{J} = \mathbf{MS}, \tag{9.59}$$

where $\mathbf{S}$ is the vector of the values of the source function (at a given frequency) at all values of optical depth τ_j, and $\mathbf{M}$ is a matrix whose elements are functions only of the τ_j.

Equation (9.115) can be written

$$S_\nu = (1 - \alpha_\nu)B_\nu + \alpha_\nu J_\nu, \tag{9.60}$$

where $\alpha_\nu = \sigma_\nu/(\kappa_\nu + \sigma_\nu)$ and in vector notation, at a given frequency

$$\mathbf{S} = (\mathbf{I} - \alpha)\mathbf{B} + \alpha\mathbf{MS}, \tag{9.61}$$

where $\mathbf{B}$ is the vector of values of B_ν, $\mathbf{I}$ is the unit matrix, and α is the diagonal matrix with elements $\alpha_{\mathbf{jj}} = \alpha_j$. This equation can be solved for $\mathbf{S}$

$$\mathbf{S} = (\mathbf{I} - \alpha\mathbf{M})^{-1}(\mathbf{I} - \alpha)\mathbf{B}, \tag{9.62}$$

which requires a matrix inversion, but all quantities on the right-hand-side are given. What has been found is the source function at a particular frequency at all the depth points τ_j, essentially by direct solution, unless one employs an iterative method to do the matrix inversion.

Once S_ν is known at all points, J_ν can be calculated, and H_ν can be obtained by straightforward integration of the second of Relations (9.54). The whole set of calculations then has to be redone for each frequency point ν_m needed to characterize the atmosphere, taking into account the fact that the optical depth scale will change with frequency. Then, at each depth point, H is calculated by integration over frequency and the discrepancy $H - H_{\mathrm{target}}$ calculated. The iteration procedure then returns to the calculation of the next temperature correction. Once the calculation has converged, one has available the run of physical variables as a function of depth as well as the information needed to self-consistently calculate the emergent intensity as a function of frequency. Much of the programming effort in such a calculation is actually in the physics of the calculation of the equation of state and in the calculation of the opacities (*not treated here*). Depending on how accurately one wishes to do the comparison with observations, one may need a very large number of frequency points in order to represent the spectrum adequately, particularly in the case of cool star atmospheres with numerous overlapping molecular band features.

9.4 TECHNIQUE FOR FLUX-LIMITED DIFFUSION IN TWO SPACE DIMENSIONS

Having given two examples of the solution of the radiative transfer equation where only one space dimension is involved, we now turn to the case where a radiative transfer solution is desired in a problem involving two space dimensions, usually in connection with the solution of the equations of hydrodynamics. To do a full frequency-dependent and angle-dependent solution for radiation transfer in such a problem, which could involve many thousands of time steps, would be costly in terms of computer resources, and often one settles for a simpler approach without full frequency dependence. As mentioned in Chapter 1, Section 1.6.4, the approach to be described gives a physically accurate solution in regions of large optical depth, but is not correct in regions of optical depth less than one.

We rewrite the Lagrangian equations of hydrodynamics (Equation 1.62), including radiation effects, but without viscosity or magnetic fields or gravity. These equations are commonly known as the equations for radiation hydrodynamics, as discussed in more detail by Mihalas and Mihalas (1984). Separate equations are written for the internal energy of the gas and that of the radiation; they are coupled by terms involving the interaction of radiation and matter through emission and absorption processes. The equations are integrated over frequency, but the method of averaging the absorption coefficient over frequency differs in the different terms of the equations

$$\begin{aligned} \frac{d\rho}{dt} + \rho\frac{\partial v_j}{\partial x_j} &= 0 \\ \rho\frac{dv_i}{dt} &= -\frac{\partial P_g}{\partial x_i} + \frac{1}{c}\epsilon_F \rho F_{\mathrm{rad},i} \quad (i = 1, 2) \\ \frac{de}{dt} + (e + P_g)\frac{\partial v_j}{\partial x_j} &= -4\pi\kappa_P\rho B + c\kappa_E\rho u, \end{aligned} \tag{9.63}$$

where e is the internal energy of the material per unit volume, P_g is the gas pressure, u is the energy per unit volume in the radiation field, B is the Planck function integrated over frequency, and $\mathbf{F}_{\mathrm{rad}}$ is the radiative flux, integrated over frequency (abbreviated in the following to $\mathbf{F}$). Thus, the last term in the second equation is the radiative force per unit volume. The equations that determine the radiation field are:

$$\rho\frac{d}{dt}\left(\frac{u}{\rho}\right) = -\nabla\cdot\mathbf{F} - \nabla\mathbf{v} : \mathbf{P} + 4\pi\kappa_P\rho B - c\kappa_E\rho u \tag{9.64}$$

$$\frac{\rho}{c^2}\frac{d}{dt}\left(\frac{\mathbf{F}}{\rho}\right) = -\nabla\cdot\mathbf{P} - \frac{1}{c}\epsilon_F\rho\mathbf{F}. \tag{9.65}$$

Here $\mathbf{P}$ is the radiation pressure tensor, defined by $\mathbf{P} = (4\pi/c)\mathbf{K}$, where $\mathbf{K}$ is defined by Equation (1.99) integrated over all frequencies, and

$$\kappa_E = \frac{1}{u}\int_0^\infty \kappa_\nu u_\nu d\nu, \tag{9.66}$$

$$\kappa_P = \frac{1}{B}\int_0^\infty \kappa_\nu B_\nu(T) d\nu, \tag{9.67}$$

$$\epsilon_F = \frac{1}{F}\int_0^\infty \epsilon_\nu F_\nu d\nu. \tag{9.68}$$

The quantity κ_P is usually known as the Planck mean. The two terms on the right-hand side of Equation (9.63) refer to, respectively, the energy lost to the gas by radiative emission processes and the energy gained by the gas by radiative absorption processes. The same two terms, with the opposite sign, appear in the radiative energy Equation (9.64), and their origin can be understood through Equation (1.107), integrated over frequency. The $\nabla\mathbf{v} : \mathbf{P}$ term in Equation (9.64) refers to the work done by the radiative forces, while the $\nabla \cdot \mathbf{F}$ term is the net rate of transport of energy out of the mass element. Equation (9.65) is obtained from integration of Equation (1.105) over frequency.

These equations apply in the co-moving (Lagrangian) frame, and it has been assumed that the material properties, including emission and absorption coefficients, are isotropic. To put the equations into the standard Eulerian coordinate system, one uses the standard transformation $d/dt = \partial/\partial t + \mathbf{v}\cdot\boldsymbol{\nabla}$, where $\partial/\partial t$ is the Eulerian derivative. As in Chapter 6, the calculations are divided into two basic parts: (1) the *advection* or *transport* step in which the $\mathbf{v}\cdot\boldsymbol{\nabla}$ term is evaluated, and (2) the *source* step, which takes care of all of the rest of the terms. Following is a discussion of the source step.

As discussed in Chapter 1, the diffusion approximation in the optically thick limit essentially involves adopting the Eddington approximation $K_\nu = \frac{1}{3}J_\nu$ and the resulting expression (Equation 1.117) for the frequency-integrated radiative flux

$$\mathbf{F} = -\frac{c}{3\kappa_R\rho}\boldsymbol{\nabla}u, \tag{9.69}$$

where κ_R is the *Rosseland* mean opacity. This expression essentially replaces Equation (9.65) and allows the entire system to be solved.*

Note that this expression is also equivalent to the time-independent form of Equation (1.105) integrated over frequency, since $\epsilon_\nu \approx \kappa_\nu$ in the case of the local thermodynamic equilibrium. Then consider Equation (9.64). In local thermodynamic equilibrium, the mean intensity $J = \frac{c}{4\pi}u = B$, so the last two terms drop out (also $\kappa_E = \kappa_P$). If we then use the principle of operator splitting, separating the radiation transport from the hydrodynamics, the velocity-dependent terms and time derivatives of the density drop out and the actual equation to be solved is:

$$\frac{\partial u}{\partial t} = -\boldsymbol{\nabla}\cdot\mathbf{F}. \tag{9.70}$$

The frequency-integrated form of Equation (1.104) gives the same result. The basic physical approximation is that in LTE the rate of emission essentially balances the rate of absorption in a given volume. Therefore, it is sufficient for many applications to simply add the two energy equations together, retain the flux divergence term, and use the single energy equation as given by Equation (9.72), below.

The radiation energy density is $u = aT_\mathrm{R}^4$ where T_R is the radiation temperature. In general, the material energy density, which can be written $e = \rho c_v T_\mathrm{M}$, where T_M is the material temperature, must be solved for separately by use of Equation (9.63). In LTE, the radiation and the matter are so closely coupled that $T_\mathrm{R} = T_\mathrm{M} = T$. Then assuming that the radiation transfer is being done as a separate step so that the density is constant and velocity-dependent terms are not included, Equation (9.63)

* In the diffusion limit, the flux mean ϵ_F reduces to the Rosseland mean.

and Equation (9.64) can be added together to give

$$\rho c_v \frac{\partial T}{\partial t} + a \frac{\partial T^4}{\partial t} = -\nabla \cdot \mathbf{F}. \tag{9.71}$$

Then use $\partial u/\partial t = 4aT^3 \partial T/\partial t$ to rewrite Equation (9.71) as

$$(1 + C_n)\frac{\partial u}{\partial t} = -\boldsymbol{\nabla}\cdot\mathbf{F}, \tag{9.72}$$

where C_n is treated as a constant and is evaluated at time step n

$$C_n = \frac{c_v \rho}{4aT^3}. \tag{9.73}$$

Thus, in the solution of the Equation (9.72), the quantity Δt is simply replaced by $\Delta t/(1 + C_n)$ and the difference Equation (9.89), discussed below, is solved. The net result is that Equations (9.63) are solved, without the terms on the right-hand side of the energy equation, but including the effects of radiative diffusion. Equation (9.64) and Equation (9.65) are not solved. An example of a solution under this approximation is given in Section 6.6.2.

In this section, we take a more general approach and retain the separate energy equations. One may wish to use the diffusion approximation in regions that are optically thin and in which local thermodynamic equilibrium does not apply. Then one must satisfy the constraint (see Chapter 1) that, in the limit $\tau \to 0$,

$$|\mathbf{F}| = cu. \tag{9.74}$$

To accomplish this, one rewrites Equation (9.69) as

$$\mathbf{F} = -\frac{c\lambda}{\kappa_R \rho}\boldsymbol{\nabla} u, \tag{9.75}$$

where λ is the flux limiter, two examples of which are given in Section 1.6.4. Whatever the form of λ, it must go in the limit $\tau \to 0$ to $|\mathbf{F}| = cu$ and in the limit $\tau >> 0$ to $\lambda = 1/3$.

Since the physical assumption that absorption processes exactly balance emission processes is no longer valid in the optically thin regions, the solution of the radiative transfer problem includes the substep given by Equation (9.70), plus an additional substep

$$\frac{\partial u}{\partial t} = 4\pi \kappa_P \rho B - c\kappa_E \rho u. \tag{9.76}$$

We first solve Equation (9.70) by the alternating direction implicit (ADI) technique. The method is similar to that described by Turner and Stone (2001). We consider a two-dimensional Cartesian (x, y) grid with uniform grid spacings in each direction, Δx, Δy, but with $\Delta x \neq \Delta y$. The method can be extended to three space dimensions in a straightforward manner. The grid setup is shown in Figure 9.3. The radiation energy density as well as the mass density ρ and opacity κ_R are defined at

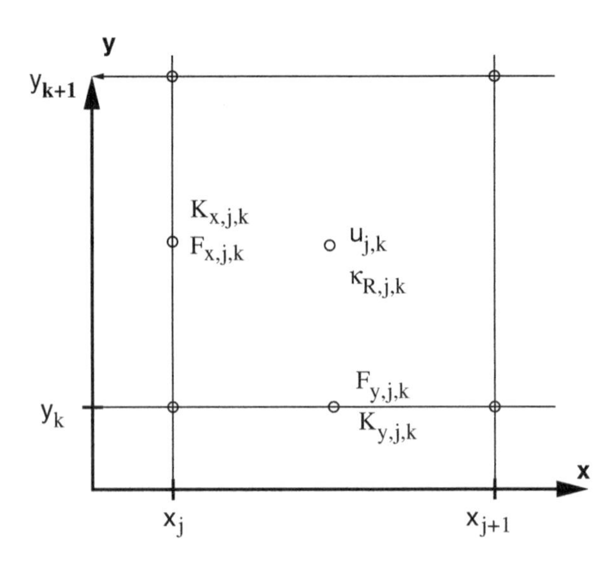

FIGURE 9.3 Grid setup for flux-limited diffusion. F_x and F_y are the components of the radiation flux in the x- and y-directions, respectively, u is the radiation energy density, and κ_R is the Rosseland mean opacity.

zone centers. The inverse of the scale height of the energy density $|\nabla u|/u$ is first calculated

$$S_{j,k} = \frac{4\left(\nabla_x^2 + \nabla_y^2\right)^{1/2}}{\bar{u}}, \tag{9.77}$$

where

$$\bar{u} = u_{j,k} + u_{j-1,k} + u_{j,k+1} + u_{j-1,k-1} \tag{9.78}$$

$$\nabla_x = \frac{u_{j,k} + u_{j,k-1} - u_{j-1,k} - u_{j-1,k-1}}{x_{j+1} - x_{j-1}} \tag{9.79}$$

$$\nabla_y = \frac{u_{j,k} + u_{j-1,k} - u_{j,k-1} - u_{j-1,k-1}}{y_{k+1} - y_{k-1}}. \tag{9.80}$$

Note that the quantity $S_{j,k}$ is defined at zone corners. The photon mean free path is defined at zone centers as

$$m_{j,k} = \frac{1}{\kappa_{R,j,k}\rho_{j,k}}. \tag{9.81}$$

The dimensionless ratio R defined by Equation (1.121), which is the ratio of the mean free path to the scale height, can be expressed in a spatially centered scheme (not necessarily always used) by

$$R_x = 0.25(m_{j,k} + m_{j-1,k})(S_{j,k} + S_{j,k+1}) \tag{9.82}$$

and the flux limiter λ_x is calculated according to

$$\lambda_x = \frac{2 + R_x}{6 + 3R_x + R_x^2}. \tag{9.83}$$

Similarly, the ratio R averaged for the y-direction is:

$$R_y = 0.25(m_{j,k} + m_{j,k-1})(S_{j,k} + S_{j+1,k}) \tag{9.84}$$

and the flux limiter λ_y is defined in an analogous way to λ_x. Now write the two components of the radiation flux

$$F_x = -K_x \frac{\partial u}{\partial x} \tag{9.85}$$

$$F_y = -K_y \frac{\partial u}{\partial y} \tag{9.86}$$

$$K_x = 0.5c\lambda_x(m_{j,k} + m_{j-1,k}) \tag{9.87}$$

$$K_y = 0.5c\lambda_y(m_{j,k} + m_{j,k-1}). \tag{9.88}$$

Note that the fluxes, as well as the Rs and Ks, are defined at the centers of cell interfaces, as illustrated in Figure 9.3.

The basic difference equation to be solved is:

$$\begin{aligned} \frac{u_{j,k}^{n+1} - u_{j,k}^{n}}{\Delta t} = & \frac{1}{(\Delta x)^2}\left[K_{x,j+1,k}\left(u_{j+1,k}^{n+1} - u_{j,k}^{n+1}\right)\right. \\ & \left. - K_{x,j,k}\left(u_{j,k}^{n+1} - u_{j-1,k}^{n+1}\right)\right] \\ & + \frac{1}{(\Delta y)^2}\left[K_{y,j,k+1}\left(u_{j,k+1}^{n+1} - u_{j,k}^{n+1}\right)\right. \\ & \left. - K_{y,j,k}\left(u_{j,k}^{n+1} - u_{j,k-1}^{n+1}\right)\right], \end{aligned} \tag{9.89}$$

where the coefficients K_x and K_y are evaluated at time step n, but the us themselves are backward (fully implicitly) time-differenced.

The basic Equation (9.89) is solved by the ADI method. A pseudo-time t^* is introduced and the following equation is relaxed to a steady state by solving it with a series of pseudo-time steps Δt_m^*. The procedure is the same as that used in Chapter 6, Section 6.6.2, but there the problem is set up in cylindrical (R, Z) coordinates and no distinction is made between the internal energy of the gas and that of the radiation

$$\frac{\partial u}{\partial t_m^*} = -\frac{\partial u}{\partial t} - \nabla \cdot \mathbf{F} \to 0\,. \tag{9.90}$$

For a given pseudo-time step Δt_m^*, the calculation is split into two parts: sweeps through the grid in the x-direction for half of Δt_m^*, followed by sweeps through the grid in the y-direction for the other half of that pseudo-time step. If we let $u_{j,k}^m$ be the energy density at the completion of one pseudo-time step; $u_{j,k}^{m+1/2}$ the energy density after the completion of the x-sweeps for the next pseudo-time step (Δt_m^*); and $u_{j,k}^{m+1}$ the energy density after the completion of the y-sweeps for that pseudo-time step, then the difference equations become

$$\begin{aligned}\frac{u_{j,k}^{m+1/2}-u_{j,k}^{m}}{\Delta t_m^*} &= \frac{1}{2(\Delta t)}\left(u_{j,k}^{n}-u_{j,k}^{m+1/2}\right)\\ &+\frac{1}{2(\Delta x)^2}\left[K_{x,j+1,k}\left(u_{j+1,k}^{m+1/2}-u_{j,k}^{m+1/2}\right)-K_{x,j,k}\left(u_{j,k}^{m+1/2}-u_{j-1,k}^{m+1/2}\right)\right]\\ &+\frac{1}{2(\Delta y)^2}\left[K_{y,j,k+1}\left(u_{j,k+1}^{m}-u_{j,k}^{m}\right)-K_{y,j,k}\left(u_{j,k}^{m}-u_{j,k-1}^{m}\right)\right]. \quad (9.91)\end{aligned}$$

Note that the solution is implicit for $u_{j,k}^{m+1/2}$ in the x-direction, but explicit (using u^m values) in the y-direction; also the Ks are always fixed at their time step n values. For the second half of the pseudo-time step the roles of the two directions are reversed and the solution becomes implicit for $u_{j,k}^{m+1}$ in the y-direction

$$\begin{aligned}\frac{u_{j,k}^{m+1}-u_{j,k}^{m+1/2}}{\Delta t_m^*} &= \frac{1}{2(\Delta t)}\left(u_{j,k}^{n}-u_{j,k}^{m+1}\right)\\ &+\frac{1}{2(\Delta x)^2}\left[K_{x,j+1,k}\left(u_{j+1,k}^{m+1/2}-u_{j,k}^{m+1/2}\right)-K_{x,j,k}\left(u_{j,k}^{m+1/2}-u_{j-1,k}^{m+1/2}\right)\right]\\ &+\frac{1}{2(\Delta y)^2}\left[K_{y,j,k+1}\left(u_{j,k+1}^{m+1}-u_{j,k}^{m+1}\right)-K_{y,j,k}\left(u_{j,k}^{m+1}-u_{j,k-1}^{m+1}\right)\right].\end{aligned} \quad (9.92)$$

The solution to the two-dimensional diffusion equation, thus, is broken down into $J+K-2$ one-dimensional problems, where J and K are, respectively, the number of grid points in the x- and y-directions. The x-sweeps are done sequentially, each one for fixed k and varying j. Then, the y-sweeps are done, each for fixed j and varying k. Thus, for a given x-sweep at a fixed $k=k1$, the unknowns are $u_{j,k1}^{m+1/2}$, $j=1,2,\cdots,J-1$. Equation (9.91), therefore, can be written

$$A_j u_{j-1,k1}^{m+1/2}+B_j u_{j,k1}^{m+1/2}+C_j u_{j+1,k1}^{m+1/2}=-E_j. \quad (9.93)$$

Exercise

Calculate the coefficients A_j, B_j, C_j, and E_j. Equation (6.106) provides a guide.

For each sweep, the set of equations can be solved by the technique described in Chapter 2, Section 2.6. Boundary conditions must be provided at both ends of each strip. Simple ones, such as a prescribed value for $u_{1,k1}$ and $u_{J,k1}$, or a zero derivative at the boundary, are easily incorporated into the scheme. An entirely analogous set of equations is solved during the y-sweeps for fixed $j=j1$ (Equation 9.92).

The time steps Δt_m^* are determined by the requirement that all regions of the grid be physically coupled, both on large and small scales. The diffusion coefficient in Equation (9.75) is $K_{\text{diff}}=c\lambda/(\kappa_R\rho)$ and the characteristic time to diffuse over a distance Δx is:

$$t_{\text{diff}}\approx\frac{(\Delta x)^2}{K_{\text{diff}}}. \quad (9.94)$$

For example, in a protostellar disk with a density of 10^{-10} g cm^{-3} and $\kappa_R \approx 1$, the diffusion time over an optically thick region of 100 AU is 10^{10} s. Assuming that $K_{\rm diff}$ is approximately constant over the grid, the minimum time step is taken to be (Black and Bodenheimer, 1975):

$$\Delta t^*_{\rm min} = \frac{(\Delta x)^2}{4K_{\rm diff}} \tag{9.95}$$

and

$$\Delta t^*_{\rm max} = \frac{(J-1)^2(\Delta x)^2}{4K_{\rm diff}}, \tag{9.96}$$

where J is the total number of grid points. The pseudo-time steps Δt^*_m are distributed logarithmically between these values (see Chapter 6, Section 6.6.2), with a total number of steps in the range 10 to 20, depending on the range of scales involved. In practice, the optimal length and number of pseudo-time steps must be determined experimentally.

Once the sequence of time steps has been gone through for both directions, the solution is checked against the original difference equation (9.89)

$$\mathrm{ERR} = \frac{|A|}{B} \tag{9.97}$$

$$\begin{aligned} A = {} & \frac{u^{n+1}_{j,k} - u^n_{j,k}}{\Delta t} - \frac{1}{(\Delta x)^2}\left[K_{x,j+1,k}\left(u^{n+1}_{j+1,k} - u^{n+1}_{j,k}\right)\right. \\ & \left. - K_{x,j,k}\left(u^{n+1}_{j,k} - u^{n+1}_{j-1,k}\right)\right] \\ & - \frac{1}{(\Delta y)^2}\left[K_{y,j,k+1}\left(u^{n+1}_{j,k+1} - u^{n+1}_{j,k}\right)\right. \\ & \left. - K_{y,j,k}\left(u^{n+1}_{j,k} - u^{n+1}_{j,k-1}\right)\right] \end{aligned} \tag{9.98}$$

and B is a normalizing factor essentially equal to the sum of the absolute values of the individual terms in A. The goal is to reduce the error to order 10^{-7}. If it is larger than that, the whole ADI is repeated, several times if necessary, until it does converge. If convergence does not occur, the number of pseudo-time steps can be increased and the actual time step Δt can be reduced.

We now consider the absorption and emission terms as expressed by Equation (9.76). This expression can be written in difference form

$$u^{n+1}_{j,k} - u^n_{j,k} = \Delta t\left(4\pi\kappa B^{n+1}_{j,k}\rho^n - c\kappa\rho^n u^{n+1}_{j,k}\right). \tag{9.99}$$

In this equation, it has been assumed that the emission coefficient j_ν can be replaced by $\kappa_\nu B_\nu$, so that the situation is close to LTE. Under these conditions, we may also assume that the frequency-independent quantities $\kappa_P = \kappa_E = \kappa$, where actually the Planck mean should be used. Equation (9.99) must be solved implicitly (Turner and Stone, 2001; Bowers and Wilson, 1991, Section 6.3) because the matter–radiation interaction time that is involved in this equation may be much shorter than the typical time step Δt, which is usually a hydrodynamic time step. Furthermore when the optical depth is large, the two terms on the right-hand side of Equation (9.99) are almost equal; a difference is being taken between large and almost equal quantities, which can lead to numerical errors.

The usual procedure is to solve this equation in conjunction with the material energy equation, written as the partial solution

$$\frac{de}{dt} = \rho c_v \frac{dT}{dt} = -4\pi\kappa\rho B + c\kappa\rho u, \tag{9.100}$$

since an imbalance in emission vs. absorption processes implies a transfer of energy from the radiation field to the matter or vice versa. This equation will be solved implicitly as well, with u and B evaluated at time $n+1$; here, T is the temperature of the material. Write

$$B(T)^{n+1} = B(T)^n + \frac{\partial B}{\partial T}\frac{\partial T}{\partial t}\Delta t. \tag{9.101}$$

Substitute Equation (9.101) into Equation (9.100)

$$\rho c_v \frac{dT}{dt} = -\kappa_{\text{eff}}\rho[4\pi B(T)^n - cu^{n+1}], \tag{9.102}$$

with

$$\kappa_{\text{eff}} = \frac{\kappa}{1 + \frac{4\pi\kappa}{c_V}\Delta t \frac{\partial B}{\partial T}}. \tag{9.103}$$

Then the radiation energy equation (9.99) can be written (using Equation 9.100 and Equation 9.102)

$$u^{n+1}_{j,k} - u^n_{j,k} = \Delta t\left(4\pi\kappa_{\text{eff}} B^n_{j,k}\rho^n - c\kappa_{\text{eff}}\rho^n u^{n+1}_{j,k}\right). \tag{9.104}$$

Once this has been solved for $u^{n+1}_{j,k}$, the material energy equation becomes

$$\rho^n c^n_V\left(T^{n+1}_{j,k} - T^n_{j,k}\right) = \Delta t\rho^n\kappa_{\text{eff}}\left(4\pi B(T)^n - cu^{n+1}_{j,k}\right), \tag{9.105}$$

from which one obtains $(T^{n+1}_{j,k})$. The quantities ρ, $\frac{\partial B}{\partial T}$, and c_V are evaluated at time step n; to simplify matters, the absorption coefficient can also be evaluated at time step n.

The $P_g\nabla\cdot\mathbf{v}$ term in Equation (9.63) can be evaluated in a separate semi-implicit step, as defined by Equation (6.71) and Equation (6.72). For simplicity, we also treat the $\nabla\mathbf{v} : \mathbf{P}$ term as a separate implicit step, in other words, by operator splitting, although it could be combined with the calculation of the absorption and emission terms (Turner and Stone, 2001)

$$u^{n+1}_{j,k} - u^n_{j,k} = \Delta t(\nabla\mathbf{v} : \mathbf{P}^{n+1}). \tag{9.106}$$

On the right-hand side, the **P**s are converted to us by use of the definition of the generalized Eddington factor

$$P^{ij} = f^{ij}u, \tag{9.107}$$

which is entirely analogous to Equation (1.108). The double-dot product on the right-hand side is then expanded

$$u^{n+1}_{j,k} - u^n_{j,k} = \Delta t\, u^{n+1}_{j,k}\left[\nabla\mathbf{v}_{11}f^{11} + \nabla\mathbf{v}_{22}f^{22} + (\nabla\mathbf{v}_{12} + \nabla\mathbf{v}_{21})f^{12}\right], \tag{9.108}$$

where

$$\nabla \mathbf{v}_{11} = (v_{x,j+1,k} - v_{x,j,k})/\Delta x$$
$$\nabla \mathbf{v}_{22} = (v_{y,j,k+1} - v_{y,j,k})/\Delta y$$
$$\nabla \mathbf{v}_{12} = (v_{y,j,k} - v_{y,j-1,k})/\Delta x$$
$$\nabla \mathbf{v}_{21} = (v_{x,j,k} - v_{x,j,k-1})/\Delta y, \tag{9.109}$$

where the velocities v_x and v_y are defined at the same locations as F_x and F_y, so that ∇v_{12} and ∇v_{21} should be zone-centered by averaging. The velocities and Eddington components are evaluated at time step n, so that Equation (9.108) is easily solved for the partial update to $u^{n+1}_{j,k}$, one zone at a time.

To get the Eddington components, it can be shown (Levermore, 1984) that the scalar Eddington factor f is related to the flux limiter λ by

$$f = \lambda + \lambda^2 R^2, \qquad R = |\nabla u|/(\kappa_R \rho u), \tag{9.110}$$

where R, again, is the ratio of the photon mean free path to the scale height of the radiation energy density. Then the Eddington components are (Levermore, 1984):

$$f^{ij} = 0.5(1 - f)\delta_{ij} + 0.5(3f - 1)s_i s_j, \tag{9.111}$$

where δ_{ij} is the Kronecker delta and the s_i and s_j are the direction cosines with respect to the direction of the gradient of the energy density ∇u.

We now summarize this simple example of the solution of the frequency-integrated equations of radiation hydrodynamics in an (x, y) coordinate system in the approximation of flux-limited diffusion. The variables to be determined include the density ρ, the two (in general, three) components of velocity, and the energy densities of matter and radiation, e and u, respectively.

The first step is the evaluation of the advection terms for all variables. How this is done for the density, material energy, and velocity is described (for the case of the Zeus code) in Chapter 6, Section 6.5.2. For the radiation energy density u, the procedure is entirely analogous to that for the material energy density. The mass fluxes are first calculated for each zone

$$\mathcal{F}^{\rho}_{x(j,k)} = v_{x(j,k)} \bar{\rho} A_{x(j,k)}$$
$$\mathcal{F}^{\rho}_{y(j,k)} = v_{y(j,k)} \bar{\rho} A_{y(j,k)}, \tag{9.112}$$

where A is the area of the cell face through which the flow is going and $\bar{\rho}$ is the interpolated value of density at that face. The methods for interpolation are described in Chapter 6, Section 6.3.2. Then each mass flux is converted to a radiation energy flux by multiplying by the interpolated *specific* radiation energy density $\bar{\epsilon}_{\rm rad} = \bar{u}/\bar{\rho}$

$$\mathcal{F}^r = \bar{\epsilon}_{\rm rad} \mathcal{F}^{\rho}, \tag{9.113}$$

which gives energy transferred per second. Then, two sweeps are made through the grid; one does advection in the x-direction and the other in the y-direction. The result is two partial updates to the internal energy density

$$
\begin{aligned}
{}^1u^{\mathbf{n+1}}_{j,k} &= u^{\mathbf{n}}_{j,k} - \frac{\Delta t}{\Delta V}\left(\mathcal{F}^r_{x(j+1,k)} - \mathcal{F}^r_{x(j,k)}\right) \\
u^{n+1}_{j,k} &= {}^1u^{\mathbf{n+1}}_{j,k} - \frac{\Delta t}{\Delta V}\left(\mathcal{F}^r_{y(j,k+1)} - \mathcal{F}^r_{y(j,k)}\right),
\end{aligned}
\tag{9.114}
$$

where ΔV is the volume of the cell. For reasons of numerical accuracy, it is worthwhile to reverse the order of these two steps every time step.

Once the advection has been performed for all variables, the source terms are evaluated in the second and third lines of Equation (9.63) and in Equation (9.64).

- For the momentum equation, the velocities in the x- and y-directions are updated according to Equation (6.70), which is an explicit calculation of the contribution of the pressure gradient and of gravity (if included). The velocity components v_x and v_y are substituted for the quantities u and w, respectively, in those equations.
- For the $P_g \nabla \cdot \mathbf{v}$ term in the material energy equation, the time-centered procedure given in Equation (6.71) and Equation (6.72) can be used. The same substitution for the velocity components applies.
- The update to the radiation energy density u is obtained by the solution of Equation (9.69) and Equation (9.70) through the ADI technique with pseudo-time steps, described in this section.
- The emission and absorption terms in the radiation energy equation and the material energy equation are taken into account implicitly through Equation (9.104) and Equation (9.105), respectively.
- The $\nabla \mathbf{v} : \mathbf{P}$ term in the radiation energy equation is evaluated implicitly by the use of Equation (9.108).
- The components of the flux $\mathbf{F}$ obtained from Equation (9.69) are used to update the velocity components in the momentum equation (second line of Equation 9.63) for the effects of radiative acceleration.

9.5 EXAMPLE: SPECTRUM OF A ROTATING, COLLAPSING OBJECT

As described in Section 9.2, it is possible in the spherically symmetric case to calculate simultaneously and self-consistently the radiation intensity, radiation temperature, source terms, and extinction coefficients at various wavelengths, and thereby derive the theoretical spectrum of a calculated density structure. This spectrum then can be compared with those of observed objects.

However, in many situations numerical hydrodynamic calculations in two or three space dimensions are required to simulate observed astrophysical objects and to derive numerically the density and temperature structure as a function of time. It is generally not feasible to include radiation transfer in such simulations with full frequency and directional dependence. Nevertheless, it is of particular interest to be able to compute the observable properties of such multidimensional structures, e.g., the overall spectrum or contour maps of equal radiation intensity at a particular wavelength. In this section, we show how this information can be obtained in a somewhat approximate manner.

Hydrodynamic collapse calculations for rotating protostars have been carried out by Yorke and Bodenheimer (1999). Starting with an assumed initial spherically symmetric density distribution and a temperature of 20 K, plus an assumed uniform rotation with a given total angular momentum, they solve the full set of $2\frac{1}{2}$-D hydrodynamic equations in cylindrical (R, Z) coordinates, assuming symmetry with respect to the equatorial plane and with respect to the rotation axis. Radiation transfer is taken into account with flux-limited diffusion, thus, a frequency-averaged opacity is used. An example of the structure of a particular model, 398,000 years into the collapse, is shown in Figure 6.14, which shows the density and temperature structure in a plane through the rotation axis. The structure has the form of a thick disk; there is an unresolved young star at the center of the grid and an infalling envelope surrounding the whole structure. A more detailed discussion of the properties of this model is given in Chapter 6, Section 6.7.3.

The spectrum of such a model can be obtained at a given moment in time through the use of the solution of the ray equation for the continuum at various frequencies, as described in Section 9.1. At each grid point in the two-dimensional structure both temperature and density are available from the hydrodynamic simulation. The simplest way to calculate the frequency-dependent properties is to assume that the source function is the Planck function $B_\nu(T)$. In the disk region of the protostar, which has temperatures less than 1000 K, the main source of opacity is dust. Rather than a frequency-averaged opacity, a frequency-dependent opacity is used, such as that calculated by Preibisch et al. (1993) for a mixture of silicate dust (evaporation temperature $T_{\rm evap} = 1500$ K), amorphous carbon ($T_{\rm evap} = 2000$ K), and ice ($T_{\rm evap} = 125$ K). These opacities depend on temperature, density, and frequency ν. In the simplest approximation, one simply solves Equation (9.1) along various rays through the structure under consideration with $S_\nu = B_\nu$ and $\epsilon_\nu = \kappa_\nu$, i.e., scattering is not considered. The other main approximation is that the temperature is determined from the hydrodynamics plus radiative transfer in the flux-limited diffusion approximation with a frequency-averaged opacity. The temperature is not recalculated to be consistent with the derived frequency-dependent radiation intensity.

The setup for the frequency-dependent calculation is shown in Figure 9.4. The (S, T, U) grid can be rotated to any desired angle θ between the U-axis and the Z-direction, which, in the example of Figure 6.14, corresponds to the vertical (rotation) axis. Thus, $\theta = 0°$ corresponds to an observer looking at the object pole-on, while $\theta = 90°$ corresponds to equator-on. For the example shown in Figure 9.5, 64 frequencies are chosen, logarithmically spaced between $\nu = 6 \times 10^{10}$ Hz (radio region) and $\nu = 3 \times 10^{15}$ Hz (UV region). A grid is set up in the (S, T) plane; it can be nonuniformly spaced to increase the resolution in the central regions. If the object is being viewed pole-on, it is symmetric and one set of, say, 121 grid points, along $T = 0$, is required. If the object is being viewed equator-on, the symmetry requires a grid in the first quadrant only, with 121 by 121 points in (S, T). At any other angle, a grid of 121 by 242 points would be needed.

Each (S, T) point thus corresponds to a line of sight through the object, with U, the coordinate along the line of sight, ranging from $-U_0$ to $+U_0$. Equation (9.1) is solved along each line of sight, starting at $(S, T, -U_0)$ (with s in that equation replaced by U). The increments dU are taken to be small enough so that the intensity varies only

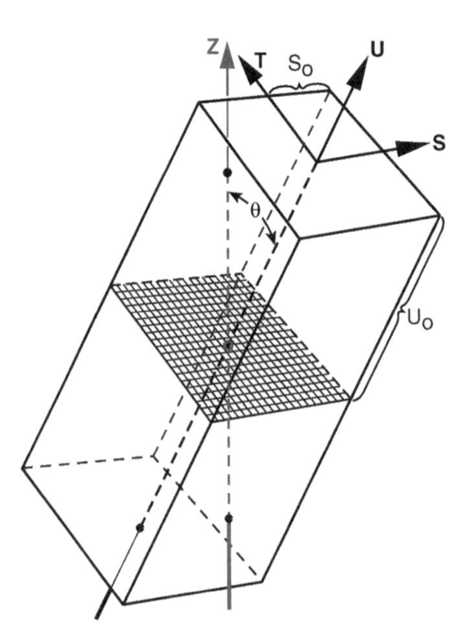

FIGURE 9.4 The (S, T, U) coordinate system, tilted by an angle θ with respect to the Z-axis of the object whose spectrum is being calculated. For the case of Figure 6.14, that axis corresponds to the vertical axis (the rotation axis), and the structure is independent of the azimuthal angle ϕ. The origin of the coordinate system is at the center of the object; the shaded plane corresponds to $U = 0$. The coordinate U runs from $-U_0$ to $+U_0$, where U_0 is chosen to be large enough so that it lies outside the region of significant emission. The observer is assumed to be located far away in the $+U$ direction.

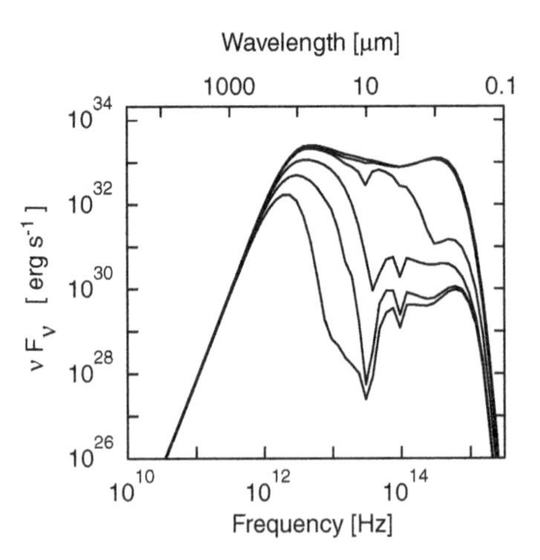

FIGURE 9.5 Theoretical spectra of the protostar of Figure 6.14 at various viewing angles, ranging from 90° (*lower curve*) to 0° (*upper curve*).

by a small amount from point to point. The solution for each increment is obtained according to Equation (9.5), Equation (9.6) or Equation (9.7), depending upon the optical depth of the increment. A boundary condition must be applied at $-U_0$, where the integration begins. If one assumes that the protostar is embedded in an ambient medium with temperature T_0, then one may use $I_\nu(S, T, -U_0) = B_\nu(T_0)$.

The result of each integration is the observable intensity I_ν at the point (S, T, U_0). To obtain the total intensity at a given frequency, one integrates over all the lines of sight that have been calculated in the (S, T) plane. Then to obtain the spectrum, one combines the information obtained at all the calculated frequencies.

Exercise

The code *trace.f* solves for the spectrum and intensity maps of an object whose density and temperature structure are given by the user. Instructions on how to use this code are given in Chapter 10, Section 10.1. The source function is assumed to be the Planck function. With the initial data supplied with the code, the intensity is calculated at 48 frequencies. The code is best applied for a cool, dusty envelope, for which opacities are supplied, around a stellar central source.

1. Take Option 1, a spherically symmetric envelope. Assume the central source is a star like the Sun. Allow the code to determine the temperature distribution in an envelope out to 1000 AU. Vary the density constant, which gives the density at the innermost zone of the envelope. Use values of 10^{-16}, 10^{-18}, and 10^{-20}. Plot the spectrum in each case. Can you explain the results? The density distribution, in the input data provided, is assumed to be proportional to $(\text{distance})^{-1}$. What happens, in the first of the above cases, if you change it to $(\text{distance})^{-1.5}$?
2. Take Option 2, an axisymmetric and flattened envelope. Again, let the code determine the temperature distribution. Take a density constant of 10^{-16}. Obtain the spectrum when viewing the object pole-on, at an angle of 45°, and equator-on. Can you explain the results? Plot contours of equal intensity when viewed equator-on at a wavelength of 10 μm. Now flatten the envelope even more than the 2:1 axis ratio given by the input data. What happens?

The situation becomes more complicated when scattering becomes important. Now the approximate source function becomes

$$S_\nu = \frac{\kappa_\nu B_\nu(T) + \sigma_\nu J_\nu}{\epsilon_\nu}, \tag{9.115}$$

so S_ν depends not only on the local temperature but on the radiation field itself. We use the zero-order moment equation (1.104)

$$\frac{1}{c}\frac{\partial J_\nu}{\partial t} + \nabla \cdot \mathbf{H}_\nu + \kappa_\nu \rho (J_\nu - B_\nu) = 0. \tag{9.116}$$

where J_ν and H_ν are defined by Equation (1.96) and Equation (1.97), respectively. Note that the energy density $u_\nu = (4\pi/c)J_\nu$ and the flux $F_\nu = 4\pi H_\nu$.

The moment equation can be solved if a relation exists between J_ν and H_ν. That relation can be chosen to be the frequency-dependent version of the flux-limited diffusion equation (1.120), which is to be used with all appropriate cautions

$$\mathbf{H}_\nu = -\frac{\lambda_\nu}{\epsilon_\nu \rho} \nabla J_\nu, \tag{9.117}$$

where the flux-limiter (Levermore and Pomraning, 1981) at a particular frequency λ_ν has the properties discussed in Chapter 1, Section 1.6.4

$$\lambda_\nu = \frac{1}{R}\left(\coth R - \frac{1}{R}\right). \tag{9.118}$$

Alternatively, the approximate expression given by Equation (1.122) can be used

$$\lambda_\nu = \frac{2 + R}{6 + 3R + R^2}; \tag{9.119}$$

the maximum error, relative to Equation (9.118) is 7%. The quantity R (Equation 1.121), which is used to determine λ_ν is now frequency-dependent; it is the ratio of the mean free path of a photon to the scale height of the energy density at ν

$$R = \frac{|\nabla u_\nu|}{\epsilon_\nu \rho u_\nu}. \tag{9.120}$$

The moment equation (9.116), plus the flux-limited diffusion equation (9.117), can be used to solve for a steady-state value of J_ν at all chosen frequencies. One method of solution is the ADI technique, using a set of pseudo-time steps to relax the first term of Equation (9.116) to zero. The procedure is similar to that described in Section 9.4, except the whole procedure has to be completed for each frequency independently. Once J_ν has been obtained at all points and at all frequencies, the source function in Equation (9.115) is known and the ray equations can be solved for each (S, T) point as above.

Examples of results, including scattering, are shown in Figure 9.5 and Figure 9.6. The emergent spectrum of the protostar is shown in Figure 9.5 as a function of the viewing angle θ. This protostar has generated a central stellar-like source with a luminosity of 7.4 $L_\odot$ and a surface temperature of about 5000 K. At 0° the spectrum is fairly flat from 100 μm to less than 1 μm. The spectrum at 90° shows a major peak near 100 μm, but also a secondary maximum in the optical. The optical maximum is caused by light emitted poleward by the central object, in which direction the optical depth is low, and then scattered by the disk material in the direction of the equator. The total energy emitted at 0° is about 30 times greater than that at 90°, indicating that it is much more likely that such objects will be detected pole-on. Intensity maps at wavelengths of 100 μm and 10 μm are shown in Figure 9.6. Intensities are strongly peaked at the position of the central source, but there are small secondary maxima off-center at 10μm, in the 75° view, caused by scattered light.

The main caution to be observed in this general approach is that the temperature distribution used to obtain the function S_ν for the frequency-dependent integrations is not necessarily the same as that which would have been obtained had the hydrodynamic calculations actually included the full frequency and directional dependence of the radiation intensity. The more optically thick the object is, the better the approximation.

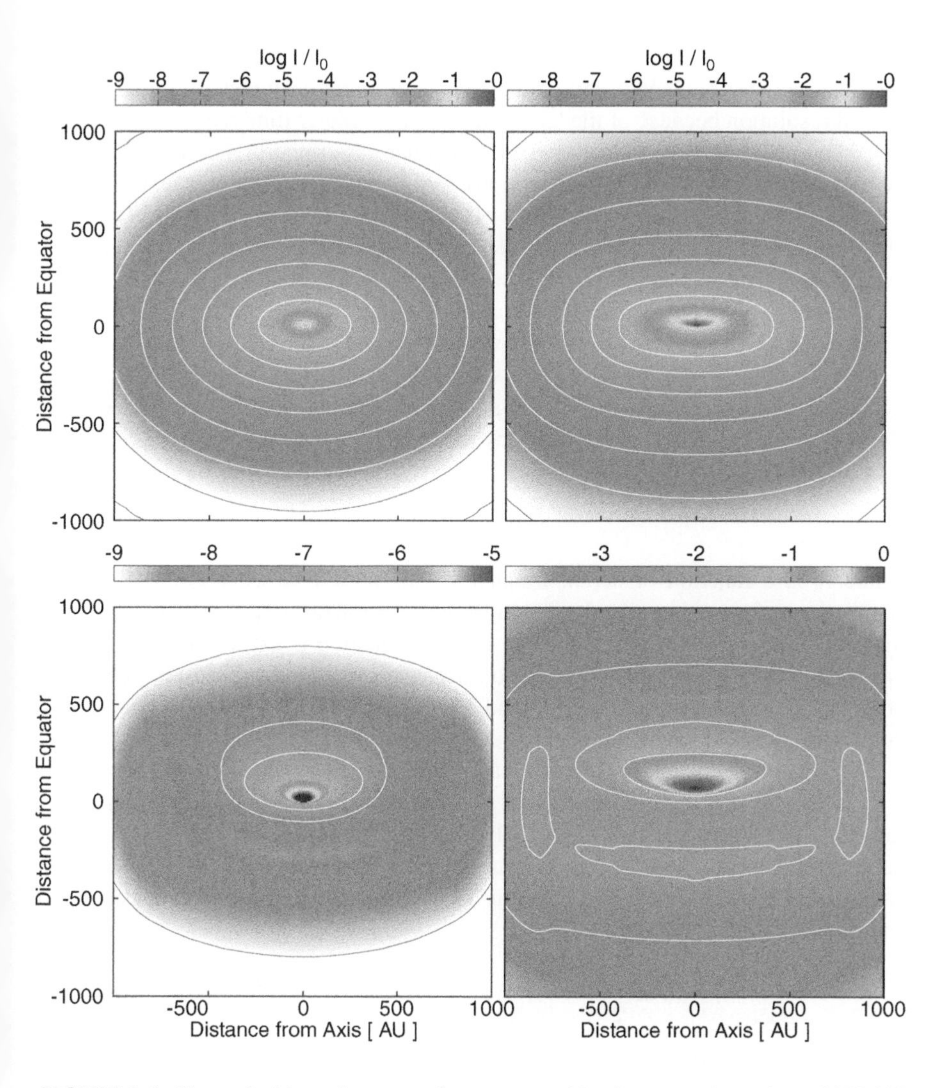

FIGURE 9.6 Theoretical intensity maps of a protostar of 1 solar mass after a time of 250,000 years after the beginning of collapse. *Upper panels:* wavelength of 100 μm at viewing angles of 60° (*left*), and 75° (*right*); *lower panels:* wavelength of 10 μm at viewing angles of 60° (*left*), and 75° (*right*). Angles are measured from the rotation axis. *Greyscale*: the logarithm of intensity of radiation normalized to the maximum value in a given panel.

9.6 EXAMPLE: 3-D CALCULATIONS OF THE SOLAR PHOTOSPHERE

An excellent test of the ability of a three-dimensional hydrodynamics code to reproduce observed physical phenomena is provided by the simulation of Stein and Nordlund (1998) of the outer layers of the Sun. In these layers, both radiation and convection contribute to the energy transport and the layers are marginally optically thick, so that the flux-limited diffusion approximation is not adequate. These authors chose to include frequency- and angle-dependent radiation transfer, although with limited resolution because of the large amount of computer time required. The grid-based code models only a small piece of the solar photosphere, measuring 6×10^8 by 6×10^8 cm in the horizontal directions and 3×10^8 cm in the vertical direction with a maximum number of grid points of 200 by 200 by 82, respectively. This box, however, is sufficiently large to include more than 10 solar granulation elements. Local thermodynamic equilibrium ($S_\nu = B_\nu$) is assumed.

The code solves the standard Eulerian equations (1.42) with the external force as a constant gravity and with an artificial viscosity added to ensure numerical stability, to mediate shock waves, and to damp short wavelength noise. The internal energy includes ionization energy. In the energy equation, the term $\nabla \cdot \mathbf{F}$ is included as in Equation (6.98). This term is replaced by the equivalent quantity

$$-\int \nabla \cdot \mathbf{F}_\nu d\nu = \int \int \rho \epsilon_\nu (I_\nu - S_\nu) d\Omega d\nu, \tag{9.121}$$

where I_ν is a function of depth, frequency, and angle (see Equation 1.100). The standard transfer equation is solved along rays at several different angles by a method first outlined by Feautrier (1964). The transfer equation is expressed as a second-order differential equation, converted to difference equations, and solved subject to boundary conditions at both the top and bottom of the atmosphere. The method is described in detail by Mihalas (1978, Section 6.3).

The Feautrier method has definite advantages in some situations, as compared with the ray-tracing method described in Section 9.2. If the optical depth is large, the quantities I_ν^+ and I_ν^- are very nearly equal, so the difference of the two, which determines the radiative flux, is inaccurate because of roundoff error, even if the calculations are done in double precision. The Feautrier technique involves averaging as follows: define

$$x_\nu = \frac{1}{2}\left(I_\nu^+ + I_\nu^-\right); \quad y_\nu = \frac{1}{2}\left(I_\nu^+ - I_\nu^-\right). \tag{9.122}$$

One then adds and subtracts the two equations (9.10) to obtain

$$\begin{aligned} \mu \frac{dy_\nu}{dr} &= -\epsilon_\nu \rho (x_\nu - S_\nu) \\ \mu \frac{dx_\nu}{dr} &= -\epsilon_\nu \rho y_\nu. \end{aligned} \tag{9.123}$$

Differentiating the second equation and using the inward-measured optical depth $d\tau_\nu = -\epsilon_\nu \rho dr$, one obtains the second-order equation

$$\mu^2 \frac{d^2 x_\nu}{d\tau^2} = x_\nu - S_\nu. \tag{9.124}$$

This equation is discretized in three variables — angle μ, depth τ, and frequency ν. For given μ, ν the resulting difference equations connect the values of x_ν at three neighboring points — τ_{j-1}, τ_j, and τ_{j+1}. The resulting system of equations is entirely analogous to Equations (2.77), except that there is no actual time dependence, and the solution for x_ν and y_ν by elimination and back-substitution is also very similar to that outlined in Chapter 2, Section 2.6. The radiative term (Equation 9.121) in the energy equation, is then obtained by integrating over angle and frequency.

The typical simulation is run for 1 hour of solar time, which takes weeks of processor time, even with course angular and frequency grids. This calculation does not, of course, resolve the full range of length scales associated with convection, a problem that applies to all numerical simulations of this type; current computers are nowhere near to being large enough and fast enough. However, it does resolve the larger scales near the photosphere of the Sun, and can be compared to observations made on that scale. The results (Figure 9.7) show excellent agreement with the observed granulation pattern in the solar photosphere; the bright regions correspond to rising warm convective elements and the darker regions correspond to cooler, sinking elements. The computed upflow/downflow velocities also agree well with observations. The three-dimensional hydro simulation gives a far better representation of solar

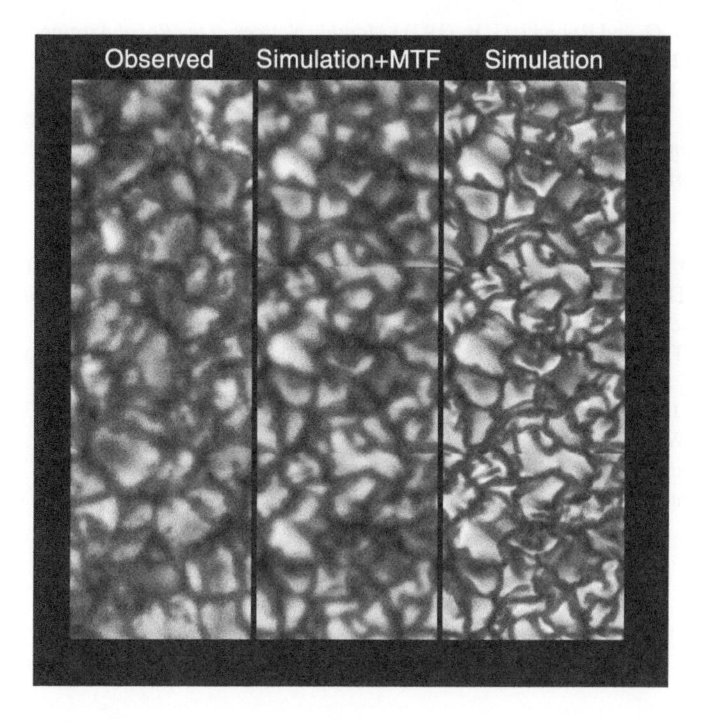

FIGURE 9.7 Comparison of the numerical model and the observations of the solar photosphere. Plotted is the emergent radiation intensity in white light as a function of position on the solar surface. *Left panel*: an image obtained with the Swedish Vacuum Solar Telescope on La Palma, Canary Islands. The area measures $18 \times 6 \times 10^8$ cm. *Right panel*: the calculated emergent intensity; three simulation images taken at 1 min intervals are placed side by side to produce the area of $18 \times 6 \times 10^8$ cm. *Center panel*: the calculated intensities smoothed to give a comparable spatial resolution to that of the observations. (From Stein, R. F. and Nordlund, A. (1998) *Astrophys. J.* **499**: 914. With permission.)

surface convection than does the one-dimensional, mixing-length theory commonly used in stellar structure calculations. The results of the three-dimensional simulations have also been used to derive spectral line profiles, also in excellent agreement with observations (Asplund et al., 2000).

REFERENCES

Asplund, M., Nordlund, A., Trampedach, R., Allende Prieto, C., and Stein, R. F. (2000) *Astron. Astrophys.*, **359**: 729.

Avrett, E. H. and Krook, M. (1963) *Astrophys. J.* **137**: 874.

Black, D. C. and Bodenheimer, P. (1975) *Astrophys. J.* **199**: 619.

Bowers, R. L. and Wilson, J. R. (1991) *Numerical Modeling in Applied Physics and Astrophysics* (Boston: Jones and Bartlett).

Feautrier, P. (1964) *C. R. Acad. Sci. Paris* **258**: 3189.

Gray, D. F. (1992) *The Observation and Analysis of Stellar Photospheres,* 2nd ed. (Cambridge: Cambridge University Press).

Hummer, D. G. and Rybicki, G. B. (1971) *Monthly Notices of the Royal Astronomical Society* **152**: 1.

Kurucz, R. L. (1969) *Astrophys. J.* **156**: 235.

Kurucz, R. L. (1970) *Atlas: A Computer Program for Calculating Model Stellar Atmospheres* (Cambridge, MA: Smithsonian Astrophysical Observatory Special Report #309).

Levermore, C. D. (1984) *J. Quant. Spectrosc. Radiat. Transfer* **31**: 149.

Levermore, C. D. and Pomraning, G. C. (1981) *Astrophys. J.* **248**: 321.

Mihalas, D. (1978) *Stellar Atmospheres,* 2nd ed. (San Francisco: W. H. Freeman).

Mihalas, D. and Mihalas, B. W. (1984) *Foundations of Radiation Hydrodynamics* (New York: Oxford University Press).

Preibisch, T., Ossenkopf, V., Yorke, H. W. and Henning, T. (1993) *Astron. Astrophys.* **279**: 577.

Stein, R. F. and Nordlund, A. (1998) *Astrophys. J.* **499**: 914.

Turner, N. J. and Stone, J. M. (2001) *Astrophys. J. Suppl.* **135**: 95.

Yorke, H. W. (1980) *Astron. Astrophys.* **86**: 286.

Yorke, H. and Bodenheimer, P. (1999) *Astrophys. J.* **525**: 330.

Yorke, H. and Shustov, B. (1981) *Astron. Astrophys.* **98**: 125.

10 Numerical Codes

On the accompanying CD-ROM, several codes are provided to illustrate the methods described in this book. This chapter provides a brief description of these codes, identifies the files containing the codes (as well as the input and output files), and gives some practical suggestions on how to run them. Users are encouraged to experiment with these codes and to modify them to suit their own requirements. Many of the codes include a large number of parameters, and the authors have not tested all possible combinations of these parameters. Thus the user may have to exercise his/her own ingenuity to get the programs to run in certain situations. Also, these programs are meant to be used as learning tools, and they do not necessarily represent the current state of the art in advanced research numerical computation.

10.1 RADIATION TRANSFER

The code *trace.f* in combination with the header file *tracedefs.h* calculates the observable spectrum and produces maps of specific intensity on the plane of the sky for an object (referred to as the "envelope") whose density and temperature structure are provided by the user. The object may be either spherically symmetric or axially symmetric; in the latter case, the densities and temperatures are provided as a function of cylindrical coordinates (R, Z). The program as provided gives a simple spherical model and a simple spheroidal model, but the user may modify the subroutine INIT to obtain different distributions of density and temperature. In normal operation, there is a central stellar source with a given luminosity and surface temperature, whose specific intensity as a function of frequency is assumed to be that of a black body. To simplify the calculation, the source function in the envelope is assumed to be the Planck function, i.e., scattering is not included.

Input: The user provides, in the file *trace.inp*,

- Names of the output files that will contain the intensity maps and the spectra that are calculated.

- Number of grid points in (R, Z) in the envelope model.
- Log ($L/L_{\odot}$), log T_{eff}, and log radius (in cm) of the central source.
- Number of grid points in the S direction in one quadrant of the radiative transfer grid (S, T), the ratio of the size of the first zone to the size of the envelope, and a factor (normally set to 1), which defines the ratio of the overall size of the radiative transfer grid to the envelope size. The radiative transfer grid is described in Section 9.5.
- Viewing angle of the observer's line of sight relative to the symmetry (Z) axis; zero degrees corresponds to a pole-on view.
- Number of frequency points and the lowest and highest frequency to be considered. The data to be provided are log (ν/c), where ν is the frequency and c is the velocity of light.
- Data for obtaining detailed information on the integration along a particular line of sight, on the standard output file.
- Maximum radius of the envelope, the maximum density (at the inner edge) of the envelope, and the ratio of equatorial axis to polar axis of the spheroid.
- Radius of the observer's beam within which the spectrum is calculated. Several beams can be used, with a radius less than the envelope size.
- Frequencies at which the intensity maps are to be produced (e.g., "4" means the fourth frequency on the list).
- Information regarding the background radiation outside the cloud — the radiation dilution factor and the temperature.
- Parameters for a model for a small disk inside the first zone of the envelope model. The quantities provided are the numerical resolution for the disk (zone size in cm), the outer radius of the disk, the power-law temperature exponent for the disk, and the power-law surface density exponent of the disk.

Output:

- The standard output file (unit 6) reprints the input data, gives the dust opacity data, prints the locations of the grid points in the (S, T) plane, and provides any line-of-sight information that the user has requested. The output observable "flux" (actually specific intensity times the element of area in the $[S, T]$ plane), integrated over that plane, in ergs per second per unit frequency interval, is given for each frequency.
- The output file for maps specified by the user gives emergent intensities over the (S, T) plane. For a spherically symmetric model or a model viewed pole-on, only the radial distribution of intensity is given. For an axisymmetric model viewed equator-on, a 2-D map of intensities is given in the first quadrant of the plane. For other angles, the number of S points is NS and the number of T points is $2 * NS - 1$. The intensities are given in the first and fourth quadrants of the plane. These data can be plotted by the user with a contour routine.
- The output file for spectra specified by the user (e.g., *flux.log*) gives the log of the flux as a function of the log of the frequency (in Hz). The units of

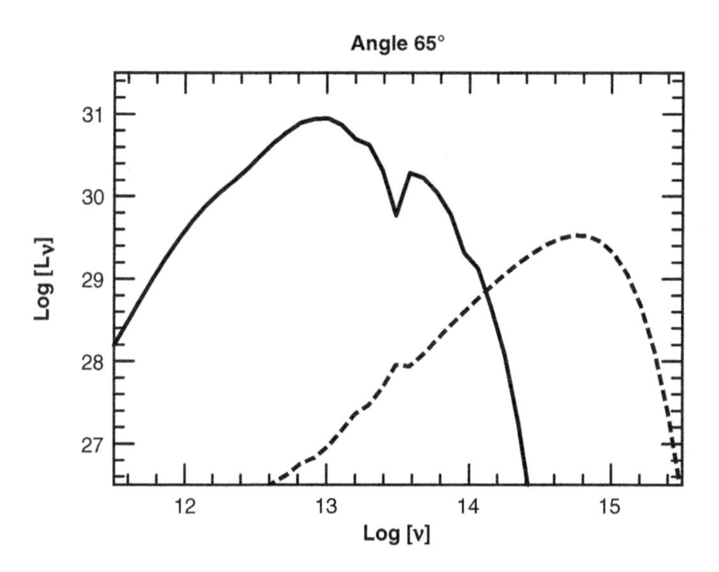

FIGURE 10.1 Example of results from *trace.f.* The spectrum of a central star with $T_{\mathrm{eff}} = 10^4$ K and log $L/L_{\odot} = 1.56$, surrounded by a spheroidal dusty envelope with axis ratio 2:1 and an equatorial size of 5×10^{16} cm. The observer's line of sight is inclined by 65° to the axis of symmetry. The vertical axis gives the energy radiated per unit time per unit frequency interval in arbitary units. *Solid line:* optically thick envelope calculated with an envelope maximum density of 10^{-16} g cm^{-3}, a density distribution that drops off as $1/R$, where R is the distance to the center, and an inner disk with radius 7×10^{14} cm. An absorption feature arising from silicate dust is noticeable. *Dashed line:* optically thin envelope calculated with the same density distribution but with a maximum density of 10^{-21} g cm^{-3} and an inner disk with radius 3×10^{14} cm.

the "flux" are ergs/s/Hz. These results are not overwritten, but are added to when new spectra are calculated, as long as the file name remains the same. Also, at the end of the standard output file, spectra are given in the same format, which take into account the beam sizes (one spectrum per beam size) that are specified in the input. Also calculated is the equivalent temperature of the spectrum, known as the bolometric temperature (Myers and Ladd, 1993)

$$T_{\mathrm{bol}} = 1.25 \times 10^{-11} \bar{\nu} \quad \text{where} \quad \bar{\nu} = \frac{\int \nu F_\nu d\nu}{\int F_\nu d\nu} \tag{10.1}$$

is the flux-weighted mean frequency. The constant is chosen so that a black-body radiator at T_{eff} has $T_{\mathrm{bol}} = T_{\mathrm{eff}}$. An example of a calculated spectrum is shown in Figure 10.1.

10.2 STELLAR EVOLUTION

The purpose of the one-dimensional Lagrangian stellar evolution code STELLAR is to calculate the evolution for a range of masses and compositions, starting from the point in pre-main-sequence evolution where the star first comes into hydrostatic equilibrium, continuing through the main sequence, and ending well into the post-main-sequence phase where helium-burning nuclear reactions have set in. The equations that are solved are discussed in Chapter 5, Sections 5.1, 5.2, and 5.3. The boundary conditions are described in Section 5.4. The four basic differential equations are expressed as the difference equations (5.39) to (5.42). The four basic variables that are solved for as a function of mass fraction and time are the pressure P, the radius r, the luminosity L, and the temperature T. The Henyey method is used to solve these equations on a grid of N points, as described in Section 5.5.

An evolutionary sequence is specified by the mass and chemical composition (X, Y, Z), where X is the fractional hydrogen abundance by mass, Y is the fractional helium abundance, and $Z = 1 - X - Y$. In STELLAR, the metals Z are subdivided into carbon, oxygen, nitrogen, and other elements. Given these parameters, the evolutionary sequence is started in the pre-main-sequence phase when the star is fully convective. A separate program *polytr.F* (with *parm.h*) is run to produce an equilibrium polytropic model with index $n = 1.5$ (other indices are also possible), which provides starting estimates for the basic dependent variables. The starting values of mass, composition, radius, and effective temperature are given in the input file *polytr1.inp* for the case of 1 $M_\odot$. The file README gives the commands to run the polytrope program and the program STELLAR itself. The program STELLAR is compiled by use of *makefile*, which is easily modified to change compilers; f77 and f90 are both allowed. The program segments needed are: *addsub.F, atmos.F, gi.F, henyey.F, invstate.F, nucrat.F, opacity.F, stellar.F, parm.h*, and *var.h*. To run the program, the equation of state tables *eostable* and *eospointer* are needed, as are the opacity tables *opac.cool, GN93hz, hyd.cond*, and *hel.cond*. To compile STELLAR, give the command "make stellar".

Once the polytrope output *polyout* has been produced, STELLAR is run with the input file *pmsstar1.start*, again for 1 $M_\odot$. This file gives an explanation of the various parameters that are needed to run STELLAR, e.g., "NMOD" is the number of time steps to be calculated in a given run. For the initial run, "NMOD" is typically specified as 100 (not more). For subsequent runs, the input file *pmsstar1.inp* is used and the input parameters may have to be modified as the evolutionary run proceeds. The main output file is specified on the command line; in the example in README, it is *pmsstar1.lst*. Every "NRIT" time steps, a detailed output is produced on this file, giving P, r, L, T (in centimeter-grams-seconds [cgs] units), as a function of the mass interior M, at every grid point. The density "RHO," the composition parameters, and the ratio of radiation pressure to total pressure $(1 - \beta)$ are also given. In the second column of this output, an asterisk indicates that the zone is convective. Also every "NRIT" time steps, a record is stored on the specified binary output file (example: *pmsstar1.mod*); these models are used to restart a run. At times between major outputs, a few summary lines of model data are given.

The following list gives a brief description of each of the elements of the program.

- The file *atmos.F* provides the surface boundary condition, as discussed in Section 5.4. The luminosity is assumed to be constant in the atmosphere, i.e., there is no energy generation, but the mass interior to a given point is calculated correctly with depth. Input to the atmosphere consists of the current estimates of the total luminosity and total radius. Output values at the bottom of the atmosphere include the pressure, temperature, density, and radius. These quantities serve as the boundary conditions for the interior calculation. The program simply integrates the equations of hydrostatic equilibrium and convective or radiative transfer inward, starting at optical depth $\tau = 0.001$. The radiation transfer is frequency-averaged by use of a Rosseland mean opacity (see *opacity.F*, below). In convection zones, a mixing-length theory is used to calculate the actual temperature gradient; the parameter is L/H, the ratio of the mixing length to the local pressure scale height. This quantity is specified by the user in the input files *.start* or *.inp*. The convective temperature gradient (Equation 5.37) is calculated in the subroutine "nabla," which is included in the file *gi.F* and which is also used in the interior calculation. The location of the bottom of the atmosphere is automatically adjusted in mass as the star evolves, so as to keep the temperature at that point within a given temperature range, "Atmx" to "Atmn," specified by the user in the input files. In normal operation, the actual temperature at the base of the atmosphere will be the average of these values. Another constraint on the bottom of the atmosphere is that the fractional radius contained within the atmosphere should be less than a value specified by the user, say, 8%. This constraint applies especially during extreme red giant phases, when otherwise the atmosphere could become very deep.
- The file *stellar.F* reads the input files, controls the calculation of the evolution, produces output data, and adjusts the time step "dtime" (units: seconds) so that the maximum fractional change in the whole set of dependent variables, per time step, falls within the range "chgmax" and "chgmin" (example: 0.10 and 0.03).
- The file *henyey.F* calculates the matrix elements in Equation (5.66) and solves the matrix for the corrections $\delta \mathbf{x}^{\mathbf{j}}$ to the dependent variables at all grid points. It adds the corrections to the variables, then recalculates the matrix elements and solves the matrix again, continuing iteratively until the maximum relative change for the last iteration among all the dependent variables falls below a given tolerance. The tolerances "epsP," "epsR," "epsL," and "epsT" are provided by the user in the input file and typically are $\approx 10^{-4}$. If no convergence is achieved after "itmx" iterations, the time step is repeated with a smaller value of "dtime" until it converges or until "dtime" falls below "dtmn," the specified minimum value, at which point the program stops.

- The file *gi.F* provides the subroutines of *henyey.F* that actually calculate the functions G_i^j and their derivatives with respect to the dependent variables, as described in Section 5.5.
- The file *addsub.F* adds or deletes grid points. The criteria for adding or deleting points are read from the input files. Basically, the variables L, P, M_r, and hydrogen abundance X are examined at all grid points. If, for example, between two neighboring grid points the relative pressure change is more than 4%, then a new grid point is inserted between the two. Or, if luminosity differences or composition differences are too large, a point is inserted. On the other hand, if the relative pressure change is less than 0.5% and the relative luminosity difference and the relative composition difference are also less than their minimum values, then the grid point is deleted. For example, in the input files the quantities "dPmx" and "dPmn" are, respectively, the relative differences required for addition or deletion of a point. A separate criterion is applied for the mass of a zone dM_j. If dM_j/M_j is greater than an input quantity "dzmax," then a point is added. On the other hand, if dM_j/M_j would, after insertion, be less than the quantity "dzmin," then the zone is not added.

  Other subroutines determine whether and how much mass is to be added or subtracted from the star during a given time step, and do the recalculation of the values of the zone masses. The mass flux (positive for addition of mass, negative for subtraction) is specified by the user at the end of the input data files *.start* and *.inp*, in *cgs* units.
- The file *invstate.F* provides the density, given the temperature, pressure, and composition, as well as other required thermodynamic quantities. The physical effects included (see Chapter 5, Section 5.6.1) are ideal gas, radiation pressure, electron degeneracy, and nonideal effects as tabulated by Saumon et al. (1995). All quantities are found by interpolation in tables (file *eostable*). Five tables are provided, with hydrogen mass fractions ranging from 0 to 0.8 and with metal abundance fixed at .02. In each table, the temperature ranges from 125 to 10^9 K. For each temperature, a range of pressures is provided, and for each pressure the table includes entries for density ρ, thermodynamic quantities α and δ (Equation 5.82 and Equation 5.10), and the specific heat c_V. From these quantities, one can obtain c_P and ∇_{ad} from Equation (5.81) and Equation (5.79). The range of temperatures and pressures is sufficient for calculations of evolutionary tracks of stars, from 0.1 to 20 $M_\odot$. For cores of evolved stars, additional tables are provided for compositions of zero hydrogen and heavy element mass fractions of 0.2, 0.4, 0.6, 0.8, and 1.0.
- The file *nucrat.F* contains two parts, the first calculates nuclear reaction rates and energy generation rates, and the second calculates the change in composition arising from nuclear reactions, including the effects of convective mixing, if applicable (Equation 5.21 and Equation 5.24). The nuclear reactions for hydrogen burning include separate calculations for $^1H + {}^1H$, $^2D + {}^1H$, and $^3He + {}^3He$. The astrophysical cross section factors for these reactions are taken from Bahcall (1989, Table 3.2). Thus, the PPI chain (see

Chapter 5, Section 5.6.3) is not assumed to be in equilibrium, which allows the program to calculate the main-sequence evolution of very-low-mass stars. An approximate correction is made for the PPII and PPIII chains. For the CNO cycle, the abundances of ^{12}C, ^{14}N, and ^{16}O are followed at all layers of the star. A simple reaction network is used to calculate their abundance changes with time as well as the energy generation. The cross section factors are taken from Bahcall (1989, Table 3.4). Thus, the CNO cycle for hydrogen burning is not assumed to be in equilibrium. Helium burning by the triple alpha process and the reaction $^{12}C + ^4He$ are included above a temperature of 5×10^7 K; the energy generation rates are given by Equation (5.124) and Equation (5.127), respectively. Also, above that temperature neutrino loss rates (Kippenhahn and Weigert, 1990, Section 18.6) by the plasma process, the photo process, and the pair production process, are included. The basic rates of energy loss per gram per second are taken from Beaudet et al. (1967), and the corrections of Munakata et al. (1985) are applied.

- The file *opacity.F* provides Rosseland mean opacities for the stellar interiors radiative diffusion equation. The opacities are provided in tables, as a function of temperature, density, and composition. The basic set of tables (file GN93hz) is a selection of OPAL tables (Iglesias and Rogers, 1996). These provide opacities for 70 temperature values in the range $3.75 \leq \log_{10} T \leq 8.70$ and 19 R values in the range $-8.0 \leq \log_{10} R \leq +1.0$, where $R = \rho/T_6^3$ and $T_6 = 1. \times 10^{-6} T$ [degrees]. The opacity file in STELLAR includes seven tables for solar composition $Z = 0.02$, with X values of 0.0, 0.1, 0.2, 0.35, 0.5, 0.7, and 0.8. These are sufficient for stellar evolution up to the beginning of helium burning. An additional seven tables are included for zero hydrogen and increasing abundances of carbon and oxygen, which should be adequate for evolution up to the beginning of carbon and oxygen burning (which are not included in the nuclear rates).

  The lower temperature cutoff at log T = 3.75 has long been a problem with the OPAL opacities. A new set of tables is now available (Ferguson et al., 2005) for temperatures in the range $2.7 \leq \log_{10} T \leq 4.50$ for the same range in R as above. In the overlapping temperature region, these opacities, which include the effects of molecules and grains, agree well with those of OPAL. Tables are available for an X range of 0 to 0.9 and a Z range of 0 to 0.08; the abundances of elements within Z are scaled solar values. (The present version of STELLAR contains just one table at low temperatures, the solar composition table of Alexander and Ferguson [1994].)

  In the temperature-density region where electron degeneracy is important, electron conduction must be taken into account in the energy transport. The corresponding opacity has been tabulated by Hubbard and Lampe (1967). STELLAR interpolates between two tables, one with a pure hydrogen composition, and one with a pure helium composition. The total opacity is then calculated from Equation (5.19).

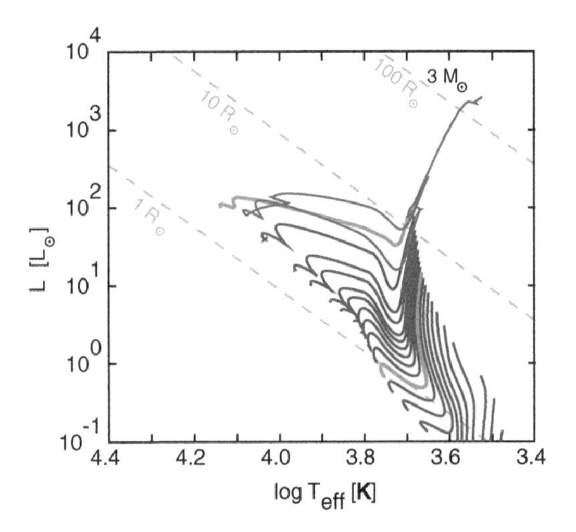

FIGURE 10.2 ***A color version of this figure follows page 212*** Evolutionary tracks in the (log T_{eff}, log $L/L_\odot$) diagram. The blue and red curves are pre-main-sequence tracks computed by D'Antona and Mazzitelli (1994). The upper and lower red curves are computed for 3 and 1 $M_\odot$, respectively. The orange curve for 3 $M_\odot$ was produced by the code STELLAR, and includes both pre-main-sequence and post-main-sequence phases.

- The file *lstmod.F* is a separate utility program that reads the stored binary models (example: *pmsstar1.mod*) and converts them into formatted files. Before running this program, copy the *.mod* file onto the file *fort.9*.

An example of the output of the code, an evolutionary track in the Hertzsprung–Russell diagram for a star of 3 $M_\odot$, as compared with pre-main-sequence tracks calculated by D'Antona and Mazzitelli (1994), is shown in Figure 10.2.

10.3 ONE-DIMENSIONAL LAGRANGIAN HYDRO

The code *lh1.f* (together with the file *commons.f*) simulates the explosion of a supernova in the interstellar medium. It solves the one-dimensional adiabatic hydrodynamic equations according to the difference scheme described in Chapter 6, Section 6.2, on a Lagrangian grid. The code is specifically designed for the study of shocks and it includes artificial viscosity. The Exercise near the end of Section 6.2 provides a guide on how to use the code and analyze the results. Several sets of input data are provided in the file *lh1.dat*; they all refer to the same problem, but will run for different amounts of time to illustrate the development of the various features of the flow. Output is provided to the screen as well as to two files, *lh1.out* and *lh2.out*, which provide, respectively, the physical parameters of the final model and the thermal and kinetic energies as a function of time. Plot routines (*pltmod* and *plte*, respectively) are provided for each of these files. The initial condition deposits a large amount of thermal energy into the grid and, at later times, some of this thermal energy is

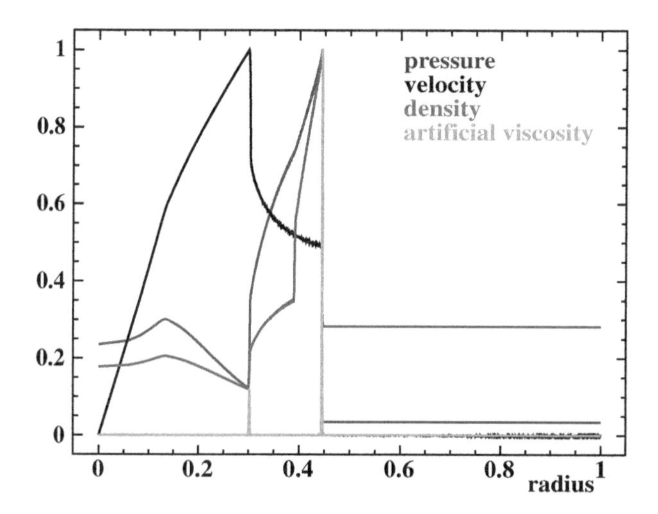

FIGURE 10.3 *A color version of this figure follows page 212* Illustration of the results from the code "lh1.f" after 16,000 time steps.

converted into the kinetic energy of expansion. An example of the results is shown in Figure 10.3. At this stage of the expansion, there is a strong outward-moving shock at a radius of 0.45 (normalized) units and an inward moving weaker reverse shock at a radius of 0.3 units. The shock locations are pinpointed by the high values of the artificial viscosity there. The actual units in the program are set up so that the unit of length is the outer radius of the grid, the unit of density is that of the ambient medium outside the hot bubble (e.g., a typical density of the interstellar medium is 10^{-24} g cm^{-3}) and the unit of internal energy per unit mass is that of the ambient medium (e.g., $1.5 \times 8.31 \times 10^7 \times 100$ K ergs per gram for neutral hydrogen).

10.4 ZEUS: 3-D HYDRODYNAMICS

The version of ZEUS3D that is provided on disk does hydrodynamics in one, two, or three space dimensions with the option of using the following coordinate systems: (x, y, z), (z, r, ϕ), or (r, θ, ϕ). Gravity is not included. Magnetic effects are included, and the magnetic field is calculated according to the constrained transport method. The magnetic effects have not been extensively tested by the authors.

When the file *Zdistr.tar* is unpacked (*tar xvf Zdistr.tar*) it creates a directory *Zdistr*. The file README.txt in that directory contains the basic instructions on how to use this version of ZEUS. The subdirectory /ZEUS3D/doc contains the ZEUS3D user manual itself.* The source code is found in /ZEUS3D/src/zeus34, and specific problems already set up and ready to run are found in /ZEUS3D/examples. The subdirectory /ZEUS3D/bin contains the script which is executed in order to build the executable version of ZEUS itself.

* ZEUS3D User Manual: © The Board of Trustees, University of Illinois (1994).

To run the specific examples that are provided, no changes to the program or input data are needed. Just follow the directions in the file /Zdistr/README.txt. It is of particular importance to make sure the environment variables are correctly set every time you use this package, as described in Section 3.2 of the README file. To change the input parameters or the code or to set up a new problem, one should consult the ZEUS3D and EDITOR manuals, which appear in /ZEUS3D/doc. The user will need to be resourceful and/or persistent in order to overcome the problems that ZEUS may present. The code provided is set up for the Linux operating system.

10.5 N-BODY CODES

Restricted 3-body: The code *toomre.f* calculates galaxy–galaxy encounters using the restricted 3-body method described in Toomre and Toomre (1972). For reasons of economy, they simply assume that each galaxy consists basically of a central point mass, and the disks are represented by a set of massless test particles each of which moves in accordance with the gravitational forces associated with the two point masses. Thus, the self-gravity of the disk is not included. The orbit integrations are performed with the Bulirsch–Stoer method with relatively low accuracy, in three space dimensions. The six input orbital elements of the two-galaxy orbit, as well as the masses of the galaxies, are provided near the top of the main program; these can be varied by the user to obtain different types of encounters. There is no additional input file. Output appears in the files *gal1.pos* and *gal2.pos*, which give the (x, y, z) positions of the two central point masses as a function of time. At certain specific times, which can be specified by the user, a snapshot file (*snap001*, etc.) is produced that gives the (x, y, z) positions of all particles. An example of the output is Figure 10.4, which is obtained by running the code as provided. The file *macro.total* is used with the plot routine "mongo" to obtain this plot. As indicated at the head of the program, a few routines from *Numerical Recipes* will have to be provided by the user.

Bulirsch–Stoer few-body integrator: The *integrator.f* code, along with the input data file *input*, integrates the Year 2000 solar system forwards in time using the Bulirsch–Stoer method. The code as provided is set up to run nine bodies plus the central mass. To run a different number of bodies, one should change the parameter "nb" in the main routine, and also in the subroutines "derivs" and "Energysum." The code as provided opens up data files for nine possible planets. For other "nb," delete or insert more "open" statements. It is also necessary to add or delete lines with the additional orbital elements for the additional planets in the *input* file. These extra lines would go in exact analogy to the lines for nine planets for which the code is initially configured.

In the input data file, the user specifies the time (in years) at which the simulation should end, the number of time steps between one-line writes to each of the output files, the mass of the central star (in solar masses), and the orbital elements at the start of the integration for all objects (nine, in this case). The orbital elements are (1) the period or semimajor axis, (2) the mean anomaly (M, in degrees) or the time of periastron passage (in Julian days) or the mean longitude ($\mathrm{M} + \varpi$, in degrees), (3) the starting time in Julian days, (4) the eccentricity, (5) the longitude of periastron (ϖ, in degrees), (6) the inclination i, and (7) the longitude of the ascending node (Ω, in degrees). The planet masses are then specified in units of 10^{27} g and an accuracy

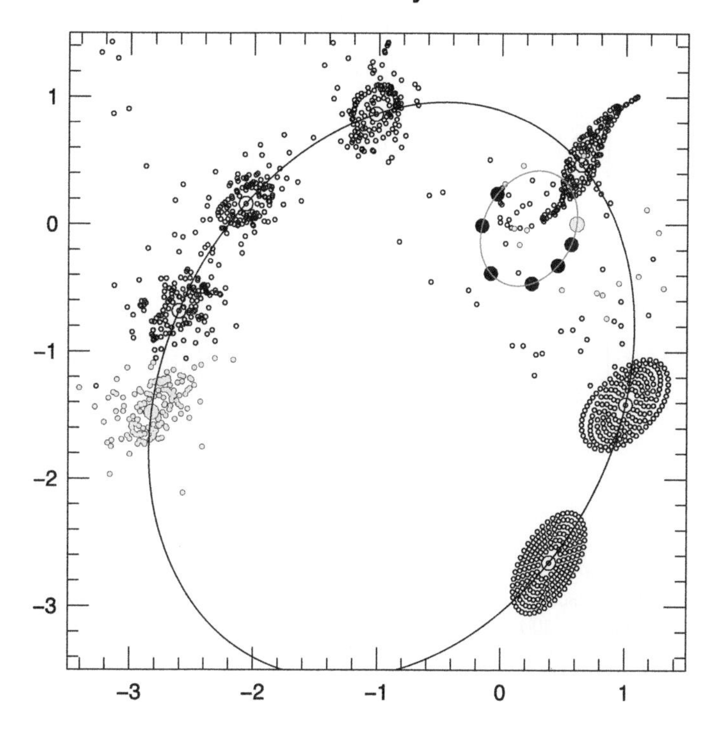

FIGURE 10.4 Galaxy–galaxy encounter as calculated with the code "toomre.f." The two galaxies start off with a mass ratio of 5 to 1. The calculation is done in a three-dimensional Cartesian coordinate system, and positions in the (x, y) plane are plotted. The more massive galaxy moves along the inner ellipse; the less massive, along the outer path. *Large filled dots:* seven successive positions of the more massive galaxy during the encounter; *large open circles:* the corresponding seven successive positions of the lower-mass galaxy; *small open circles:* the positions of the disk particles; *grey symbols*: final positions of the two systems. The high-mass galaxy does not have a disk, but the program does allow one to be added. The initial plane of the disk is inclined to that of the orbit by $\pi/3$ radians and the orbital planes of the two galaxies are inclined by $\pi/6$ radians. Note the "tidal tail" of disk particles that occurs as a result of the close encounter.

criterion is provided, which should generally be in the range 10^{-13} to 10^{-14}. The integrations can be carried out in either astrocentric or Jacobian coordinates. The time step length is specified as a fraction of the orbital period of the innermost planet. The value of this parameter should be around 0.1.

The output files, one for each planet, are labeled *planet.n* where "n" runs from one to the total number of planets. The seven columns of this output contain, respectively, the time in years, the semimajor axis in AU, the eccentricity, the inclination, the longitude of periastron (degrees), the longitude of the ascending node (degrees), and the mean anomaly (degrees). An example of output from the program is shown in Figure 10.5, where the mean anomaly as a function of time for a 1000-year integration

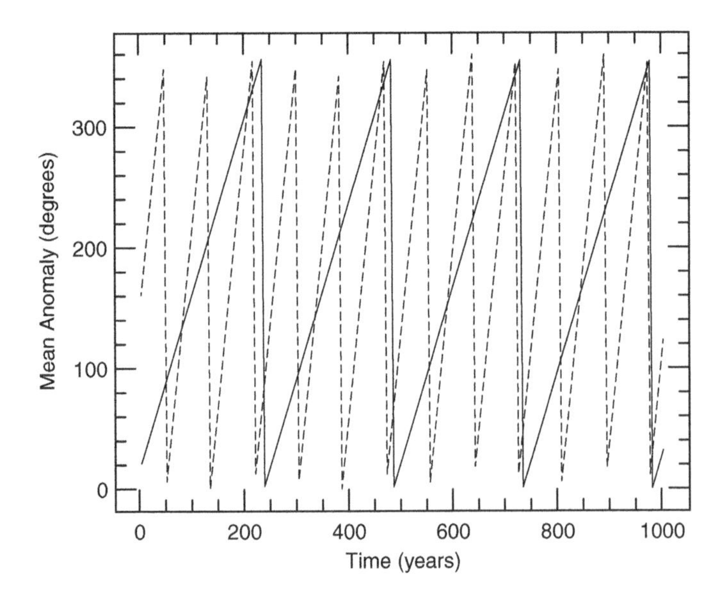

FIGURE 10.5 Example of the integration of the orbits of the eight planets of the solar system plus Pluto with a Bulirsch–Stoer code over a period of 1000 years, starting from the state of the planets on January 1, 2000. The mean anomaly as a function of time is plotted for Uranus (*dashed line*) and for Pluto (*solid line*). Note that the mean anomaly increases linearly with time from 0° to 360° in one period, then jumps back to 0°.

is shown for Uranus and Pluto. This program requires the same *Numerical Recipes* routines as does *toomre.f*.

Symplectic integration: The code *symplec.f* with input *input.symplec* uses the Wisdom–Holman (1991) algorithm to perform symplectic integration. As described in Section 3.6, this program is useful for calculating the orbits of N planets around a star, where N is of the order of 10 or less, for very long time periods, up to billions of years. The code could be used for studying the long-term stability of the solar system. This version of the code does not provide an accurate description of close encounters.

The configuration provided integrates the orbits of two planets around a single star. The input required includes the total integration time required (in years), the size of the time step (in days), the frequency for writing output (40,000 means that one line is written to each output file every 40,000 time steps), and the mass of the central star in solar masses. The initial orbital elements of the two planets are then specified, including the periods, either the mean anomaly or the time of periastron passage, the starting time, the eccentricities, the argument of perihelion, the positions of the ascending nodes, and the planet masses (in Jupiter masses) or their velocity amplitudes. The times are given in Julian days, but the basic time unit of the program is years. Additional planets can be added by adding the appropriate lines in *input.symplec* and modifying the program, in the section which reads the input data, to accommodate the additional input. The parameter "nb" must also be changed. The data in the

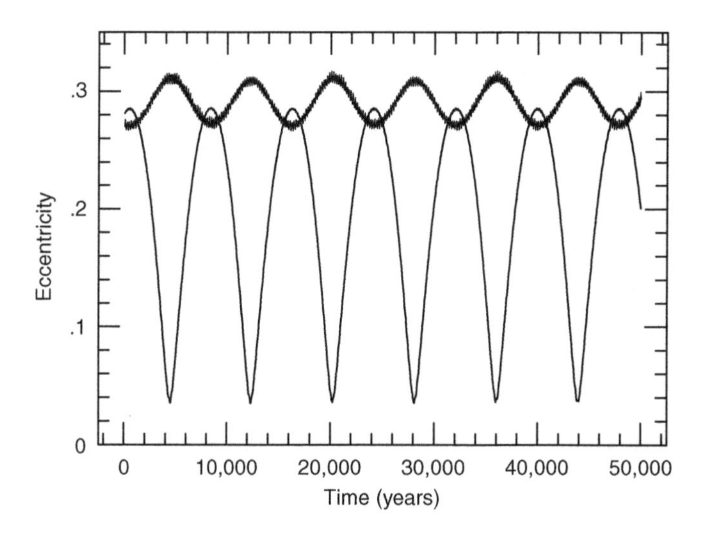

FIGURE 10.6 The eccentricities of the orbits of the two outer planets around the star Upsilon Andromedae, assuming that the planets have the minimum masses consistent with the radial velocities measured for the star, and that the starting values of the eccentricities are the present observed values, 0.28 and 0.27, for the inner and outer planet, respectively. Note the short-period oscillations in the eccentricity of the outer planet (*upper curve*).

existing input file correspond to the outer two planets in orbit about the star Upsilon Andromedae, the first extrasolar multiple planetary system discovered.

An output file is opened for each planet (*planet.2* and *planet.3* in this case). Each line for *planet.2* provides the time, the periastron (in AU) of the current orbit, the apastron (in AU), the eccentricity, the inclination, the argument of periastron, the longitude of the ascending node, the mean anomaly, the minimum distance of the planet and star (in AU), the minimum distance between the two planets (in AU), and an accuracy check. Similar quantities are provided for *planet.3*. The output for *planet.2* is also written to the screen. The accuracy parameter included in the input data is not used. The typical time step to be used is of order 1/100 the shortest period of the system. One can check the accuracy by repeating the calculation at a shorter time step. Figure 10.6 shows the results for the eccentricities of the two planets as a function of time for the first 50,000 years. This code requires no routines from *Numerical Recipes.*

***N*-body integration with large *N*:** The code *nbodyfft.f* combined with *fftdefs.h* does large-N integrations using the FFT-scheme in three space dimensions, as described in Section 7.1.2, to compute the gravitational potentials. The code as provided integrates the positions and velocities of 10^5 particles by means of a fourth-order Runge–Kutta procedure. The gravity is softened by a parameter ϵ. The code, thus, is an example of a "particle–mesh" method. It is currently configured to run the Hohl (1971) bar instability problem. This problem shows how an initially axially symmetric disk of stars deforms, after a few rotation periods, into a two-armed spiral structure, which is later dominated by a central bar-like structure surrounded by an axisymmetric disk. The program can also be run as a general-purpose N-body code. There is no input file; the setup of parameters and initial conditions is done in the main program. The output is a series of snapshots at specified time intervals (*snap000*, *snap001*, etc.) that provide the positions in (x, y, z) of all the particles. Several routines from *Numerical Recipes*, as indicated at the head of the program, are to be provided by the user. The setup provided is a two-dimensional problem in (x, y) coordinates. To convert it to run in three space dimensions, simply change the parameters "n3" and "ngridz" to, for example, 256 (or any power of 2), comment the line "dz = 0" in the subroutine "set_gravity3d," and modify the initial model, which is set up near the beginning of the program.

10.6 SMOOTHED PARTICLE HYDRODYNAMICS

TREESPH is a smoothed particle hydrodynamics code written by Lars Hernquist and Neil Katz (Hernquist and Katz, 1989). It involves a combined user-specified system of gas and/or collisionless particles. The code employs a variable smoothing length and adopts a tree algorithm for both the nearest neighbor search and the computation of the gravitational potential. Extensive comments regarding the use of the code are included at the start of the source list. Users of the code are urged to read these thoroughly.

The main program and all of the subroutines are contained in a single file, *sph.f*. The common blocks, array declarations, and numerical parameters for program flow and operation are contained in the header file *treedefs.h*. The file TREEPAR contains all of the physical parameters (such as the required evolution time, thermodynamic parameters, etc.). These are generally self-explanatory and detailed descriptions are contained in the comment lines of the source routine itself. Additionally, the user provides a model input file, named TREEBI. The first line gives the total number of bodies (particles), the number of collisional particles, and the number of star particles (zero in the example given). The second line gives the number of dimensions (generally 3). The third line gives the starting time, generally zero. Then one specifies the masses, positions, and velocities, and, optionally, other variables that define the specific initial condition of the system, such as the densities. An example, corresponding to two self-gravitating gas spheres prior to a collision, is included here to show the required format. On operation, TREESPH produces sequentially numbered snapshot files starting with *SNAP001*, each of which shows the state of the system at a particular time, with the same structure as the TREEBI file. After the masses, positions, and velocities, additional physical variables are listed. Additional output includes (1) a status file, *TREEOUT*, which shows the number of steps completed, (2) a dump file *SYSDUMP*, which gives the state of the system at the end of the run and contains more information than the *SNAP* files, and (3) a logfile *TREELOG*, which details the global variables at each output interval.

The code complies with f77-O sph.f -o sph, or the corresponding command in f90. There may be machine-specific changes that have to be made, especially for the timing routine. For example, in f90 the program segment "FUNCTION second" will have to be activated. Users may wish to reformat the *TREELOG* file to make it easier to read. An example of the output produced by running the code with the example of input that is provided is shown in Chapter 4, Figure 4.1.

REFERENCES

Alexander, D. and Ferguson, J. (1994) *Astrophys. J.* **437**: 879.
Bahcall, J. N. (1989) *Neutrino Astrophysics* (Cambridge: Cambridge University Press).
Beaudet, G., Petrosian, V., and Salpeter, E. (1967) *Astrophys. J.* **150**: 979.
D'Antona, F. and Mazzitelli, I. (1994) *Astrophys. J. Suppl.* **90**: 467.
Ferguson, J. W., Alexander, D. R., Allard, F., Barman, T., Bodnarik, J. G., Hauschildt, P. H., Heffner-Wong, A., and Tamanai, A. (2005) *Astrophys. J.* **623**: 585.
Hernquist, L. and Katz, N. (1989) *Astrophys. J. Suppl.* **70**: 419.
Hohl, F. (1971) *Astrophys. J.* **168**: 343.
Hubbard, W. and Lampe, M. (1969) *Astrophys. J. Suppl.* **18**: 297.
Iglesias, C. and Rogers, F. (1996) *Astrophys. J.* **464**: 943.
Kippenhahn, R. and Weigert, A. (1990) *Stellar Structure and Evolution* (Berlin: Springer).
Munakata, H., Kohyama, Y. and Itoh, N. (1985) *Astrophys. J.* **296**: 197.
Myers, P. C. and Ladd, E. F. (1993) *Astrophys. J.* **413**: L47.
Saumon, D., Chabrier, G., and van Horn, H. (1995) *Astrophys. J. Suppl.* **99**: 713.
Toomre, A. and Toomre, J. (1972) *Astrophys. J.* **178**: 623.
Wisdom, J. and Holman, M. (1991) *Astron. J.* **102**: 1528.

Index

F

G

H

LIMITED WARRANTY

Taylor & Francis Group warrants the physical disk(s) enclosed herein to be free of defects in materials and workmanship for a period of thirty days from the date of purchase. If within the warranty period Taylor & Francis Group receives written notification of defects in materials or workmanship, and such notification is determined by Taylor & Francis Group to be correct, Taylor & Francis Group will replace the defective disk(s).

The entire and exclusive liability and remedy for breach of this Limited Warranty shall be limited to replacement of defective disk(s) and shall not include or extend to any claim for or right to cover any other damages, including but not limited to, loss of profit, data, or use of the software, or special, incidental, or consequential damages or other similar claims, even if Taylor & Francis Group has been specifically advised of the possibility of such damages. In no event will the liability of Taylor & Francis Group for any damages to you or any other person ever exceed the lower suggested list price or actual price paid for the software, regardless of any form of the claim.

Taylor & Francis Group specifically disclaims all other warranties, express or implied, including but not limited to, any implied warranty of merchantability or fitness for a particular purpose. Specifically, Taylor & Francis Group makes no representation or warranty that the software is fit for any particular purpose and any implied warranty of merchantability is limited to the thirty-day duration of the Limited Warranty covering the physical disk(s) only (and not the software) and is otherwise expressly and specifically disclaimed.

Since some states do not allow the exclusion of incidental or consequential damages, or the limitation on how long an implied warranty lasts, some of the above may not apply to you.

DISCLAIMER OF WARRANTY AND LIMITS OF LIABILITY

The author(s) of this book have used their best efforts in preparing this material. These efforts include the development, research, and testing of the theories and programs to determine their effectiveness. Neither the author(s) nor the publisher make warranties of any kind, express or implied, with regard to these programs or the documentation contained in this book, including without limitation warranties of merchantability or fitness for a particular purpose. No liability is accepted in any event for any damages, including incidental or consequential damages, lost profits, costs of lost data or program material, or otherwise in connection with or arising out of the furnishing, performance, or use of the programs in this book.

For Product Safety Concerns and Information please contact our EU representative GPSR@taylorandfrancis.com Taylor & Francis Verlag GmbH, Kaufingerstraße 24, 80331 München, Germany

T - #0014 - 160425 - C0 - 234/156/19 [21] - CB - 9780750308830 - Gloss Lamination